Seed Physiology
Volume 2

Seed Physiology

Volume 2

Germination and Reserve Mobilization

Edited by

David R. Murray

*Biology Department
The University of Wollongong
New South Wales*

ACADEMIC PRESS

(Harcourt Brace Jovanovich, Publishers)
Sydney   Orlando   San Diego   Petaluma   New York
London   Toronto   Montreal   Tokyo
1984

ACADEMIC PRESS AUSTRALIA
Centrecourt, 25-27 Paul Street North
North Ryde, N.S.W. 2113

United States Edition published by
ACADEMIC PRESS INC.
Orlando, Florida 32887

United Kingdom Edition published by
ACADEMIC PRESS, INC. (LONDON) LTD.
24/28 Oval Road, London NW1 7DX

Copyright © 1984 by
ACADEMIC PRESS AUSTRALIA

All rights reserved. No part of this publication may be reproduced or transmitted in any form or by any means, electronic or mechanical, including photocopy, recording, or any information storage and retrieval system, without permission in writing from the publisher.

Printed in Australia

National Library of Australia Cataloguing-in-Publication Data

Seed physiology. Volume 2. Germination and
reserve mobilization.

  Includes bibliographies and index.
  ISBN 0 12 511902 X.

  1. Seeds – Physiology. 2. Germination. I. Murray,
  David R. (David Ronald), 1943– .

582'.0333

Library of Congress Catalog Card Number: 84-71959

Contents

List of Contributors	ix
Preface	xi
Contents of Volume 1	xiii

1. Structural Aspects of Dormancy — 1

V. N. TRAN AND A. K. CAVANAGH

I.	Introduction	1
II.	The Concept of Dormancy	2
III.	Dormancy, Seed Viability and Longevity	5
IV.	Structural Factors Influencing Coat-Imposed Dormancy	8
V.	Mechanical Strength of the Seed Coat	24
VI.	Treatment to Break Dormancy	31
VII.	Concluding Remarks	38
	References	40

2. Metabolic Aspects of Dormancy — 45

J. D. ROSS

I.	Introduction	45
II.	Gene Repression	46
III.	Hormonal Changes	48
IV.	Phytochrome Control of Metabolism	61
V.	Respiration	63

	VI.	Mobilization of Reserves	66
	VII.	Conclusion	69
		References	73

3. Early Events in Germination   77

E. W. SIMON

	I.	Introduction	77
	II.	The Phase of Imbibition	78
	III.	Metabolism	94
	IV.	Concluding Discussion	110
		References	111

4. Mobilization of Polysaccharide Reserves from Endosperm   117

ANNE E. ASHFORD AND FRANK GUBLER

	I.	Introduction	117
	II.	Starch Hydrolysis in Cereal Endosperm	118
	III.	Cell Wall Mobilization	146
	IV.	Future Prospects	156
		References	157

5. Mobilization of Nitrogen and Phosphorus from Endosperm   163

MICHAEL J. DALLING AND PREM L. BHALLA

	I.	Introduction	163
	II.	Mobilization of Nitrogen	164
	III.	Mobilization of Phosphorus	182
	IV.	Conceptual Model of Nitrogen and Phosphorus Mobilization	193
		References	196

6. Mobilization of Oil and Wax Reserves   201

RICHARD N. TRELEASE AND DIANE C. DOMAN

	I.	Introduction	202
	II.	Oil and Wax Composition of Mature Seeds	202
	III.	Developmental Patterns of Glyoxysomes and Their Enzymes	207
	IV.	Lipolysis of Triacylglycerols	217
	V.	Lipolysis of Wax Esters	223
	VI.	β-Oxidation of Fatty Acids	224
	VII.	Metabolism of Acetyl Coenzyme A	228
	VIII.	Control of Lipid Mobilization	237
	IX.	Summary and Perspectives	240
		References	242

7. Axis–Cotyledon Relationships during Reserve Mobilization — 247

DAVID R. MURRAY

I.	The Influence of Cotyledons on Axis Development	247
II.	Regulation of Reserve Mobilization from Cotyledons by the Axis	261
III.	Conclusions	276
	References	277

Plant Species Index — 281

Subject Index — 287

Contributors

Numbers in parentheses indicate the pages on which the authors' contributions begin.

ANNE E. ASHFORD (117), Botany Department, The University of New South Wales, P.O. Box 1, Kensington, New South Wales 2033, Australia.

PREM L. BHALLA (163), Plant Sciences Section, School of Agriculture and Forestry, The University of Melbourne, Parkville, Victoria 3052, Australia.

A. K. CAVANAGH (1), School of Engineering and Architecture, Deakin University, Victoria 3217, Australia.

MICHAEL J. DALLING (163), Plant Sciences Section, School of Agriculture and Forestry, The University of Melbourne, Parkville, Victoria 3052, Australia.

DIANE C. DOMAN (201), Department of Botany and Microbiology, Arizona State University, Tempe, Arizona 85287, United States of America.

FRANK GUBLER (117), Botany Department, The University of New South Wales, P.O. Box 1, Kensington, New South Wales 2033, Australia.

DAVID R. MURRAY (247), Biology Department, The University of Wollongong, New South Wales 2500, Australia.

J. D. ROSS (45), Department of Botany, The University of Reading, Whiteknights, Reading RG6 2AS, England.

E. W. SIMON (77), Botany Department, The Queen's University of Belfast, Belfast BT7 1NN, Northern Ireland.

V. N. TRAN (1), School of Engineering and Architecture, Deakin University, Victoria 3217, Australia.

RICHARD N. TRELEASE (201), Department of Botany and Microbiology, Arizona State University, Tempe, Arizona 85287, United States of America.

Preface

The authors of the seven chapters comprising this Volume have confronted some of the major unanswered questions about seed dormancy, germination and early post-germination development of the seedling. Dormancy is considered first: structural constraints to germination are described by V. N. Tran and A. K. Cavanagh (Chapter 1), then metabolic barriers preventing germination are discussed by J. D. Ross (Chapter 2). A comprehensive account of the physical and biochemical events that follow imbibition of water by dry seeds is given by E. W. Simon (Chapter 3). In the remaining chapters, attention is concentrated on the cell biology and metabolic regulation of reserve mobilization in seeds and seedlings. Anne E. Ashford and Frank Gubler review the mobilization of polysaccharide reserves from endosperm (Chapter 4). Then a complementary account of the mobilization of nitrogen and phosphorus from external storage tissues is given by Michael J. Dalling and Prem L. Bhalla (Chapter 5). In Chapter 6, Richard N. Trelease and Diane C. Doman review the mobilization of lipid reserves in seed tissues. A useful feature of this chapter is the inclusion of jojoba (*Simmondsia chinensis*), a species unique in its storage of wax esters as a major reserve. In conclusion, I review the subject of embryonic axis–cotyledon interaction, considering mainly those species where the cotyledons are adapted for the storage of reserves (David R. Murray, Chapter 7).

The authors have thoroughly reviewed the literature, as well as emphasized new information and techniques. Important questions for future investigation are clearly identified. Both this Volume and its companion ('Seed Physiology Volume 1. *Development*') will provide a valuable resource for advanced students, teachers and researchers in plant physiology, biochemistry, agronomy and related disciplines.

I thank all those who provided information and material for presentation, and copyright owners, who readily gave permission for the reproduction of data already published. I also thank the staff of Academic Press for their assistance in bringing this project to fruition.

My own Chapter in this Volume would not have been possible without the enthusiastic collaboration of my former students. It is a pleasure to record my appreciation of the efforts of Megan Mowchanuk (née Collier), Karl Gordon, Janet Crump, Bill Ashcroft, Mark Peoples, Stephen Waters, Franklin Vairinhos and Maria Vidovic.

Finally, I must thank all the contributors to both Volumes for their patience with the editorial process, and for the high quality of their contributions.

Contents of Volume 1

1. **The Seed and Survival**

   DAVID R. MURRAY

2. **The Carbon and Nitrogen Nutrition of Fruit and Seed — Case Studies of Selected Grain Legumes**

   J. S. PATE

3. **Accumulation of Seed Reserves of Nitrogen**

   DAVID R. MURRAY

4. **Accumulation of Seed Reserves of Phosphorus and Other Minerals**

   J. N. A. LOTT

5. **The Synthesis of Reserve Oligosaccharides and Polysaccharides in Seeds**

   N. K. MATHESON

6. **Synthesis of Storage Lipids in Developing Seeds**

   C. R. SLACK AND J. A. BROWSE

7. **Toxic Compounds in Seeds**

   E. A. BELL

CHAPTER **1**

Structural Aspects of Dormancy

V. N. TRAN and A. K. CAVANAGH

I.	Introduction	1
II.	The Concept of Dormancy	2
	A. Dormancy and Its Effects	2
	B. Dormancy Types	3
III.	Dormancy, Seed Viability and Longevity	5
	A. Viability of Old Seeds	5
	B. Dormancy of Dry *versus* Fully Imbibed Seeds in Storage	7
IV.	Structural Factors Influencing Coat-Imposed Dormancy	8
	A. Seed Morphology and Terminology	9
	B. Seed Coat Structure	10
	C. Role of Specialized Features of the Seed Coat	14
	D. Nature and Depth of Impermeability	22
V.	Mechanical Strength of the Seed Coat	24
	A. The Seed Coat under Compression	26
	B. The Seed Coat under Tension	28
	C. Thin Shell Theory Applied to the Seed Coat	29
VI.	Treatment to Break Dormancy	31
	A. Natural Breakdown of Impermeable Seeds	31
	B. Artificial Treatments	33
VII.	Concluding Remarks	38
	References	40

I. INTRODUCTION

Dormancy imposed by the seed coat has interested agriculturalists, horticulturalists and botanists since at least the third century B.C. Then, the Greek writer Theophrastus stated that the seeds of certain pulses were 'hard' and require soaking in 'nitre' prior to sowing if they were to germinate (Evenari, 1980/1981). He also suggested that soil and climatic conditions could control the degree of 'hardness' developed in seeds. Little additional work was done until the 17th and 18th centuries when anatomical examinations of seeds were made by Grew (1682), Malpighi (1687) and Gaertner (1791). Their work laid the foundation for the extensive studies of Chalon (1874), Mattirolo and Buscalioni (1892), Pammel (1899), Maisel (1909), Netolitzki (1926), Pitot (1936) and Corner (1951, 1976) who have contributed much to what we currently know about the structure of seeds. Nobbe (1876) introduced the term 'hard seed' and suggested that the waxy cuticle of such seeds was responsible for their impermeability, an idea which is now known to be incorrect. Since then, much has been written on the topic of coat-imposed dormancy: Crocker (1906, 1916), Barton and Crocker (1948), Porter (1949), Barton (1965a, 1967), Quinlivin (1971), Ballard (1973), Rolston (1978) and Werker (1980/1981). Yet certain structural aspects are not fully understood. As stated by Werker (1980/1981), 'It is surprising how vast the literature is on seed dormancy caused by the embryo envelopes and how meagre is the information concerning the structure and histochemistry of these envelopes'.

In this chapter, we review progress that has been made in recent years towards the understanding of the coat-imposed dormancy and emphasize the lack of experimental information which exists. We also attempt to clarify some terminology, such as the use of the word 'strophiole', which has caused confusion in the past, and to describe the function of various structural features of seeds. Finally, we consider the effects on the seed coat of various treatments designed to enhance germination and indicate the lack of knowledge which exists regarding their mechanism(s) of action.

The majority of the work discussed refers to leguminous and malvaceous genera. These are usually agriculturally important and also contain the largest number of species with dormant seeds. However, it should be appreciated that seed coat-imposed dormancy is widespread in the plant kingdom (Harrington, 1972; Ballard, 1973).

II. THE CONCEPT OF DORMANCY

A. Dormancy and Its Effects

Though the term 'dormancy' is sometimes used to describe the condition of any seed not in the process of germinating, in this chapter we will follow conventional practice and define dormancy as a state in which a viable seed will not germinate when placed in conditions normally considered to be adequate for germination, that is, when provided with suitable temperature, adequate moisture and oxygen (Roberts, 1972). The causes of dormancy are many and varied (Roberts, 1972; Mayer and Poljakoff-Mayber, 1975;

Copeland, 1976; Nikolaeva, 1977): impermeability of the seed coat to water and gases, immaturity of the embryo, special requirements for temperature or light, presence of inhibitors, and mechanical restriction to embryo growth and development or to radicle extension in germination. Dormancy in a given case may be due to only one cause, or to several of the above factors operating together.

Dormancy in seeds provides a means by which germination is delayed until favourable conditions for growth and establishment in the field are met. It thus plays an important role in the survival of plant species as it distributes germination in time and allows seeds to overcome adverse sowing conditions such as heavy frosts, dry weather or excessive moisture. Dormancy also gives other advantages; for example, in cotton, (*Gossypium hirsutum*), impermeable seeds maintain seed quality under adverse conditions of harvest (damp seasons) and storage (high humidity) (Rolston, 1978). The degree of dormancy varies, even within a given batch of seeds. For instance, Kondo (cited by Barton, 1965a) found that after 14 years soaking in water, 6% of the seeds of *Robinia pseudo-acacia* were still impermeable. The pattern of breakdown of dormancy in 869 seeds of *Albizzia julibrissin* soaked in water over a 10 year period is shown in Figure 1. Thirty-nine percent germinated in the first year, 23% in the second, 11% in the third and 23.5% in the fourth. After five years, nearly 97% had swollen and germination continued spasmodically for the next five years, one seed remaining impermeable after 16 years.

The condition of seed dormancy also poses problems. The most notable is finding a means of breaking it if a satisfactory germination is to be achieved (Crocker and Barton, 1953). Another difficulty is the ability of such seeds to remain ungerminated in soil for many years, to appear as unwanted plants in later crops, even after years of intensive cultivation. An interesting example of this problem was provided by Sherry (1971) for the wattle plantations of Natal. The land concerned, originally native grassland, was ploughed and planted with black wattle (*Acacia mearnsii*) in 1884. In 1900, the area was cleared of wattles and sown with maize (*Zea mays*). It was cropped and weeded annually until 1944 and was then left fallow. Within two years, wattle seedlings appeared among the maize stubble, reaching a density in excess of 1400 plants per hectare, a remarkably high residual figure.

B. Dormancy Types

Harpur (1957) and Roberts (1972) recognized three types of dormancy — innate, induced and enforced — though other classifications have been

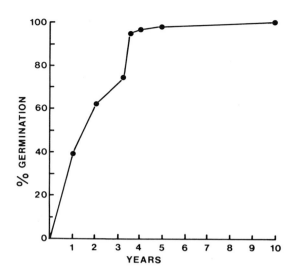

Fig. 1. Germination of *Albizzia julibrissin* seeds soaked in water for up to 10 years (data of Moore and Moore, 1962).

proposed, for example, Nikolaeva (1977). Innate dormancy (sometimes also known as primary dormancy) prevents germination during development and maturation of the seed on the mother plant and usually also for some time after shedding and harvesting (Karssen, 1980/1981a). Seeds with impermeable coats represent one form of innate dormancy. Even when a seed has lost its innate dormancy, it may fail to germinate following exposure to one or more unsuitable conditions, for example, excessive moisture, too much light, too high or too low a temperature. Such induced or secondary dormancy is characterized by persistence of the dormant condition even after the seeds have been returned to a situation that was originally favourable for germination. By contrast, seeds with enforced or imposed dormancy can germinate immediately upon removal of the environmental limitation. The latter is usually applied to seeds buried in soil which will not germinate, even if fully imbibed, until exposed to light or brought close to the surface and a more favourable temperature regime. Enforced dormancy can persist for hundreds or possibly thousands of years, as shown by the longevity of the buried seeds listed by Harrington (1972). The mechanisms underlying secondary dormancy are still not fully understood but are assumed to be similar to those operating with dormancy in general (Mayer and Poljakoff-Mayber, 1975). Factors such as temperature, absence of light or oxygen, presence of volatile or allelopathic inhibitors and moisture conditions have all been shown to contribute (Karssen, 1980/1981a, b).

III. DORMANCY, SEED VIABILITY AND LONGEVITY

It is sometimes suggested that seed viability and longevity are causally related to innate dormancy — the greater the dormancy which exists, the longer the viability of the seed. However, Roberts (1972) studied the problem using 6 cultivars of rice (*Oryza sativa*) as test material and concluded that at least for this case, such a relationship did not exist. Nevertheless, seeds can retain viability for hundreds or even thousands of years and it is likely that some form of enforced dormancy must be preventing germination, especially in cases where seeds are fully or partially imbibed for long periods of time.

A. Viability of Old Seeds

The life of seeds and their viability after extended periods of storage are subjects of scientific interest and commercial importance. Yet the information available is limited to seeds of a few hundred species and little is known about the potential life and viability of the seeds of most plant species.

Beal in 1879 and Duvel's in 1902 set up comprehensive experiments to determine longevity of seeds buried at different depths in soil. Results were reported after 100 years for Beal's experiment by Kivilaan and Bandurski (1981) and after 39 years for Duvel's by Toole and Brown (1946). Ødum (1965) gave germination results for seeds taken from archeological excavations in Denmark and these showed that life spans in excess of 1700 years were possible. Ewart (1908), Becquerel (1907 and 1934), Turner (1933) and Schjelderup-Ebbe (1936) (cited by Barton, 1965b) reported the results of laboratory germination experiments with seeds of known age taken from herbarium specimens and seed collections. In the last case, the seeds were stored under conditions of ambient temperature and humidity, that is, with no environmental control. There still remains the need to determine seed longevity under optimum conditions of controlled temperature and humidity, such as was attempted by Went and Munz (1949).

1. Storage in Air
When stored dry with little or no environmental control, leguminous and malvaceous seeds retain their viability exceptionally well. Thus Harrington

(1972) lists 90 species from these families as capable of germinating after 50 years storage, while a few remain viable for around 150 years, for example, *Cassia multijuga* at 158 years and *Albizzia julibrissin* at 147 years. More recent observations on seeds recovered from bricks of adobe buildings in California and North Mexico have been published by Spira and Wagner (1983). The following seven species yielded at least one viable seed (age in years in parentheses): Papilionaceae — *Medicago (hispida) polymorpha* (200), *Melilotus indicus* (183), *Trifolium* sp. (193); Malvaceae — *Malva parviflora* (183 to 200); Chenopodiaceae — *Chenopodium album* (143), *C. murale* (183); Poaceae — *Hordeum (murinum) leporinum* (200). The highest proportion of viable seeds (summing all samples) was shown by *Medicago (hispida) polymorpha*, with 68% (17/25) germinating (Spira and Wagner, 1983). The water-impermeable coat of such seeds served the important functions of protecting the embryo and preserving its moisture content, thus slowing down degenerative changes to cells during storage.

2. Storage in Soil

The conditions provided by burial in soil appear to be particularly favourable for long-term survival of many seeds, including a high percentage without a water impermeable coat. Harrington (1972) lists almost 200 species capable of germinating after 20 years burial, representing 49 families. Data from Ødum (1965) show that at least 19 species were viable after more than 500 years burial. Those with outstanding longevity include the 10 000 years for *Lupinus arcticus*, 1000 years for *Nelumbo nucifera* found in cold peat soil in Manchuria and 1750 years for *Chenopodium album* and *Spergula arvensis*. The latter group contained seeds which were buried under conditions of extreme cold and low oxygen, and at least in some cases, could have involved seeds which were fully or partially imbibed. It is still not clear whether seeds will retain viability longer when fully imbibed or in a dry state.

3. Storage in Water

Although a low seed moisture content is known to be vital for long-term viability, Shull (1914) demonstrated that permeable seeds of *Juncus bufonius*, *J. tenuis* and *Sium circutaefolium* were still germinable after 7 years submersion. The life of impermeable seeds in water is even longer — up to 16 years for *Albizzia julibrissin* (Moore and Moore, 1962), 14 years for *Robinia pseudo-acacia* (Kondo, 1929) and at least 10 years for *Acacia melanoxylon* stored in seawater (Cambage, 1928).

B. Dormancy of Dry *versus* Fully Imbibed Seeds in Storage

Relatively little is known about the reasons why some species are able to survive long periods of storage with little impairment of viability. Berjak and Villiers (1972) and Villiers and Edgecumbe (1975) (quoted by Heydecker, 1977) were the first to show that at thermodormancy inducing temperatures, lettuce seed (*Lactuca sativa*) survived better in the imbibed than the air dry state. While some extremely long-lived buried seeds have a water-impermeable seed coat (e.g., *Nelumbo nucifera*, *Lupinus arcticus* and many leguminous and malvaceous seeds from the Duvel experiment [Toole and Brown, 1946]), others, such as *Viburnum blattaria* (100 years buried), *Chenopodium album* and *Spergula arvensis* (1750 years buried), were probably fully imbibed for most of the time.

Osborne (1977) points out that loss of viability can be quite variable in dry non-dormant seeds, for example, rye (*Secale cereale*) with a life of 3-4 years and wheat at 30-40 years. The large seeds of most crop plants in the Duvel experiment did not survive one year's burial (Tool and Brown, 1946) while seeds of some tropical species, for example, *Pothos longipes* and *Schefflera actinophylla*, deteriorate within a few days (Elliot and Jones, 1980). On the other hand, seeds of *Canna compacta*, which lack a water-impermeable seed coat, were germinated when 620 years old and produced normal offspring (Osbornem, 1977). Osborne (1977) maintains that while old seeds appear externally unchanged, on imbibition they germinate more slowly than those of newer harvests and exhibit what is called loss of vigour. She attributes these changes to a number of biochemical factors — impaired respiratory activity, reduced ATP levels, impaired and leaky cell membranes and delayed syntheses of nucleic acid and protein when water is again supplied. Species which deteriorate rapidly in storage have a number of characteristics — they usually exhibit little dormancy, (Osborne, 1977), cannot be dried without loss of viability and cannot be frozen without injury (Harrington, 1970). Though similar cell degeneration occurs in seeds with coat-imposed dormancy, the rate of deterioration is much reduced, probably because the impermeable seed coat prevents any external moisture from reaching the embryo. Life spans for these seeds in excess of 50 years are not uncommon, as noted above (Section III,A,*1*).

Dormant, imbibed seeds do not germinate but apparently maintain a regular turnover of many cellular constituents and can continually repair any cytological damage they may suffer (Osborne, 1977; Heydecker, 1977). They are prevented from germinating by a metabolic block (Chapter 2). Osborne (1977) suggests that dormancy may need to be broken before viability is lost.

IV. STRUCTURAL FACTORS INFLUENCING COAT-IMPOSED DORMANCY

The coat of impermeable seeds is usually mechanically hard and was held responsible for water exclusion by early gardeners. Hence there arose the practice of boiling hard seeds to 'soften' them (Bowie, 1832). Though the term 'soft' is still employed today to describe impermeable seeds which have become permeable either naturally or as a result of treatment, its use in this context is a misnomer. Seeds gradually soften only after they imbibe water and commence to swell. It is found that treatment ruptures a small, specialized area of the seed coat (the lens of leguminous seeds, or a region near the chalazal discontinuity of malvaceous seeds) and this permits water absorption and consequent germination. Treatments, using dry or moist heat, have little effect on the seed coat itself (Tran and Cavanagh, 1980).

Experimental observation shows that during storage, coat-imposed dormancy breaks down and given sufficient time, most seeds will eventually take up water (Fig. 1). Until the work of Hamly (1932), little was known about the mechanism of such breakdown and it was usually assumed that the seed coat had deteriorated in some way. Hamly believed that water exclusion was a property of the outer layers of the coat, that is, the suberized walls and caps of the Malpighian cells, but, more importantly, he showed that impermeability was lost when the highly stressed cells at the 'strophiole' (lens) separated, thus forming a 'strophiolar cleft' and permitting water entry. He found that rupture of the lens tissue could occur during storage or as a result of heating or by shaking the seeds in a bottle (impaction). Subsequent studies by Aitken (1939), by Gladstones and Quinlivin (reviewed by Quinlivin, 1971) and Ballard and co-workers (reviewed by Ballard, 1973) confirm the important function of the 'strophiole' (lens) in papilionoid legumes. However, a number of important questions still remain to be answered: (i) what causes impermeability at seed maturity: is it a chemical or mechanical process? (ii) what is the mechanism of water loss in the maturing seed and after impermeability develops? (iii) what layer(s) of the seed coat are responsible for impermeability and what is the depth to which impermeability extends? and (iv) what part of the seed coat is affected by treatments which break coat-imposed dormancy? In the remainder of this section, we will discuss these problems. The major factors which influence coat-imposed dormancy are the structure and nature of the testa or seed coat. Field and environmental conditions and heredity can also influence seed permeability (Aitken, 1939; Lebedeff, 1947; Kyle and Randall, 1963; Baciu-Miclaus, 1970; Rolston, 1978) but it is not known whether these factors exert their influence through purely chemical or mechanical means.

A. Seed Morphology and Terminology

The seed develops from the fertilized ovule of the flower. A typical seed consists of an outer seed coat or *testa*, an inner seed coat or *tegmen* (both derived from the ovular integuments) and an embryo. Endosperm may or may not be present (Section 1.I,C of Volume 1). In seeds lacking endosperm, reserve materials are usually stored in the cotyledons or seed leaves of the embryo.

The external morphology of leguminous seeds has been discussed by several authors (Boelcke, 1946; Isely, 1955; Kopooshian and Isely, 1966; Cavanagh, 1981; Gunn, 1981) but as yet there is no standardized terminology. A prominent feature is the *hilum* or abscission scar left when the seed detaches from the *funiculus* or funicle, which is the elongated stalk which formerly attached the ovule to the placenta of the ovary. The hilum may be large and characteristic as in many papilionoid seeds or it may be hidden under remnants of funicular tissue (the *epihilum* of Lackey, 1981). It is closed by one or two layers of elongated cells and is not normally permeable. Close to the hilum is the *micropyle*, the former passage for the pollen tube between the integuments of the ovule. It is mostly a sealed or plugged opening, sometimes regarded as a breathing hole or pore (Polhill, 1976), frequently inconspicuous as in mimosoid and caesalpinioid seeds but occasionally prominent and brightly coloured.

On the opposite side of the hilum is a mound of tissue commonly referred to as the *strophiole* (Hamly, 1932) but also known as a *boss* (Miller, 1967) or *lens* (Gunn, 1981). The name strophiole is frequently misapplied and has been used in the literature in several senses: (i) as a hilar outgrowth on the seed (Gaertner, 1791); (ii) as a region of structural weakness in the outer layer of the seed coat of some leguminous seeds (Hamly, 1932); and (iii) as a swelling of the funiculus (Bentham, 1864; Fahn, 1974).

The word 'strophiole' is derived from 'strophiolum' which was originally defined by Gaertner (1791) as a 'spongy, glandular, porous outgrowth on the ventral (hilar) side of the seed' (Zimmermann, 1936). Such outgrowths are prominent on seeds of species such as *Asarum*, *Aristolochia*, *Peperomia*, and *Ricinus*. Thus, on the basis of priority, we consider the term should be restricted to the description of hilar outgrowths. Its use in a structural sense, which apparently originated with Hamly (1932), or as a name for a funicular swelling, is not appropriate. Gunn (1981) proposed the name lens as an alternative and we have followed his proposal. The relationship between the hilum, micropyle and lens in a typical *Acacia* seed is shown in Figure 2; for other species, see Pitot (1935, 1936), Zimmermann (1936), Corner (1951), Miller (1967), Polhill (1976) and Chapter 1 of Volume 1. The function of the lens in papilionoid and mimosoid legumes is further discussed in Section IV.C.*3*.

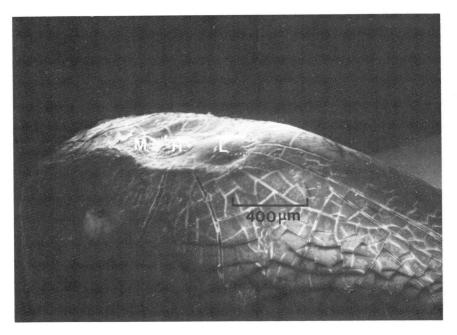

Fig. 2. SEM of the hilum area of a seed of *Acacia suaveolens*. H = hilum, L = lens, M = micropyle (from Cavanagh, 1980).

B. Seed Coat Structure

During growth of the seed, the one or two integuments of the ovule undergo significant histological changes. In most cases, the inner integument is lost and the seed coat is derived from the outer integument alone. This occurs in leguminous seeds where the so-called palisade layer (Malpighian cells) is derived from the outer epidermis of the outer integument. The mesophyll tissue, which forms the body of the testa, develops from the middle cell layer or layers of the outer integument (Corner, 1951). Additional cell layers which may or may not be present include (hypodermal) hour-glass or osteosclereid cells, inner hour-glass cells and endosperm (Fig. 3A). By contrast in some seeds, both integuments contribute, for example, in

Fig. 3. (A) Main macroscopic features of papilionoid seeds. (B) Anatomical features of papilionoid seeds: (1) diagrammatic longitudinal section of hilar region; (2) the same in transverse section; (3) diagrammatic transverse section of part of seed; and (4) diagrammatic transverse section of small part of hilar region (zone shown in 2) (from Polhill, 1976).

1. *Dormancy—Structural* 11

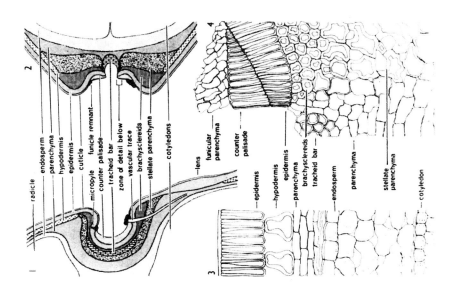

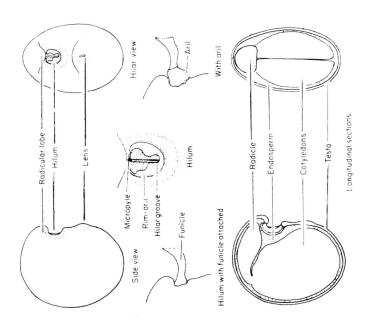

Gossypium (Reeves, 1936b; Bhojwani and Bhatnagar, 1979), in *Abutilon* (Reeves, 1936a; Winter, 1960) in *Sida spinosa* (Egley and Paul, 1981, 1982) and in *Cuscuta campestris* (Hutchinson and Ashton, 1979). The palisade layer in these seeds is derived from the outer epidermis of the *inner* integument. Inside the palisade is a thin layer of parenchyma cells equivalent to the mesophyll of leguminous seeds. The outer integument develops into two layers in mature seeds, the outer layer containing the single celled, thin walled hairs or cotton fibres in the case of *Gossypium* species.

The structure most implicated in impermeability is the palisade layer of the seed coat. This has a thin layer of cuticle which consists of waxy hemicellulose or pectinaceous substances deposited over the testa surface but not over the hilum (Gunn, 1981). A sub-cuticular layer, described by Corner (1951) as a 'mucilage stratum', is sometimes conspicuous. As its name implies, it may become gelatinous or mucilaginous when wet. The tops of the Malpighian cells of the palisade layer are buried in this material which comprises several lamellae. It consists mainly of pectin in its outer part and hemicellulose in the inner section and becomes very hard and hydrophobic during the last stages of seed maturation (Werker, 1980/1981). Both the cuticle and the sub-cuticular layer have been considered to be the cause of impermeability (White, 1908; Rees, 1911; Khudaire, 1956; Werker *et al.*, 1973; Gutterman, 1978). This view is not supported by the findings of Ballard (1973) and Tran and Cavanagh (1980) and the fact that shallow scratches which penetrate the cuticle do not break impermeability (Werker, 1980/1981).

The so-called palisade layer is composed of elongate cells known variously as Malpighian, macrosclereid, epidermal or prism cells. The cells are columnar and oppressed and arranged with their long axis perpendicular to the surface of the seed. In some domestic lines of rough-coated cowpeas the Malpighian cells have very thin walls and collapse as the seed dries out. Their alignment is not always clear in these cases (Lush and Evans, 1980). Length variations are enormous even within a single seed, especially around the lens and hilum areas, and range from as little as 5–10 μm in some cultivated lines of cowpeas (*Vigna unguiculata* subsp. *unguiculata*) (Sefah-Dedeh and Stanley, 1979) to 400–500 μm in seeds with exceptionally thick coats, for example, *Mucuna urens* and *Physostigma venenosum* (Rowson, 1952). In cross section, the Malpighian cells are usually polyhedral with 5 or 6 sides. The lower sections show only slight secondary wall thickening while in the upper half, the cell lumen is frequently almost closed by twisted internal ridges (Polhill, 1976) or fluted or flange-like thickenings (Miller, 1967). The lumen top of individual cells can be open and in contact with the overlying cuticle or in some species such as the mimosoid *Prosopis farcata* and members of the papilionoid genera *Melilotus* and *Trifolium* the cells are closed by dome-like caps. While many workers believe the caps are the main source of water exclusion, particularly when suberized (Hamly, 1932) or pectinized (Werker *et al.*, 1973), this does not explain how cells with open lumens maintain impermeability. It is likely that in such cases (and perhaps in many other seeds) impermeability extends to the base of the Malpighian cells, as shown for *Gleditsia triacanthos* (Cavazza, 1950), *Phaseolus vulgaris* (Baciu-Miclaus, 1970) and *Acacia farnesiana* (Tran and Cavanagh, 1980).

Running through the Malpighian cells in their upper half is the *light line* (or *linea lucida* of early authors — Mattirolo and Buscalioni, 1892; Pammel, 1899). Although sometimes even today considered as a source of impermeability (Esau, 1977), studies by Hamly (1935), Cavazza (1950), Stein von Kamienski-Jancke (1958) and Scott *et al.* (1962) have shown beyond doubt that the light line is an optical effect and has no structural significance. It is caused by differences in refraction of light at particular locations along the cell, variously ascribed to changes in chemical composition (Reeves and Valle, 1932; Hamly, 1935; Stein von Kamienski-Jancke, 1958) or to twisting of internal thickenings of the secondary wall (Gunn, 1981) or to a change in orientation of the microfibrils in this region (Scott *et al.*, 1962). The light line lies close to the outer ends of the Malpighian cells and is only occasionally deeper than 50 μm from the surface, even in seeds with very thick coats (Rowson, 1952).

Cells which constitute the inner seed coat include the hour glass, osteosclereid or hypodermal pillar cells, and a layer of crushed or collapsed cells known variously as mesophyll, nutrient tissue or crushed parenchyma. The osteosclereid cells are variable in shape (Corner, 1951; Rowson, 1952; Miller, 1967; Vassal, 1973, 1975) but are most commonly flask or dumb-bell shaped with very thick walls and large air spaces between them. Though Corner (1951) ascribed no structural role to the osteosclereids, others such as Kramer and Szczesniak (1973) claim that they support the Malpighian cells so that compressive and concussive forces are absorbed by the deformed Malpighian cells. The osteosclereid cells flatten under pressure (McKee *et al.*, 1977).

The mesophyll cell layer frequently constitutes more than half the thickness of the seed coat and was shown by Hamly (1932) to be permeable. In piercing experiments with seeds of *Coronilla varia*, McKee *et al.* (1977) claimed that both the osteosclereid and mesophyll layers as well as at least part of what they termed the 'inner integument' needed to be punctured or made permeable for seeds to imbibe. There is obvious disagreement in the literature concerning the relative depth of impermeability and this topic will be discussed in more detail in Section IV,D,2.

C. Role of Specialized Features of the Seed Coat

1. The Hilum and Its Function

Hyde (1954) was the first to show that the hilum of several papilionoid legumes functioned as a hygroscopically activated one-way valve permitting moisture to be lost from the drying seed during the last stages of maturation. A fissure (the hilar groove, Fig. 4) develops during drying and can open and close in response to changes in the relative humidity surrounding the seed. With high humidity, the fissure closes and opens as the humidity is reduced (Fig. 5). The seed tends to acquire and retain a moisture content in equilibrium with the lowest relative humidity to which it has been exposed (Hyde, 1954). This is usually less than 10% in dry, impermeable seeds. The hilar groove remains closed when the seed is placed in a moist

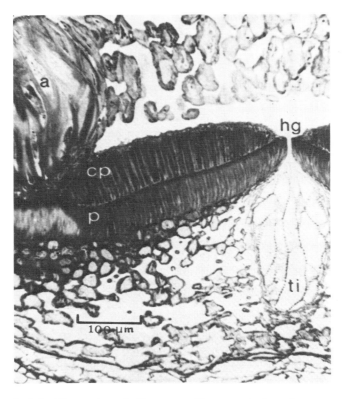

Fig. 4. Detail of the hilum region of wild cowpea, *Vigna unguiculata* subsp. *mensensis*. a = aril, hg = hilar groove, p = palisade layer, cp = counter palisade layer, ti = tracheid island (bar) (from Lush and Evans, 1980).

environment because the counterpalisade tissue of the hilum swells and presses the margins together, thus providing a very effective seal.

The hilar and sub-hilar regions of papilionate legumes are relatively complex (Fig. 3B) and have a greater variety of cells than exist elsewhere in the coat. The hilum itself is usually sunken, frequently elongated and underlain by the *tracheid bar* (Lersten, 1982) whose function is still not understood. Cuticle is absent, terminating on the outer rim of the hilum, while a funicular remnant is frequently present, sometimes giving rise to a tongue of tissue (the *hilar tongue*; Berg, 1979). An *epihilum* (Lackey, 1981) may obscure the hilum completely. The single layer of Malpighian cells of the remainder of the coat is replaced by the two layers of palisade and

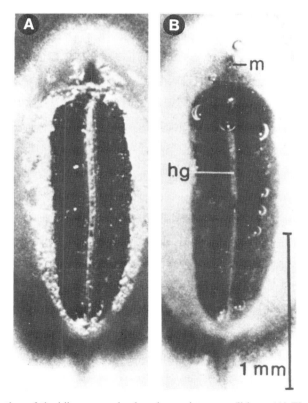

Fig. 5. The action of the hilar groove in changing moisture conditions. (A) The hilar groove opens when a seed with high moisture content is transferred from high to low relative humidity. (B) The groove closes when the relative humidity is raised. Seed of *Vigna unguiculata* subsp. *dekindtiana*. hg = hilar groove; m = micropyle (from Lush and Evans, 1980).

counterpalisade cells, the latter being derived from the funiculus. A layer of cuticle separates the two, although this may be absent or the cells poorly developed in permeable seeds (Stienswat et al., 1971). There are no osteoscleried cells, their place being taken by several types of parenchyma cells whose function is not known. The mesophyll tissue of the seed extends under the hilum up to and surrounding the tracheid bar. Despite its complexity, this general configuration of tissues in the hilar region is remarkably consistent throughout many of the papilionoid legumes (Polhill, 1976).

In other genera and species the hilum is a simple round or elongate scar, occasionally partly or wholly covered by funicular remnant. Cuticle is absent and the Malpighian cells extend across a single layer continuous with that of the rest of the coat. Osteoscleried cells are absent from beneath the hilum, the Malpighian cells bearing directly on the thick layer of mesophyll tissue. The hilum does not possess a tracheid bar and presumably it does not function as a hygroscopic valve. The manner in which water is lost from the drying seed is not yet understood. Preliminary results given in Table I appear to indicate that at least for some mimosoid seeds, a significant amount of moisture is lost through the lens.

Traversing the hilum of all seeds is the vascular bundle, a strand of conducting tissue through which water and nutrients pass during develop-

Table I. Moisture escape pathway in impermeable *Acacia* seeds during desiccation and drying at 40°C

Species	Lens status	Moisture loss (%)			Collection date
		Silica gel[a]	H_2SO_4[a]	40°C[b]	
A. accola	Free	1.7	1.3	1.0	1971
	Blocked	1.0	0.7	0.45	
A. baileyana	Free	—	6.5	5.8	1979
	Blocked	—	1.6	1.2	
A. farnesiana	Free	2.0	—	3.0	1975
	Blocked	1.2	—	1.8	
A. saligna	Free	4.3	—	6.2	1979
	Blocked	0.9	—	1.2	
A. sclerosperma	Free	2.8	—	—	1974
	Blocked	1.5			

[a] at 10 days
[b] at 2 days
Unpublished data of Tran (1981).

ment of the seed. For those seeds possessing a lens, a vascular bundle invariably continues through the mesophyll tissue immediately beneath the lens. Dell (1980) believes that it might assist in distributing water during imbibition, but Tran's experiments (Fig. 6) failed to confirm this claim. Contrary to Dell's observations, the results show that impermeable seeds drilled through the lens do not swell faster than those drilled at the opposite end. There is no evidence that the vascular bundle serves any function in the mature seed. The reason it is not permeable at the hilum remains a mystery even in cases where there is no evidence of collapsed or blocked cells in its xylem vessels (e.g., *Albizzia lophantha*, Dell, 1980).

2. The Micropyle

It is now generally held that the micropyle is sealed in impermeable seeds. Robbertse (1973) found in African acacias that the micropyle was plugged at an early stage of the seed's development with spongy endocarp material. This later disintegrated and the contents formed a cuticle-like layer over the interior of the micropylar cavity. Observations of micropylar impermeability seem to be confined to seeds of wild species. The evidence from a limited number of investigations is that some domestic lines of beans (*Phaseolus vulgaris*) (Preston and Scott, 1943; Kyle and Randall, 1963) and cowpeas (Sefa-Dedeh and Stanley, 1979; Lush and Evans, 1980) possess partly or fully open micropyles which can play a significant role in water uptake. Even in these cases, however, the response is not universal. For instance, water absorption was positively associated with the size of the micropylar orifice in Great Northern beans but not in Red Mexican beans. In the latter seeds, water was absorbed predominantly through the lens (raphe)* and hilum despite the presence of an open micropyle (Kyle and Randall, 1963).

3. The Lens

The lens in papilionoid legumes is a bulge or mound of tissue, near the hilum but on the side opposite the micropyle. In some species, for example, members of the genus *Phaseolus*, it is situated at the end of the raphe ridge while the lens (strophiole) of *Melilotus* as described by Hamly (1932) is an elevation centred in a narrow longitudinal depression. The lens may be devoid of pigmentation or it may be a darker or lighter hue than the

---

* *Raphe.* The raphe is a ridge along the seed formed by that part of the funiculus which was fused to the ovule (Fahn, 1974). It is prominent in bean and pea seeds and usually appears as a dark line running between the hilum and the lens; some authors (e.g., Kyle and Randall, 1963; Bravato, 1974) have considered it to be synonomous with lens.

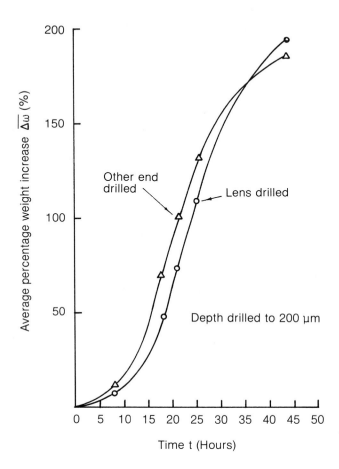

Fig. 6. Average water uptake of seeds of *Albizzia lophantha* drilled at the lens (O) and at the opposite end (△) to a depth of 200 μm.

remainder of the coat. Its colour and degree of prominence can be utilized in seed identification (Miller, 1967).

The most characteristic feature of the papilionoid lens is the degree of elongation of the Malpighian cells which may be 2-3 times the length of the cells in other parts of the coat. In subclover (*Trifolium subterraneum*), the lens cells are up to 120 μm long and 6-8 μm wide, whereas the typical size is 30-40 μm × 10 μm (Aitken, 1939). The lens Malpighian cells appear to be in a state of considerable stress and can separate between adjacent cells when seeds are shaken against a hard surface, forming a split or fissure

(Hamly's strophiolar cleft) which can reach almost to the cuticle (Fig. 7). It is through this fissure that water enters both naturally permeable seeds and those made permeable by treatment (e.g., Hagon and Ballard, 1970; Gunn, 1972; Ballard *et al.*, 1976; Lesins and Lesins, 1979).

The instability of the cells is also shown by the fact that they frequently separate during microsectioning (e.g., Zimmermann, 1936 (his 'Riss') and Miller, 1967).

The lens in mimosoid seeds differs on several counts from that in papilionoid species. It is usually relatively small and inconspicuous, the Malpighian cells are considerably shorter than elsewhere in the coat and the cells are not in a state of stress. Pitot (1936) was the first to distinguish between the two, referring to and illustrating the former as a 'plage subhilaire'. Later, Tran and Cavanagh (1979) and Dell (1980) were able to show that water entry occurred through this location in heat-treated seeds, though the mechanism did not involve the development of a fissure in the lens. Following heating, the lens becomes raised, golden and permeable (Fig. 8A), the Malpighian layer separating from the mesophyll layer. In extreme cases, the lens tissue may rupture (Fig. 8B). In *Albizzia lophantha*, the entire lens tissue is removed. Dell (1980) found that under the lens, the osteosclereid cells are replaced by thin-walled parenchyma cells. These cells break across their radial walls, permitting the lens to lift off. As similar thin-walled cells confined to the area of the lens have been found in *Acacia* seeds (Hanna, 1984), it may well be that these special cells provide a point of structural weakness in the coat of mimosoid seeds. The proper role of these cells is awaiting detailed investigation.

4. The Chalazal Discontinuity

Seeds of the family Malvaceae do not possess the lens of leguminuous seeds even though they have an in-built point of weakness in the coat and a structural break in the palisade cells layer known variously as the chalazal slit (Winter, 1960; La Croix and Stanisforth, 1964; Egley and Paul, 1981) or chalazal* pore (Pearson, 1939; Simpson *et al.*, 1940). Winter (1960) has shown in velvet leaf (*Abutilon theophrasti*) that at the chalazal end of the ovule, the inner and outer epidermises of the inner integument are discontinuous, the outer epidermis being absent from a slit-shaped region approximately 60 μm wide by 315 μm long. As the seed matures, the chalazal slit

---

* *Chalaza.* The chalaza is often at the opposite end of the seed to the hilum and is the area of fusion of the integuments and the nucellus of the ovary. In many mature seeds, the chalaza is indistinct, appearing only as a slight thickening or spreading of the vascular bundle (Corner, 1951; Robbertse, 1973) while in others, for example, *Crotalaria* sp. (Corner, 1951) it may coincide with the lens. In the latter case, it is sometimes considered as synonomous with lens (Berggren, 1963).

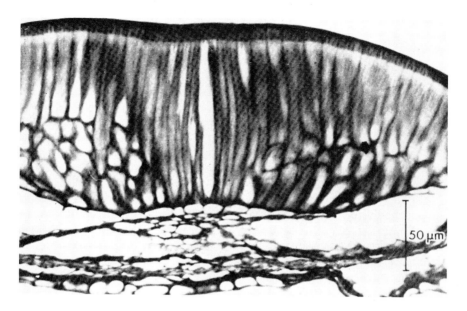

Fig. 7. Split in the lens of seeds of *Trifolium subterraneum* extending to cuticle (from Hagon and Ballard, 1970).

reaches its final dimensions of 60 μm × 740 μm. By contrast, in species of *Gossypium* the chalazal discontinuity is a pore 240–280 μm in diameter. In hard seeds, the chalazal discontinuity is closed by a 'chalazal plug' of dense, parenchymatous material connected to a 'chalazal cap'. In dormant seeds, the cap and the plug adhere tightly to the palisade layer and provide an effective seal. The plug is connected to the palisade through compact narrow 'pseudo-palisade' cells (Winter, 1960) or 'subpalisade cells' (Egley and Paul, 1981, 1982) which rupture under stress. The plug-cap connection in dormant seeds can be disrupted by hot water, ethyl alcohol or by piercing with needles (La Croix and Stanisforth, 1964; Christiansen and Moore, 1959) rendering the seed permeable. In many cultivated varieties of cotton,

Fig. 8(A) Raised and golden lens (S) in permeable microwave irradiated *Acacia* seed (after Tran and Cavanagh, 1980). (B) SEM of ruptured lens in microwave irradiated *Acacia* seed.

1. Dormancy—Structural 21

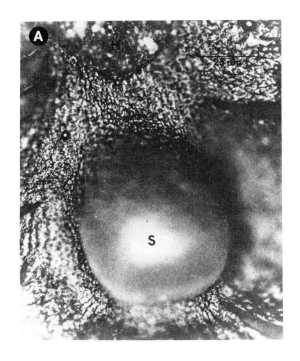

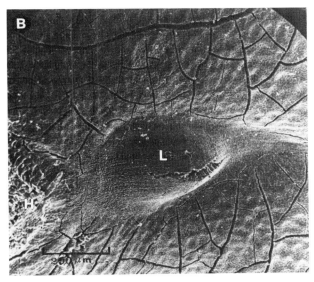

the cap is poorly organized and does not adhere tightly to the palisade cells; such seeds absorb water freely through the chalazal slit or chalazal pore.

The role of the subpalisade cells was unclear until the work of Egley and Paul (1981, 1982) with *Sida spinosa*. They found in after-ripened permeable seeds, the first sign of imbibition was the development of a raised 'blister' in the chalazal area caused by the separation of the palisade of the seed coat from the underlying tissue. Seeds without a raised blister did not imbibe. Similar blisters have been previously observed in cotton (Christiansen and Moore, 1959) and velvetleaf (La Crox and Stanisforth, 1964) where the presence of blisters correlated well with loss of impermeability. Egley and Paul (1981, 1982) found that the single layer of subpalisade cells, which are unique to the area of potential blister formation, were involved in the sequence of events leading to water uptake by the seed. The lateral walls of these cells have thin and poorly developed sections at the junction with the palisade layer and these sections provide predetermined weak sites which break, permitting palisade separation and subsequent water imbibition.

While there is strong evidence that the mechanism of water entry is the same in certain malvaceous and mimosoid seeds, there are several aspects which require further investigation: (i) what causes the palisade layer to separate from the underlying cells? (ii) how widespread is the presence of weak subpalisade cells in seeds with coat-imposed dormancy? and (iii) what is the mechanism by which hard seeds not possessing a lens or chalazal discontinuity become permeable?, for example, certain *Parkia* species (Corner, 1951).

D. Nature and Depth of Impermeability

1. Nature of Impermeability

Many authors have speculated on how impermeability develops in the drying seed and it may well be that a number of mechanisms can operate in different families. We now know that seeds become impermeable in the last stages of maturation and that the degree of impermeability is related to the seeds moisture content (Quinlivin, 1971) or to genetics (Rolston, 1978) or to the presence of pigments in the seed coat (Marbach and Mayer, 1974; Section 1.III,C of Volume 1; Werker *et al.*, 1979; Egley *et al.*, 1983), though the last is not universally accepted (Slattery *et al.*, 1982). Closure of all openings in the coat and impregnation of the cell walls with water-repellent substances are claimed to be necessary requirements for impermeability (Werker, 1980/1981). The hydrophobic substances detected by staining techniques and assumed to produce impermeability include cutin, lignin, pectin, suberin and perhaps callose and hemicellulose.

Raleigh (1930) and Corner (1951) believed that impermeability was due to contraction and presumably closer packing of the cells in the palisade layer of the coat as the seed dried out. This is a purely mechanical process and the idea has apparently received little support. The majority of the mechanisms listed by Werker (1980/1981) rely on some type of chemical substance to prevent water entry. The major difficulty with the latter hypothesis is that in the few cases where structure and histochemistry of the seed coats of various species have been compared (Coe and Martin, 1920; Raleigh, 1930; Watson, 1948; Stienswat *et al.*, 1971), it has not been possible to detect any chemical or physical difference between permeable and impermeable seeds. Moreover, there is disagreement in the literature concerning the substances actually present in the cells of the seed coat. Aitken (1939) found suberin in the outer ends of the Malpighian cells of *Trifolium subterraneum* whereas Slattery *et al.* (1982) detected only hemicellulose and no suberin. Additional information on the development of impermeability as a function of the stage of maturation was recently provided by Egley *et al.* (1983) for seeds of *Sida spinosa*. Seeds at 'stage 3' (15 to 16 days post anthesis) already possessed impermeable coats, even though the coats were pale brown in colour, and soft, with seed moisture content 115% of dry matter content. Between 'stage 3' and 'stage 4' (17 to 18 days post anthesis), the seeds lost water rapidly (water content 14.3% of dry matter), and became fully impermeable, mechanically hard and dark brown in colour. Evidently none of these parameters is by itself a reliable indicator of the development of impermeability.

Egley *et al.* (1983) employed a variety of histochemical and immunocytochemical tests to follow the course of conversion of soluble to insoluble (i.e., water repellent) compounds in the cells of the seed coat. They implicated peroxidase [EC 1.11.1.7] as an enzyme responsible for pigment formation, rather than polyphenol oxidase [EC 1.14.18.1] as in the seed coats of *Pisum elatius* (Marbach and Mayer, 1974; Section 1.III,C of Volume 1). The highest peroxidase activity coincided with the stage at which maximum conversion of soluble phenolic compounds to insoluble polyphenolic compounds was found ('stage 2', 13 to 14 days post anthesis). The most intense peroxidase activity was localized in the outer regions of the palisade (Malpighian) cells, with little activity evident elsewhere in the seed.

Although Egley *et al.* (1983) did not speculate on the mechanisms involved, their observations add weight to the view that the palisade layer is responsible for impermeability of the seed coat. We are forced to conclude that knowledge of the nature of impermeability is currently incomplete and the whole problem requires further study.

2. *Depth of Impermeability*

It is perhaps not surprising, given the current state of uncertainty regarding the nature of impermeability, that knowledge concerning the depth to which impermeability extends is also incomplete. The present view is that the Malpighian cell layer prevents water entry, and it is likely that the whole cell in many species is impermeable (Tran and Cavanagh, 1980). Implicit in such a view is the assumption that the cells below the Malpighian cells are permeable, although the results of McKee *et al.* (1977) call this into question. They found that in *Coronilla varia* the coat needed to be punctured to below the mesophyll cells to produce 'substantial' permeability. A relatively simple method of determining depth of impermeability involves drilling or grinding in successive steps to known depths below the surface and observing the seed's subsequent swelling in water. The drilling technique of Tran and Cavanagh (1980) is especially useful as accurate control can be maintained over depth and diameter of the drilled hole. The results from a limited number of investigations (Table II) clearly indicate that impermeability extends well below the light line, probably at least to the base of the Malpighian cells, although additional information is required on the relative permeability of the inner seed coat layers.

V. MECHANICAL STRENGTH OF THE SEED COAT

The mechanical strength of the seed coat is considered by Muller (1914) and Crocker *et al.* (1946) to restrict germination. Muller (1914) measured the strength of a number of rings or strips obtained from seed coats. Crocker *et al.* (1946) obtained the internal pressure necessary to rupture the seed coat of black walnuts (*Juglans nigra* L.), hickory nuts (*Carya ovata* K.) and butternuts (*Juglans cinerea* L.) directly. Tran and Cavanagh (1980) gave the compression force at fracture of a number of *Acacia* seeds. If enough seeds are compressed, it is possible to construct fracture load and deformation histograms which give an indication of the strength and elastic property of the seed coat (Tran, 1979; Tran and Cavanagh, 1980). The use of compression tests on agricultural produce has been extensively studied with a view to determine the minimum force causing damage due to mechanical handling.

Recently we measured the tensile strength of the seed coat by modifying a small balance sensitive to within 1 g. Thin strips of seed coat (about $200 \times 800 \times 4000$ μm) were carefully removed from a seed by using a fast hobbyist grinder and a dissecting knife. These were mounted in small rectangular aluminium blocks with quick acting glue before testing. Tensile force was then applied to each strip via the aluminium blocks. The stress (force/area) or the mechanical strength of each strip was calculated from the applied force, measured in newtons, N, and the cross sectional area, m^2,

Table II. Measured depths of impermeability

Species	Malpighian cells		Depth of damage to cause significant permeability, μm	Comments	Reference
	Total length μm	Depth to LL μm			
Gleditsia triacanthos	135	28	Base of Malpighian cell	—	Cavazza (1950)
Phaseolus lunatus	—	—	Base of Malpighian cell	Base of cells lignified. Fats impregnated between cells	Baciu-Miclaus (1970)
Trifolium subterraneum	45	6–7	13–24	Achieved 70%–85% germination in 20 days	Ballard (1973)
Prosopis farcata	100–120	35–50	Not given	Damage to half depth of cells gave erratic results	Werker et al. (1973)
Coronilla varia	45–50	15	> 80	Pierced to 62 μm → 43% germination. Pierced to 82 μm → 93% germination	McKee et al. (1977)
Acacia farnesiana	168	88	> 155	Seeds unswollen after 5 weeks	Tran and Cavanagh (1980)

LL = Light Line

at breaking. The unit of stress is expressed in pascals (Pa = 1 N/m²) or Megapascals. In the following sections, the results obtained are given and their implications discussed in detail.

A. The Seed Coat under Compression

Each *Acacia* seed was compressed between parallel plates (Arnold and Mohsenin, 1971) in a 5 tonne compression machine at a slow deformation rate of 0.8 mm min^{-1} so that the laws of statics still applied. Fracture was indicated by a sharp and distinct sound. The compressive load and deformation were recorded at breaking. It was not possible to calculate the compressive stress in this study as the seed is curved and the surface area of contact changes under load.

The force necessary to fracture an *Acacia* seed varies from species to species. For instance, on the average, an *Acacia longifolia* seed can withstand up to 7 kg of compressed load whereas 10 kg is the limit for a seed of *Acacia sophorae*. *Acacia* seeds are quite elastic as evidenced by the linear relationship between load and deformation (Fig. 9). The deformation varies linearly with compressive load up to the fracture point as shown (Fig. 9C). Seeds of *A. longifolia* and *A. sophorae* could be compressed up to 65% of the average fracture deformation without rendering them permeable to water (Tran, 1979) while all fractured seeds imbibed but never germinated.

The elastic property of a seed is related to the ratio load, L, to deformation, D. It is interesting to find that *A. longifolia* and *A. sophorae* have about the same L/D (0.34–0.37). Since the cotyledons tend to crumble early, the elastic property of an *Acacia* seed is predominantly that of the seed coat. It should be noted, therefore, that for two seeds having the same L/D, the seed with a thicker coat is considered to be stronger. But seed shape also played an important role. Referring to Figure 9, when a compressive load is applied to a seed, its effect at any point on the seed coat can be resolved into a tensile component, F_t, and a compressive component, F_c. On comparing components F_t in Figures 9A and 9B, it can be seen that the tensile component is greater for a seed with a larger radius of curvature. It should be remembered that the tensile component is responsible for rupturing the seed coat across its length as shown in Figure 9D. Therefore, before attempting to make a comparison between the thickness of seed coats and the degree of impermeability, two other important factors, the elastic property (L/D) and the seed shape, must also be considered. A meaningful comparison can only be made among seeds having the same shape and the same L/D. For this reason, authors such as

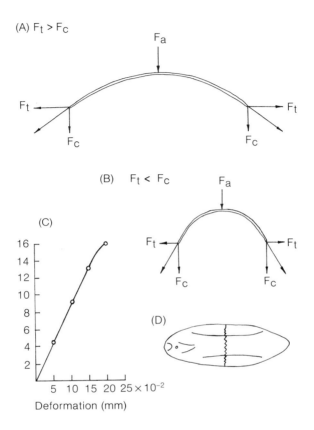

Fig. 9. *Acacia* seeds under compression. (A) and (B) show effects of seed shape on the resultant components of an applied load F_a; F_t = tensile component, F_c = compressive component: in (A) $F_t > F_c$, in (B) $F_t < F_c$, (C) Load-deformation curve for a seed of *A. sophorae*. (D) Mode of fracture of *Acacia* seeds under compression (from Tran and Cavanagh, 1980).

Zimmermann (1936), Aitken (1939) and Watson (1948) found no obvious relationship between seed coat thickness and degree of impermeability. On the other hand, Tran (1979) and Tran and Cavanagh (1980) showed that an inverse relationship was apparent between the seed coat thickness and the degree of impermeability (as measured by swelling) for several *Acacia* species (Table III).

An alternative method of comparing the degree of impermeability among seeds of different species having the same shape and elastic property is by applying the concept of maximum deformation energy (Tran and Cavanagh, 1980). Maximum deformation energy is given by the area under

Table III. Seed coat thickness and swelling percentage of *Acacia* seeds[a]

Species	L/D (N/m)	Seed dimensions (mm)	Seed coat thickness (μm)	T_M (μm)	Swelling (%)
A. longifolia	0.37	1.90 × 2.30 × 4.24	165	87	25
A. sophorae	0.34	2.57 × 3.47 × 4.55	242	104	4
A. retinodes	0.45	1.57 × 2.65 × 4.60	150	90	75
A. pycnantha	0.45	1.74 × 2.98 × 5.10	182	110	52
A. buxifolia	0.49	1.69 × 2.60 × 4.72	160	78	75
A. calamifolia	0.48	1.92 × 3.20 × 5.04	233	120	52
A. brachybotrya	0.54	2.22 × 4.05 × 5.60	285	150	24
A. melanoxylon	0.56	1.34 × 2.65 × 4.58	192	93	56

[a] Swelling percentages were recorded at 7 days. L/D, load/deformation; t_M, thickness from cuticle to base of Malpighian cells.
Data of Tran (1979).

a load and deformation curve similar to the curve shown in Figure 9C. When the load is proportional to deformation, maximum deformation energy is simply given by half the product of the load and deformation at fracture. Table IV gives some typical results of a comparison between *A. longifolia* and *A. sophorae* seeds obtained from several sources. It can be seen that *A. sophorae* has a higher maximum deformation energy and a slower rate of germination than *A. longifolia*, indicating an apparent correlation between deformation energy, strength of the seed coat and degree of impermeability.

B. The Seed Coat under Tension

Muller (1914) measured the tensile strength of strips of *Ricinus* coats and rings of *Corylus* and *Pinus* shells. From the tensile breaking strength, he calculated the internal pressure necessary to break the intact coats. He also found that the average breaking strength of a dry *Corylus* shell was more than double the average strength exerted by a growing embryo. The shell strength, however, was greatly reduced by imbibition of water. In fact, this weakening was sufficient to bring the breaking strength of a shell below that which could be exerted by a growing embryo. Crocker et al. (1946) found that the shells of black walnuts broke with lower pressure when dry than when wet and the average breaking pressure decreased when the walnuts were stored in a moist soil at 11°C, 17°C, 23°C and 28°C. But significantly

Table IV. Comparative study of maximum deformation energy and swelling percentage for different seed lots 10 days after hot water treatment. Swelling took place at 25°C in the dark.

Seed lot[a]	Moisture content (%)	Swelling percentage (control)[a]	Swelling percentage 96°C hot water[b]	Fracture load (L) (kg)	Deformation (D) (10^{-2} mm)	Maximum deformation energy (J)
L(1)	8.2	2	88	6.9 ± 2.1	20.6 ± 3.2	71.0 × 10^{-2}
L(2)	8.1	6	83	7.2 ± 2.7	18.7 ± 3.2	67.3 × 10^{-2}
L(2)[c]	—	0	47[c]	8.7 ± 2.0	23.2 ± 4.1	100.9 × 10^{-2}
L(3)	7.0	6	45	7.8 ± 3.4	21.9 ± 5.0	85.4 × 10^{-2}
S)1)	7.7	9	27	10.0 ± 3.7	28.8 ± 6.8	144.0 × 10^{-2}
S(2)	7.2	8	10	10.6 ± 3.6	30.1 ± 6.2	159.5 × 10^{-2}
S(3)	7.2	7	7	11.07 ± 3.2	35.6 ± 7.5	197.0 × 10^{-2}

[a] L, *Acacia longifolia*; S, *Acacia sophorae*. (1) Forestry Commission of New South Wales; (2) Forest Commission, Victoria; (3) Nindethana Seed Service, Western Australia.
[b] At least two replicates of 100 seeds except where otherwise indicated.
[c] Unaffected seeds after microwave treatment, one replicate of 100 seeds.
Data condensed and revised from Tran and Cavanagh (1980).

the average breaking pressure did not greatly alter with storage at 6°C. While these findings may have some bearing on the impermeability of the seed coat and hence the longevity of buried seeds, they do not necessarily indicate that the strength of the seed coat has a dominant effect in restricting an imbibed or growing embryo.

The results of our experiments show that the tensile strength of the seed coat of *Albizzia lophantha* is 29.6 MPa, of *Acacia farnesiana* 26.6 MPa, of *A. sophorae* 28.7 MPa and of *A. longifolia* 17.3 MPa. It is interesting to note that *A. sophorae* has a stronger seed coat than *A. longifolia* and *A. fornesiana*. We have also confirmed the findings of Muller (1914) concerning imbibed seed coats. The tensile strength of the imbibed seed coat of *A. longifolia* is only 2.7 MPa and that of *A. sophorae* is 7 MPa after 30 minutes wetting.

C. Thin Shell Theory Applied to the Seed Coat

When examining the rupture of the lens (strophiole) of some 20 species of *Acacia* seeds under microwave and heat treatments, Tran (1979) considered that modification at the lens is predominantly mechanical and is caused by internal pressure and the different rates of expansion in a seed. The observed ruptured, raised or swollen lens and a network of cracks on a treated seed coat are evidence of mechanical stresses. Tran (1979) postulated that a seed coat behaves structurally like a thin shell. By approximating a seed to an ellipsoid of revolution, he applied the theory of thin shells used in structural analysis (Roake and Young, 1975) to compute the hoop stress and the meridian stress which would develop all over the seed coat. It can be shown that the hoop stress is given by

$$\sigma_t = pr_2(1 - r_2r_1)/t \qquad (1)$$

and the meridian stress is given by

$$\sigma_m = pr_2/t \qquad (2)$$

where p is the internal pressure caused by internal vapour pressure and different rate of expansion in the seed, t is the thickness of the seed coat and r_1, r_2 are the radii of principal curvature at a point on the surface of revolution. Computation shows that both σ_t and σ_m are tensile in all *Acacia* seeds examined (Fig. 10). Therefore, it is conceivable that under uniform or localized heating, the tensile stresses developed could cause the weak area of a seed coat such as the lens or chalazal discontinuity to erupt and become permeable to water.

Lush and Evans (1980) put the theory to an approximate test by assuming each seed to be a sphere to simplify the calculation of r_1 and r_2. Accordingly the ratio r/t is a direct measure of the meridian stress, σ_m. The radius, r, was obtained from the cube root of seed weight and the thickness, t, was estimated from the specific weight of the seed coat where it was not available (Lush and Evans, 1980). The ratio r/t ranged from 0.21–0.51 in wild legumes. Higher values of r/t were found for all domesticates except in an accession of *Lupinus angustifolius* (narrow leafed lupin) and one of *Cicer arietinum* (chick pea); these seeds were normally permeable. From their own data and from Tran's (1979) and Zimmermann's (1936) data, Lush and Evans found that coats of legume seeds which have a low r/t ratio were likely to be impermeable to water. Therefore they concluded that it was not seed coat thickness alone but thickness relative to seed curvature (i.e., shape) that was the important measure of the degree of impermeability. It must be pointed out here that the elastic property of the seed coat has been implicitly assumed to be similar in this evaluation.

A more accurate although difficult test for the theory would be by direct measurement of the tensile strength, the radius of curvature and the seed coat thickness at the lens or chalazal discontinuity.

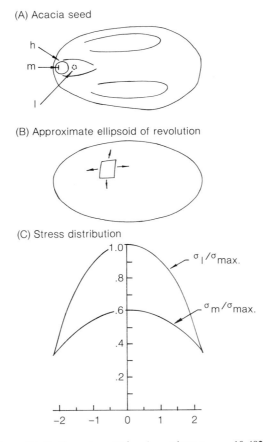

Fig. 10. Tensile stress distribution σ_t/σ_{max} and σ_m/σ_{max} where σ_{max} = 10.482 for 1 unit pressure for an *Acacia* seed 4.4 mm long and 2.5 mm wide. h = hilum, l = lens, m = micropyle (from Tran, 1979).

VI. TREATMENT TO BREAK DORMANCY

A. Natural Breakdown of Impermeable Seeds

There have been no comprehensive investigations of the mechanisms of natural breakdown of hard seeds even though most artificial treatments are probably based on speculative observations of the effects of natural agents on the seed. When impermeable seeds become permeable naturally, it is usually assumed that the seed coat has been damaged in some way, for instance, by fire, by abrasion due to wind and water movement over the ground, by microbiological attack in the soil, by high and fluctuating temperatures, or by passage through the digestive tracts of birds and animals. The effects of some of these on *Acacia* have been reviewed recently by Cavanagh (1980).

Simple observation of affected seeds without comprehensive anatomical investigation has often led to false conclusions being drawn. Thus as fire or heat produces a network of cracks, for example, in *Acacia* (Fig. 2; Cavanagh, 1980), in *Albizzia julibrissin* (Gogue and Emino, 1979) in *Coronilla varia* (Brant *et al.*, 1971) and *Rhus ovata* (Stone and Juhren, 1951), it is usually considered that water enters the hard coat at these points. However, sections through cracked seed coats reveal that the cracks frequently do not penetrate to the base of the Malpighian cells and thus water entry through them is unlikely. It is now well established in *Acacia* (Brown and Booysen, 1969; Tran, 1979; Cavanagh, 1980), *Albizzia* (Dell, 1980), *Abutilon theophrasti* (Winter, 1960; La Croix and Stanisforth, 1964), *Crotalaria* (Egley, 1979; Saha and Takahashi, 1981), *Gossypium* (Christiansen and Moore, 1959) and in many papilionoid legumes (Ballard, 1973; Ballard *et al.*, 1976) that heat causes a rupture of the seed coat at the lens or chalazal discontinuity and that this is the first site of water entry.

Fluctuating temperatures in the field have been shown by Quinlivin (1971) to render papilionoid legumes permeable by causing a fracture at the lens tissue, but an attempt by Preece (1971) to achieve similar softening in *Acacia aneura* was unsuccessful. Seeds of *Acacia karroo* (Robbertse, 1974) and *Acacia bidwillii* (A. K. Cavanagh, unpub. data) collected off the ground had damaged lens tissue and were permeable, giving up to 80% germination. It is assumed that the action of soil bacteria is responsible for such damage. Gogue and Emino (1979) found by germinating seeds of *Albizzia julibrissin* in soil containing various fungi that *Rhizoctonia* sp. in particular was able to raise germination from 8.7% in controls to 37.1% in 90 days. Scanning electron microscope (SEM) photographs showed the fungi had altered the seed surface in random areas and had not necessarily attacked specialized sites such as the hilum or lens.

The role of birds and animals in distributing seed and enhancing its germination has been known for many years (Krefting and Roe, 1949; Janzen, 1971). Once again, there is little information concerning the effects on the seed coat of passage through the digestive tract though it is usually supposed that in birds at least, the grinding action of grit in the gizzard is equivalent to mechanical scarification. Also, stomach acids are thought to corrode the coat and allow fluid penetration. Krefting and Roe (1949) showed that of 16 species of seeds fed to captive birds, 8 had marked improvement in germination after passage through the birds' intestines. Weight losses of up to 20% were recorded as compared with controls and this was attributed to seed coat thinning. In a recent SEM study, Smith (1981) found that the thickness of the seed coat was important in protecting the seed from damage and that smaller seeds were more likely to be degraded due to longer retention in the gizzard. The major damage suffered by the seed appeared to be mechanical scarification and random seed coat thinning though the author did not examine the area of hilum–lens–chalazal discontinuity.

More detailed information is available on the role played by animals in dispersing and germinating seed (Table V). Once again, there is little direct knowledge of the effects of such digestion on the seed coat. Cavanagh (1980) speculated that because germination enhancement was relatively low in most cases (frequently less than 40%), the seed probably suffered relatively little damage in the mouths of larger animals, such as elephants (*Loxodonta africana*) and cattle, though observation by Halevy (1974) with gazelles and Jarman (1976) with impala (*Aepyceros melampus*) indicated that they destroyed most of the seed consumed. Seed held for considerable periods of time in the intestines of animals, for example, six days in the case of cattle (Wickens, 1969), would be subject to possible enzymic action and degradation. Consequently, it is likely that coat damage and chemical degradation occur and that both are responsible for enhanced germination observed after animal ingestion.

B. Artificial Treatments

The methods used to render 'hard' seeds permeable have been described by Porter (1949) and Barton (1965a) and recently reviewed by Rolston (1978). Essentially, most methods are concerned with damaging the seed coat, or a specific area of the seed coat, and thus facilitating water entry. The availability of water can initiate the process of germination, although according to Khan (1977), it is likely that practices which weaken or damage the coat will also bring about other changes, for instance, increased sensitivity to light and temperature, permeability to gases, and removal of inhibitors. Similarly, it is often believed that the mechanical strength of the coat or of surrounding covers such as pericarp, glumes and lemmas acts as a restraint to germination and radicle penetration, or prevents adequate expansion and development of the embryo. Ballard (1973) cautions against too-ready acceptance of the idea that mechanical restriction is responsible for delayed germination, pointing out that: (i) seed coat properties have rarely, if ever, been measured; (ii) the expanding radicle can generate considerable force

Table V. Per cent germination after ingestion by birds and animals

Species	Dispersal agent	% germination Control	% germination After passage	% Bruchid damage	Reference
Acacia giraffae	Eland[a]	45%–49%	60%–66%	—	Leistner (1961)
	Elephant[b]	92%	78%	—	
Acacia albida	Cattle	85% at 90 days	85% at 90 days	—	Wickens (1969)
Acacia raddiana	Dorcas gazelle[c]	1%	13% ± 4%	—	Halevy (1974)
	Dorcas gazelle	4%	21%	72%	
Acacia tortilis subsp. *spirocarpa*	Impala[d]	—	8%	26%	
	Thompsons gazelle[e]	—	12%	31%	
	Dikdik[f]	—	11%	45%	Lamprey et al. (1974)
	Elephant[b]	—	19%	22%	
	Dorcas gazelle[c]	—	13.2%	—	
Acacia cyclopis	Various	9%	30%	—	Glyphis et al. (1981)
Amaranthus palmeri		95%	11%	—	
Rhus glabra	Starling[g]	0%	20%	—	Smith (1981)
Abutilon theophrasti		1%	10%	—	

[a]*Taurotragus oryx*; [b]*Loxodonta africana*; [c]*Gazella dorcas*; [d]*Aepyceros melampus*; [e]*Gazella thompsonii*; [f]*Rhynchotragus kirkii*; [g]*Sturnus vulgaris*.

over the few μm^2 of its tip; (iii) in many cases, the embryo may also be in a state of deep physiological dormancy, so that it is not always apparent whether coat or embryo properties are playing a dominant role in preventing germination.

1. Effects of Treatment on the Seed Coat

Except in a few instances, there has been little critical evaluation of how the treatment acts on the coat. Most experiments have been concerned with enhancing percentage germination rather than with understanding the basic mechanisms involved, consequently our knowledge of effects is much poorer than our knowledge of results. Additionally, most studies have been made on unsorted material containing both permeable and non-permeable seeds. A more realistic assessment of treatment effectiveness would be obtained by working with samples containing only known impermeable seeds obtained by soaking beforehand (McKee *et al.*, 1977; Tran, 1979; Tran and Cavanagh, 1980).

Seed coat treatments can be divided into 'wet' and 'dry' treatments:

Wet
Chemical action — sulphuric acid, alcohols, acetone, oxidizing agents.
Thermal action — hot or boiling water, liquefied gases.

Dry
Mechanical action — manual and mechanical scarification, impaction, high pressure.
Thermal action — dry heating, using radiant or electromagnetic waves, field temperature fluctuations.

a. Wet treatments

In general, alcohols, acetone and organic solvents have been relatively ineffective or have caused high seed mortality (Barton, 1947; Brown and Booysen, 1969). Verschaffelt (1912) believed that water entry was facilitated by penetration of alcohol through cracks in the seed coat or through the pleurogram* of mimosoid seeds. Results of other studies have been inconclusive (Barton, 1947; Cavazza, 1951) though Brown and Booysen's work with *Acacia* indicates that the hilum and/or lens is affected in some mimosoid seeds. More aggressive reagents such as methanol, corrode random holes through the coat and sulphuric acid is also believed to behave similarly, for example, in *Coronilla varia* (Brant *et al.*, 1971) although the counter palisade tissue of the hilum of lupins, however, can be hydrolysed,

---

* *Pleurogram.* The pleurogram is a characteristic groove on both sides of many mimosoid seeds whose function is unknown. It is not normally permeable though Hyde investigated its function as a one-way valve during seed desiccation (Gunn, 1981).

presumably permitting water entry (Burns, 1959). In recent years, sulphuric acid has fallen into disuse, mainly because of the dangers of handling it, but also because it offers few advantages over other techniques such as soaking in hot water. Horn and Hill (1974) reported that *Lupinus cosentinii* required up to 7 hours soaking to become permeable, and as much as 12 hours may be required in other species with very thick coats. Brant *et al.* (1971) showed after SEM examination that sulphuric acid exposed the lumens of the Malpighian cells in *Coronilla varia* and considered that such damage allowed the seeds to become permeable. However, they did not determine the actual depth of impermeability. Freezing treatments with liquid air, liquid nitrogen and liquid oxygen have been successful in reducing impermeable seeds in several legumes (Barton, 1947; Brant *et al.*, 1971) but not in others (Al-Saadawi and Abdul-Wahab, 1977; Liu *et al.*, 1981), whereas freezing in ice has no effect (Brown and Booysen, 1969). Brant *et al.* (1971) speculated that differential expansion and contraction of the seed coat in both boiling water and liquid nitrogen caused cracks in the palisade layer to the base of the Malpighian cells, thus permitting water entry.

Hot, and boiling water have long been known as effective agents for reducing impermeability (Bowie, 1832; Thornber, 1903). It is usually considered that boiling 'softens' the seed coat, allowing random water uptake, whereas in reality, boiling, like dry heat or microwave heating, has little effect on the properties of the coat (Fig. 11) and renders the seed permeable through action on the lens (Tran and Cavanagh, 1980). Sufficient evidence is now available to indicate that the lens and the chalazal discontinuity are widespread in the plant kingdom and these features perform the specific role of providing predetermined points of weakness which rupture under treatment (Sections III, IV).

b. Dry treatments

Microscopic examination of scarified seeds shows that, as expected, the seed coat is damaged to varying degrees by the process. Any mechanical impact might be expected to separate the highly stressed cells at the lens of papilionoid legumes (Section IV,C,3) but this is less effective with other legumes (Barton, 1947; A. K. Cavanagh, unpubl. res., 1980). The role of alternating temperature in breaking impermeability has been known since at least the work of Nakajima in 1933 (cited by Ueki and Suetsugu, 1958). Quinlivin (1971) showed in parallel field and laboratory trials that impermeable seeds which become permeable naturally do so at the lens area. The critical factor which determined the rate of breakdown was the maximum daily temperature to which the seeds had been exposed. There is almost no knowledge of the part played by temperature fluctuation in reducing impermeability of other types of seeds.

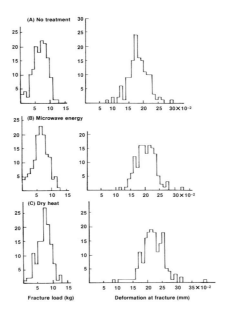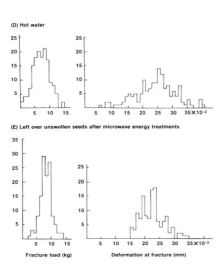

Fig. 11. Effects of treatment on fracture load and deformation histograms of *Acacia longifolia*; (A) no treatment, (B) microwave energy, (C) dry heat, (D) hot water, (E) unaffected seeds after microwave energy treatment (from Tran and Cavanagh, 1980).

Nelson (1965) reviewed the use of various radiation and electrical processes in enhancing germination but such studies have been characterized by an almost complete lack of a basic understanding of the physiological and structural effects of the treatments. The situation has been partly remedied by Tran (1979, 1981) in his work with microwave energy and its effect on *Acacia* seeds. Nelson (1976) believes that radio and microwave frequency treatments produce only a rapid heating effect whereas others consider that non-thermal effects may be present (Wayland *et al.*, 1972). Adverse effects observed include molecular breakdown of enzyme structure due to resonance (Crawford, 1977) and partial protein denaturation and inactivation of key enzymes (Hooper *et al.*, 1978). Ballard *et al.* (1976) established the effectiveness of high frequency heating on papilionoid legumes and demonstrated with serial photography that the lens was the first site of water entry. Treated seed retained viability and a high level of germination after up to 14 years of controlled storage (Nelson *et al.*, 1976). However, the treatment, though simple and inexpensive, is not effective on all impermeable seeds (Nelson *et al.*, 1978).

VII. CONCLUDING REMARKS

The work reviewed in this chapter reveals gaps in our knowledge of several aspects of coat-imposed dormancy. The water-impermeable coat protects the embryo from adverse storage and environmental conditions and actively promotes seed longevity. However, we do not know whether impermeability is due to mechanical processes (e.g., shrinkage and closer packing of the cells in the seed coat as the seed matures) or to chemical effects (e.g., impregnation of the cell walls with hydrophobic substances). A combination of both may exist, and the operative mechanisms may differ among various species. There is a need for an accurate histological examination of the cells of the coats of impermeable seeds to update our knowledge of the constituents which are present. We also need to know whether chemicals which are present are responsible for impermeability and the depth to which impermeability exists. The strength and other mechanical properties of the coat need to be measured for many more species to determine if mechanical constraints to embryo expansion or radicle extension prevent germination. Furthermore, it will be necessary to assess the change in coat properties during the course of imbibition and germination as limited data indicate that the process of imbibition leads to weakening of the coat and this may remove any mechanical restrictions that are operating.

Our current understanding of seed coat structure suggests that at least in many species of the Leguminosae and Malvaceae, water entry is prevented by the near continuous layer of Malpighian cells. Potential openings such as the hilum, micropyle, lens and chalazal discontinuity are sealed in impermeable seeds but the latter two form a region of structural weakness which can be disrupted either naturally or by several specific treatments, thus rendering the seed permeable.

There is also evidence from a small number of studies that the lens in mimosoid seeds and an area near the chalazal discontinuity in malvaceous seeds are underlain by a single or double layer of thin-walled, sub-palisade cells whose walls break readily under stress. The palisade layer (of Malpighian cells) can then lift, exposing the underlying permeable tissue. Further research is needed to determine how widely distributed these cells are in other impermeable seeds and to clarify what causes the palisade cell layer to lift in the first place. No work has been done on the means by which impermeable seeds, not containing a structural weakness, become permeable. In papilionoid seeds, the hilum functions as a hygroscopically activated, one-way valve which permits drying out of the seed during maturation. Limited work with several mimosoid seeds indicates that during drying a significant amount of moisture can be lost through the lens. But what is the mechanism by which seeds which do not possess a papilionoid hilum or a mimosoid lens dry out and become impermeable?

Although we have only mentioned the problem in passing, there have been few attempts to correlate the degree of impermeability known to be related to genetic and environmental effects with the structure and biochemistry of the seed coat. For instance, why do some lines of domestic legumes contain high percentages of impermeable seeds and what is the mechanism by which such 'hardseededness' can be increased or bred out of a particular strain? Are predominantly permeable seeds from cultivated species chemically or structurally different to seeds from predominantly impermeable lines, or for that matter, are they both different to wild seeds?

The mechanism(s) of natural breakdown of the seed coat and the effects of passage through birds and animals are two virtually unexplored areas. The available reports of the effects of artificial treatments on the seed coat are sometimes vague and frequently contradictory. There is a need for co-operative research between workers in various fields in order to understand these problems and to minimize unnecessary 'proliferation of speculation and hypotheses' (Werker, 1980/1981).

ACKNOWLEDGEMENTS

The authors are grateful to Mrs Coral Milligan for her able typing of the manuscript and to Phil Juler for reading an early draft..

REFERENCES

Aitken, Y. (1939). *Proc. Roy. Soc. Vic.* (NS) **51**, 187-212.
Al-Saadawi, I. S., and Abdul-Wahab, A. S. (1977). *Bull. Coll. Sci. Univ. Baghdad* **18**, 27-39.
Arnold, P. C., and Mohsenin, N. N. (1971). *Trans. ASAE* **14**, 74-84.
Baciu-Miclaus, D. (1970). *Proc. Int. Seed Test. Assoc.* **35**, 599-617.
Ballard, L. A. T. (1973). *Seed Sci. and Technol.* **1**, 285-303.
Ballard, L. A. T., Nelson, S. O., Buchwald, T., and Stetson, L. E. (1976). *Seed Sci. and Technol.* **4**, 257-274.
Barton, L. V. (1947). *Contrib. Boyce Thompson Inst. Plant. Res.* **14**, 355-362.
Barton, L. V. (1965a). *In* 'Encyclopaedia of Plant Physiology', XV(2) (W. Ruhland, ed.), pp. 727-745. Springer-Verlag, New York.
Barton, L. V. (1965b). *In* 'Encyclopaedia of Plant Physiology', XV(2) (W. Ruhland, ed.), pp. 1058-1085. Springer-Verlag, New York.
Barton, L. V. (1967). 'Bibliography of Seeds.' Columbia University Press, New York.
Barton, L. V., and Crocker, W. (1948). 'Twenty Years of Seed Research at the Boyce Thompson Institute for Plant Research', pp. 36-42, 70-75. Faber and Faber, London.
Becquerel, P. (1907). *Ann. Sci. Nat. Bot.*, Ser. IX, **5-6**, 193-311.
Becquerel, P. (1934). *C.R. Acad. Sci.* (Paris) **199**, 1662-1664.
Bentham, G. (1864). 'Flora Australiensis.' Vol. II, p. 302: L. Reeve and Co., London.
Berg, R. Y. (1979). *Norwegian J. Bot.* **26**, 229-254.
Berggren, G. (1963). *Sv. Bot. Tidskr.* **57**, 377-395.
Berjak, P., and Villiers, T. A. (1972). *New Phytol.* **71**, 135-144.
Bhojwani, S. S., and Bhatnagar, S. P. (1979). 'The Embryology of Angiosperms' (3rd edn), pp. 207-210. Vikas Publishing House, New Delhi.
Boelcke, O. (1946). *Darwiniana* **7**, 240-321.
Bowie, J. (1832). *Gardeners Magazine* **8** (36), 5-9.
Brant, R E., McKee, G. W., and Cleveland, R. W. (1971). *Crop Sci.* **11**, 1-6.
Bravato, M. (1974). *Acta Botanico Venezuelica* **9**, 317-361.
Brown, N. A. C., and Booysen, P. de V. (1969), *Agroplantae* **1**, 51-60.
Burns, R. E. (1959). *Plant Physiol.* **34**, 107-108.
Cambage, R. H. (1928). *J. Proc. Roy. Soc. N.S.W.* **62**, 152-167.
Cavanagh, A. K. (1980). *Proc. Roy. Soc. Vic.* **91**, 161-180.
Cavanagh, A. K. (1981). *Bull. Int. Group Study Mimosoideae* No. 9, 61-68.
Cavazza, L. (1950). *Ber. Schweiz. Bot. Ges.* **60**, 596-610.
Cavazza, L. (1951). *Nuovo. Giorn. Bot. Ital.* **58**, 393-397.
Chalon, J. (1874). *Mem. Soc. des Sci. des Arts et des Lettres du Hainaut* **10**, 55-118.
Christiansen, M. N., and Moore, R. P. (1959). *Agronomy J.* **51**, 582-584.
Coe, H. S., and Martin, J. N. (1920). 'Sweet Clover Seed.' U.S.D.A. Bull. No. 844.
Copeland, L. O. (1976). 'Principles of Seed Science and Technology', pp. 121-148. Burgess Publishing Co., Minneapolis, Minnesota.
Corner, E. J. H. (1951). *Phytomorphology* **1**, 117-150.
Corner, E. J. H. (1976). 'The Seeds of Dicotyledons' Vols I and II. Cambridge University Press, Cambridge.
Crawford, A. E. (1977). *Seed Sci. and Technol.* **5**, 671-676.
Crocker, W. (1906). *Bot. Gaz.* **42**, 265-291.
Crocker, W. (1916). *Am. J. Bot.* **3**, 99-120.
Crocker, W., and Barton, L. V. (1953). 'Physiology of Seeds', pp. 114-139. Chronica Bot., Waltham, Massachusetts.

Crocker, W., Thornton, N. C., and Schroeder, E. M. (1946). *Contrib. Boyce Thompson Inst.* **14**, 173–201.
Dell, B. (1980). *Amer. J. Bot.* **67**, 556–563.
Egley, G. H. (1979). *Weed Sci.* **27**, 355–361.
Egley, G. H., and Paul, R. N. (1981). *Am. J. Bot.* **68**, 1056–1065.
Egley, G. H., and Paul, R. N. (1982). *Am. J. Bot.* **69**, 1402–1409.
Egley, G. H., Paul, R. N. Jr, Vaughn, K. C., and Duke, S. O. (1983). *Planta* **157**, 224–232.
Elliot, W. R., and Jones, D. L. (1980). 'Encyclopedia of Australian Plants Suitable for Cultivation' Vol. I, p. 201. Lothian Publishing Co., Melbourne.
Esau, K. (1977). 'Anatomy of Seed Plants' 2nd edn, p. 466. John Wiley and Sons, New York.
Evenari, M. (1980/1981). *Isr. J. Bot.* **29**, 4–21.
Ewart, A. J. (1908). *Proc. Roy. Soc. Vic.* **21**, 1–203.
Fahn, A. (1974). 'Plant Anatomy' 2nd edn, p. 534. Pergamon Press, London.
Gaertner, J. (1791). 'De Fructibus et Seminibus Plantarum' Vol. II (1791), p. 325. Schr. Tübingen.
Glyphis, J. P., Milton, S. J., and Siegfried, W. R. (1981). *Oecologia* **48**, 138–141.
Gogue, G. J., and Emino, E. R. (1979). *J. Amer. Soc. Hort. Sci.* **104**, 421–423.
Grew, N. (1682). 'The Anatomy of Plants'. Rawlings, London. Reprint edition, 1965, Johnson Reprint, New York.
Gunn, C. R. (1972). *In* 'Alfalfa Science and Technology' (C. H. Hasen, ed.), pp. 677–687. American Society of Agronomy Inc., Madison.
Gunn, C. R. (1981). *In* 'Advances in Legume Systematics' Part 2 (R. M. Polhill and P. H. Raven, eds), pp. 913–925. Proceedings of International Legume Conference, Kew 1978. Kew Botanic Gardens and Ministry of Agriculture, Fisheries and Food, London.
Gutterman, Y. (1978). *J. Arid Environ.* **1**, 141–144.
Hagon, M. W., and Ballard, L. A. T. (1970). *Aust. J. Biol. Sci.* **23**, 519–528.
Halevy, G. (1974). *Isr. J. Bot.* **23**, 120–126.
Hamly, D. H. (1932). *Bot. Gaz.* **39**, 345–375.
Hamly, D. H. (1935). *Bot. Gaz.* **96**, 755–757.
Hanna, P. J. (1984). *New Phytol.* **96**, 23–30.
Harpur, J. L. (1957). *In* 'Proceedings of Fourth International Congress on Crop Protection' Vol. I, pp. 415–420. Braunschweig, Hamburg.
Harrington, J. F. (1970). *In* 'International Biological Program Handbook' No. 11, pp. 501–521.
Harrington, J. F. (1972). *In* 'Seed Biology' (T. T. Kozlowski, ed.), Vol. III, pp. 145–245. Academic Press, New York.
Heydecker, W. (1977). *In* 'The Physiology and Biochemistry of Seed Dormancy and Germination' (A. A. Khan, ed.), pp. 237–282. North Holland Publishing Co., New York.
Hooper, G. R., Rice, R. P., Flegler, S. L., and Putman, A. R. (1978). *J. Amer. Soc. Hort. Sci.* **103**, 173–176.
Horn, P. E., and Hill, G. P. (1974). *J. Aust. Inst. Agric. Sci.* **40**, 85–87.
Hutchinson, J. M., and Ashton, F. M. (1979). *Amer. J. Bot.* **66**, 40–46.
Hyde, E. O. C. (1954). *Ann. Bot.* **18**, 241–256.
Isely, D. (1955). *Proc. Iowa Acad. Sci.* **62**, 142–145.
Janzen, D. H. (1971). *Annu. Rev. Ecol. System* **2**, 465–492.
Jarman, P. J. (1976). *East Afr. Wildlife J.* **14**, 223–225.
Karssen, C. M. (1980/1981a). *Isr. J. Bot.* **29**, 45–64.

Karssen, C. M. (1980/1981b). *Isr. J. Bot.* **29**, 65-73.
Khan, A. A. (1977). *In* 'The Physiology and Biochemistry of Seed Dormancy and Germination' (A. A. Khan, ed.), pp. 29-50. North Holland Publishing Co., Amsterdam.
Khudaire, A. K. (1956). *Physiol. Plant.* **9**, 452-461.
Kivilaan, A., and Bandurski, R. S. (1981). *Am. J. Bot.* **68**, 1290-1292.
Kondo, M. (1929). *Ber. Ohara Inst. Landiv. Forsch.* **4**, 289-293.
Kopooshian, H. A., and Isely, D. (1966). *Proc. Iowa Acad. Sci.* **73**, 59-67.
Kramer, A., and Szczesniak, A. S. (1973). 'Texture Measurement of Foods'. D. Reidel Publishing Co., Boston.
Krefting, L. W., and Roe, E. I. (1949). *Ecol. Mono.* **19**, 269-286.
Kyle, J. H., and Randall, T. E. (1963). *J. Amer. Soc. for Hort. Sci.* **83**, 461-475.
Lackey, J. A. (1981). *Bot. Gaz.* **142**, 160-164.
La Croix, L. J., and Stanisforth, D. W. (1964). *Weeds* **12**, 171-174.
Lamprey, H. F., Halevy, G., and Makascha, S. (1974). *East Afr. Wildlife J.* **12**, 81-85.
Lebedeff, G. A. (1947). *J. Agric. Res.* **74**, 205-215.
Leistner, O. A. (1961). *Koedoe* No. 4, 101-104.
Lersten, N. (1982). *Am. J. Bot.* **69**, 98-107.
Lesins, K. A., and Lesins, I. (1979). 'Genus *Medicago* (Leguminosae): A Taxogenetic Study', pp. 24-27. Dr W. Junk, The Hague.
Liu, N. Y., Khatamian, H., and Fretz, T. A. (1981). *J. Amer. Soc. Hort. Sci.* **106**, 691-694.
Lush, W. M., and Evans, L. T. (1980). *Field Crops Res.* **3**, 267-286.
McKee, G. W., Pfeiffer, R. A., and Mohsenin, N. N. (1977). *Agronomy J.* **69**, 53-58.
Maisel, A. (1909). *Bull. Soc. d'Hist. Nat. d'Autun.* **22**, 52-128.
Malpighi, M. (1687). *'Opera Omnia'* New edition, Leiden.
Marbach, I., and Mayer, A. M. (1974). *Plant Physiol.* **54**, 817-820.
Mattirolo, O., and Buscalioni, L. (1892). *Mem. della Reale Accad. delle Scienze di Torino Ser. 2* **42**, 223-318 and 359-445.
Mayer, A. M., and Poljakoff-Mayber, A. (1975). 'The Germination of Seeds' (2nd edn), pp. 46-67. Pergamon Press, London.
Miller, R. H. (1967). 'Crotalaria Seed Morphology, Anatomy and Identification'. USDA Tech. Bull. 1373.
Moore, C. E., and Moore, A. W. (1962). *J. Tennessee Acad. Sci.* **37**, 17-18.
Muller, G. (1914). *Jahrb. Wiss. Bot.* **54**, 529-644.
Nakajima, Y. (1933). *Proc. Crop Sci. Jap.* **5**, 443-459.
Nelson, S. O. (1965). *In* 'Electromagnetic Radiation in Agriculture', pp. 60-63. Proc-165, American Society of Agricultural Engineers, St Joseph, Michigan.
Nelson, S. O. (1976). *J. Microwave Power* **11**, 271-277.
Nelson, S. O., Heckert, R. M., Stetson, L. E., and Wolf, W. W. (1976). *J. Seed Technol.* **1**, 31-43.
Nelson, S. O., Bovey, R. W., and Stetson, L. E. (1978). *Weed Sci.* **26**, 286-291.
Netolitzki, F. (1926). *In* 'Handbuch der Planzenanatomie' Bld 10, Lief 14. (K. Linsbauer, ed.). Gebr. Borntraeger, Berlin.
Nikolaeva, M. G. (1977). *In* 'The Physiology and Biochemistry of Seed Dormancy and Germination' (A. A. Khan, ed.), pp. 51-74. North Holland Publishing Co., Amsterdam.
Nobbe, F. (1876). 'Handbuch der Samenkunde'. Wiegandt, Hempel und Parey, Berlin.
Ødum, S. (1965). *Dansk Botanisk Arkiv.* **24**(2), 1-70.

Osborne, D. J. (1977). *In* 'The Physiology and Biochemistry of Seed Dormancy and Germination' (A. A. Khan, ed.), pp. 319-333. North Holland Publishing Co., Amsterdam.
Pammel, L. H. (1899). *Trans. Acad. Sci. St Louis* **9**, 91-274.
Pearson, N. L. (1939). *J. Agric. Res.* **58**, 865-873.
Pitot, A. (1935). *Bull. Soc. Bot. France* **82**, 311-314.
Pitot, A. (1936). 'Isolement et Chute de la Graine a Maturite chez les Legumineuses', pp. 67-68, 83. University of Montpelier, France.
Polhill, R. M. (1976). *In* 'Botanical Systematics' Vol. I (V. H. Heywood, ed.), pp. 143-368. Academic Press, London.
Porter, R. H. (1949). *Bot. Rev.* **15**, 221-344.
Preece, P. B. (1971). *Aust. J. Bot.* **19**, 39-49.
Preston, P. D., and Scott, L. I. (1943). *Proc. Leeds Phil. and Lit. Soc.* **4**, 123-134.
Quinlivin, B. J. (1971). *J. Aust. Inst. Agric. Sci.* **37**, 283-295.
Raleigh, G. J. (1930). *Bot. Gaz.* **89**, 273-294.
Rees, B. (1911). *Proc. Roy. Soc. Vic.* (NS) **23**, 393-414.
Reeves, R. G. (1936a). *Am. J. Bot.* **23**, 291-296.
Reeves, R. G. (1936b). *Am. J. Bot.* **23**, 394-405.
Reeves, R. G., and Valle, C. C (1932). *Bot. Gaz.* **93**, 259-277.
Roake, R. J., and Young, W. C. (1975). 'Formulas for Stress and Strain', p. 453. McGraw-Hill, New York.
Robbertse, P. J. (1973). *Tydskr. Natuurwet.* **13**, 72-95.
Robbertse, P. J. (1974). *J. Sth. Afr. Bot.* **40**, 269-273.
Roberts, E. H. (1972). *In* 'Viability of Seeds' (E. H. Roberts, ed.), pp. 321-359. Chapman and Hall, London.
Rolston, M. P. (1978). *Bot. Rev.* **44**, 365-396.
Rowson, J. M. (1952). *J. R. Microsc. Soc.* (series III) **72**, 46-55.
Saha, P. K., and Takahashi, N. (1981). *Ann. Bot.* **47**, 423-425.
Schjelderup-Ebbe, T. (1936). *Skr. Norske Vidensk. Akad. I. Mat. Nat. Kl.* **1935**, 1-178.
Scott, F. M., Bystrom, B. G., and Bowler, E. (1962). *Am. J. Bot.* **49**, 821-833.
Sefa-Dedeh, S., and Stanley, D. W. (1979). *Cereal Chem.* **56**, 379-386.
Sherry, S. P. (1971). 'The Black Wattle (*Acacia mearnsii* de Wild.)'. University of Natal Press, Pietermaritzburg.
Shull, G. H. (1914). *Plant World* **17**, 329-337.
Simpson, D. M., Adams, C. L., and Stone, G. M. (1940). 'Anatomical Structure of Cotton Seed Coat as Related to Problems of Germination'. U.S.D.A Technical Bulletin, 734.
Slattery, H., Atwell, B. J., and Kuo, J. (1982). *Ann. Bot.* **50**, 373-378.
Smith, L. B. (1981). *Scanning Electron Microscopy* **III**, 545-552.
Spira, T. P., and Wagner, L. K. (1983). *Am. J. Bot.* **70**, 303-307.
Stein von Kamienski-Jancke (1958). *Mikroskopie* (Wien) **12**, 357-392.
Stienswat, W., Pollard, L. H., and Campbell, W. R. (1971). *J. Amer. Soc. Hort. Sci.* **96**, 312-315.
Stone, E. C., and Juhren, G. (1951). *Am. J. Bot.* **38**, 368-372.
Thornber, S. (1903). Some practical suggestions concerning seed germination. *Timely Hints for Farmers* No. 50, Univ of Arizona Agric. Expt. Station, Arizona.
Toole, E. H., and Brown, E. (1946), *J. Agric. Res.* **72**, 201-210.
Tran, V. N. (1979). *Aust. J. Plant Physiol.* **6**, 277-287.
Tran, V. N. (1981). *J. Microwave Power* **16**, 277-281.

Tran, V. N., and Cavanagh, A. K. (1979). *J. Microwave Power* **14**, 21-27.
Tran, V. N., and Cavanagh, A. K. (1980). *Aust. J. Bot.* **28**, 39-51.
Turner, J. H. (1933). *Kew Bull. Misc. Inform.* **6**, 257-269.
Ueki, C., and Suetsugu, I. (1958). *Proc. Int. Seed Testing Assoc.* **23**, 69-72.
Vassal, J. (1973). *C.R. Acad. Sci. Paris, Series D* **276**, 3029-3032.
Vassal, J. (1975). *Boissiera* **24**, 285-297.
Vershaffelt, E. (1912). *Rec. Trav. Bot. Need* **9**, 401-435.
Villiers, T., and Edgecumbe, D. (1975). *Seed Sci. Technol.* **3**, 761-774.
Watson, D. P. (1948). *Ann. Bot.* (NS) **12**, 385-409.
Wayland, J. R., Davis, F. S., Young, L. W., and Merkle, M. G. (1972). *J. Microwave Power* **7**, 358-388.
Went, F., and Munz, P. A. (1949). *Aliso* **2**, 63-75.
Werker, E. (1980/1981). *Isr. J. Bot.* **29**, 22-44.
Werker, E., Dafni, A., and Negbi. M. (1973). *Bot. J. Linn. Soc.* **66**, 223-232.
Werker, E., Marbach, I., and Mayer, A. M. (1979). *Ann. Bot* (NS) **43**, 765-771.
White, J. (1908). *Proc. Roy. Soc. Vic.* (NS) **21**, 203-210.
Wickens, G. E. (1969). *Kew Bull.* **23**, 181-202.
Winter, D. M. (1960). *Am. J. Bot.* **47**, 157-162.
Zimmermann, K. (1936). *Landwirtsch Vers.-Stn.* **127**, 1-56.

CHAPTER 2

Metabolic Aspects of Dormancy

J. D. ROSS

I.	Introduction	45
II.	Gene Repression	46
III.	Hormonal Changes	48
	A. Abscisic Acid	49
	B. Gibberellins	51
	C. Cytokinins	55
	D. Ethylene	56
	E. Auxins	60
	F. Short Chain Fatty Acids	61
IV.	Phytochrome Control of Metabolism	61
V.	Respiration	63
VI.	Mobilization of Reserves	66
VII.	Conclusion	69
	References	73

I. INTRODUCTION

Dormancy is frequently defined as a state of suspended or greatly reduced physiological activity, which tends to imply that dormant seeds are metabolically inactive. This is a misleading idea, as dormant seeds certainly do have an active metabolism, although there is frequently a transitory period, which may be relatively lengthy, during which changes in metabolism may occur as dormancy is lost. It is perhaps more useful to consider dormancy as a time of suspended growth during which physiological development and differentiation can still occur.

It is often interesting to speculate on how any particular physiological phenomenon evolved; for instance, in respect to what selective pressures did dormancy arise? It is possible that it came about as a mechanism allowing the seed, or equivalent structure, to be retained for longer periods on the parent plant, thus extending the period of seed development and maturation. The necessity to inhibit the germination of such a seed while it continued to accumulate reserves could have resulted in the origin of a dormancy system as a supression of vivipary. The extension of this phase into the period subsequent to shedding and reimbibition would then carry the selective advantages of aiding both temporal and spatial dispersion of the propagules.

During the past twenty years two major concepts have dominated the approach to our investigation and understanding of dormancy. The first of these is that dormancy is controlled by gene repression and derepression (Tuan and Bonner, 1964), and the second is that dormancy, in common with other physiological phase-changes, is under control of a balance of hormonal promoter–inhibitor concentrations (Section III). These concepts will be discussed below in the light of experimental evidence, as will the opinion, often stated, that mobilization of reserve materials is exclusively a function of the germination process (Evans *et al.*, 1975; Bewley and Black, 1978).

II. GENE REPRESSION

The genetic control of development is probably the most difficult aspect of contemporary biology to investigate. The idea that certain genes are expressed only during seed development, maturation or germination is not usually doubted. However, the well known hypothesis of Tuan and Bonner (1964) that plant organs are dormant because of direct repression of some part of the genome remains to be conclusively demonstrated. Many studies have been carried out on this topic but supportive evidence remains slight.

For dormancy to be overcome in certain species, including Norway maple (*Acer platanoides*) and hazel (*Corylus avellana*), the seed must undergo a period of exposure to low temperature (stratification). In a recent study by Slater and Bryant (1982), the stratified seed of Norway maple was shown to accumulate RNA in the embryonic axis. The synthesis of RNA was measured by the incorporation of radioactively labelled precursors and most of the newly synthesized RNA was ribosomal RNA (rRNA). An earlier investigation by Wood and Bradbeer (1967) using hazel seeds had yielded similar results. However, in the *Acer* system other forms of RNA were also labelled, but no significant differences were shown between the stratified and non-stratified seed in the proportion of the [^{3}H]uridine incorporation associated with poly (A)-rich (polyadenylated) RNA. These authors were therefore unable to provide any evidence in support of gene derepression. The proportion of RNA represented by messenger RNA (mRNA) actually decreased during the chilling treatment, and they could detect no differences in proteins, labelled *in vivo*, between treatments. Slater and Bryant (1982) concluded that unmasking of any part of the genome did not occur, and the increased growth potential brought about by chilling was attributed to an increase in the protein-synthesizing machinery, mainly rRNA. It must be emphasized that these authors were comparing stratified seeds with warm-imbibed controls, and not chilled seeds with unchilled seeds after the application of exogenous gibberellic acid (GA_3). Other reports on nucleic acid metabolism following gibberellin (GA) treatment do show differences in nucleic acid labelling (Pinfield and Stobart, 1969; Jarvis and Shannon, 1981). It should be recognized that since treatment of seeds with GA normally stimulates germination and its associated different and increased metabolic activities, this experimental approach cannot yield any information on the natural means of breaking dormancy.

Other workers using different species have reached similar conclusions to those of Slater and Bryant (1982). A comparison has been made between dormant and after-ripened seeds of *Vaccaria pyramidata* (Hecker and Kohler, 1979). As no differences in the patterns of RNA or protein labelling could be shown, again it was concluded that the developmental block that controls dormancy did not reflect genetic repression. Similarly, De Klerk and Linskens (1979) using embryos of *Agrostemma githago* were unable to demonstrate any significant differences in protein labelling for 48 hour after the start of imbibition. However, after 48 hour one distinct protein did accumulate in the cotyledons of the dormant seeds. The rate of uptake of amino acids was much higher (49%) in the after-ripened embryos, although the percentage incorporation of total uptake was about the same in both the dormant and non-dormant axes. This suggests that an increase in the rate of active transport of amino acids may develop early in the process of overcoming dormancy in some species. In other species with different mechanisms of breaking dormancy the situation is equally unresolved. Lettuce (*Lactuca sativa*) seeds require light for germination. Whether in light or darkness, lettuce seeds show similar rates of RNA synthesis until 12 hours after a light stimulus, when the rate of RNA synthesis increases rapidly in illuminated seeds until it is five-fold greater than in seeds kept in the dark. However, by this time the processes of germination are well underway (Frankland *et al.*, 1971).

The difficulties of separating the metabolism associated with the breaking of dormancy from that of germination and growth make studies of this kind difficult to interpret. In addition, technical expertise and methods may not yet have attained the degree of sophistication necessary for the detection of small discrete changes in the pattern of mRNA synthesis. The combination of these two problems may be the reason for the lack of any confirmation of the Tuan and Bonner hypothesis, which remains logical, but as yet without experimental evidence.

Imbibed dormant seeds can, and do, synthesize RNA and protein and the alleviation of dormancy is associated with an increased capacity for protein synthesis. Most reports thus appear to favour a concept of control at the level of translation.

III. HORMONAL CHANGES

The concept of a balance of promoter–inhibitor hormones was first applied to seeds by Villiers and Wareing (1960) in their study of dormancy in ash (*Fraxinus excelsior*) and then later elaborated by Amen (1968) and Khan (1971, 1975). It is this concept that reflects the paradigm of developmental plant physiology during the last twenty years. The model of hormonal interactions borrowed from animal studies seemed to offer a solution to many problems, including those of seed dormancy and germination. It was quickly established that in a great number of species, application of phytohormones, in particular gibberellins, to the dormant seed brought about germination. Conversely, exogenous inhibitors, notably abscisic acid (ABA), could prevent non-dormant seeds from germinating. Thus the model was supported by a strong body of circumstantial evidence and it offered a number of assumptions that could be tested. Unfortunately, experimentation has not usually provided evidence that confirms the validity of the model.

Trewavas (1981) makes the point that many of our assumptions are ill-founded and that even classical experimental data does not support the hypothesis. For example, we can take the measurements of endogenous levels of phytohormones obtained prior to and during any physiological change. Experiments with exogenously applied growth substances show a log-linear relationship between concentration and response. However, endogenous levels appear to change relatively little, certainly not over orders of magnitude. Trewavas (1981) suggests that the log-linear relationship is an argument against control by changes in endogenous concentration. Similarly other assumptions about timing, and whether these substances are ever limiting, are called into question. An alternative approach suggested would be to search for changes in the sensitivity of tissues, or organs, to these growth substances.

The narrowing of the classical view to the subject of seed dormancy highlights the problem. The model predicts a change in the hormonal balance in response to environmental factors, which would then trigger the activation of metabolism that leads to germination. This means that inhibitory substances must predominate in the dormant seed, then in response to an environmental signal, promotory hormones are synthesized, and germination results.

A. Abscisic Acid

Perhaps the most consistent evidence we have for the involvement of growth regulators in seed dormancy comes from the studies on ABA. It has been shown frequently that maturing seeds accumulate ABA, for example in wheat (*Triticum aestivum*) (King, 1976). Even those species that do not exhibit a true physiological dormancy in the shed seed, for example, pea (*Pisum sativum*) accumulate ABA in the seed during development (Eeewens and Schwabe, 1975). Freshly harvested seeds of many species which show dormancy, such as ash, sycamore (*Acer pseudoplatanus*) and hazel do contain relatively high levels of ABA (Sondheimer et al., 1968; Webb and Wareing, 1972; Williams et al., 1973). In the case of hazel, a cold-requiring seed, most of the ABA appears in the pericarp and testa. Removal of these seed coats at the harvest stage, before the seed dries out, allows the naked embryo to germinate. The freshly harvested (and dormant) intact seed contains 3.9 nmol ABA per seed, of which 97.5% is to be found within the pericarp and testa. Removal of the coats results in 62% germination (Williams et al., 1973). That the coat inhibition is caused by soluble inhibitors, and is not a physical effect, can be shown by incubating the naked embryos with the stripped seed coats in the same dish. When this is done, the inhibition is restored. Following harvest, when the seed dries and true embryo dormancy is established, removal of the coats no longer allows germination. In order to break this innate dormancy the hazel seed must be stratified at 5°C for about 35 days, after which time germination will occur on transfer to a warmer temperature. Gibberellic acid can induce germination of the unchilled seed and this will be discussed later (Section III,B).

Radicle protrusion (equals germination) occurs in chilled naked embryos as frequently as does germination of chilled intact hazel seeds, and yet the testa still contains about 40% of its initial ABA content, while the embryo has 20% of its original content (Williams et al., 1973). The irrelevance of the presence of the testa, with its still relatively high ABA content, suggests that during either the onset of true embryo dormancy or the chilling process, the seed has become less sensitive to ABA. Indeed, remarkably high exogenous concentrations are necessary to prevent germination of the chilled seed: about 0.1 mM compared with 0.1 μM effective on the freshly harvested naked seed. Thus it would seem ABA no longer has a significant role to play in the dormancy of hazel seed after the onset of true embryo dormancy.

This data from hazel and similar observations on other species, for example, ash (Sondheimer *et al.*, 1968) and sycamore (Webb and Wareing, 1972) suggest that the initial dormancy of the newly harvested seed may be imposed by ABA but not the physiological or innate dormancy developed during dry storage. As developing seeds have often been shown to have relatively high ABA levels, such as those found in wheat (King, 1976), it may be concluded that the early stages of hazel dormancy are imposed by the seed coats through the action of ABA, and this is an extension of the vivipary-avoidance process.

That ABA does impose primary dormancy on seeds has been elegantly demonstrated by Karssen (1982) using single-gene mutants of *Arabidopsis thaliana*. These mutants showed increased transpiration rates and a lack of seed dormancy, correlated with a lowered ABA content in leaves and mature seeds. In the wild-type seeds the ABA concentrations reached 200–500 ng per g fresh weight, whereas in the mutants the concentration never exceeded 10 ng per g. The wild-type seeds developed dormancy during a period of after-ripening but the mutants did not. Mendelian backcrossing experiments confirmed that the induction of dormancy was correlated with ABA derived from the embryo and not with maternal production, thus implying lack of transport of any active ABA between coats and embryo.

At which level, and via what mechanism, ABA exerts its inhibitory influence in the developing or mature seed is an interesting line of inquiry. Studies on precocious germination of surgically excised cotton (*Gossypium hirsutum*) embryos (Ihle and Dure, 1972) and the inhibition of GA_3-induced synthesis of α-amylase [EC 3.2.1.1] in barley (*Hordeum vulgare*) aleurone cells (Ho and Varner, 1976) lead to the conclusion that the inhibition is at the level of mRNA translation. These authors conclude that the mechanism operates because of the continuous production of an ABA-induced regulator, either RNA or protein in nature. Jarvis and Shannon (1981) have demonstrated significant inhibition of poly (A)-rich RNA synthesis by ABA in hazel embryonic axes, which implies transcriptional control. However, the same workers have shown that only 30% of the mRNA in hazel embryonic axes is polyadenylated in fresh seeds, and that this proportion declines during the dormancy-inducing period of dry storage (Shannon *et al.*, 1981). In addition, they were able to show different peptide patterns when comparing labelling from *in vivo* and *in vitro* experiments, and conclude that in hazel there is also strong translational control, although they do not directly suggest that ABA is responsible for this.

B. Gibberellins

The general observation that application of exogenous GA can bring about germination of the dormant seeds of many species led to the hypothesis that GA must accumulate in the seed in response to the environmental stimulus that triggers dormancy loss. This simple hypothesis has been the starting point of a large number of investigations which have proven only one point: the situation is immensely more complicated than was originally believed.

Only in one species, hazel, has a clear sequence of events been shown to occur, although many other species seem to follow a similar pattern. Frankland and Wareing (1966) reported that the chilling treatment necessary for dormancy breakage of hazel was accompanied by a slight increase in the GA content of the seeds. They did, however, point out that the resultant measured concentrations were well below those needed to be applied exogenously to induce germination of the unchilled seed. Later, Bradbeer (1968) showed that CCC (2-choloroethyltrimethyl ammonium chloride, an inhibitor of GA biosynthesis) inhibited the germination of chilled seeds if applied after the chilling treatment as the seeds were returned to the germination temperature. It was later shown that there was a negligible change in GA concentration during the stratification treatment but there was a massive increase when the now non-dormant seeds were subsequently returned to the warmer germination temperature (Ross and Bradbeer, 1968, 1971a). All of this work had been carried out using thin layer chromatography and bioassay techniques and the identities of only two GAs (GA_1 and GA_9) were later confirmed by gas liquid chromatography–mass spectrometry (GLC-MS) (Williams et al., 1974). This latter study also demonstrated that the freshly harvested seed, prior to the onset of embryo dormancy, contained comparable levels of GA to those in the previously chilled germinating seed. As mentioned above (Section III,A) removal of the inhibitor-containing coats allows these fresh seeds to germinate.

The levels of GA in the stratified seeds remained depressed on their return to warmer temperatures if the growth retardants CCC or phosphon D [tributyl 1 (2,4-dichlorobenzyl) phosphonium chloride] were supplied (Ross and Bradbeer, 1971b). This was confirmed by Arias et al. (1976) who extended the observation to show that de novo GA biosynthesis in the embryonic axis was inhibited, and that the isolated cotyledons appeared to accumulate GA released from some bound form. In excised cotyledons the levels of GA were 60% lower than in attached cotyledons. This implies that GA is translocated from the axis to the cotyledons, which is not unexpected as the endogenous concentrations are several hundred-fold greater in the axis.

The conclusion of these reports is that while chilling relieved some unknown block to GA biosynthesis, synthesis took place on the subsequent transfer of the seed to the higher temperatures suitable for germination. Thus GA biosynthesis appears to be involved with the growth and metabolism of the germinating seed and not with the metabolism of dormancy breakage.

In an investigation of GA content during the 16 week stratification period of cherry seed (*Prunus avium*) similar results were obtained in that no qualitative or quantitative differences could be detected during after-ripening (Proctor and Dennis, 1968). In a study of the role of hormones in the germination of sycamore, Pinfield and Stobart (1972) concluded that although GA failed to induce germination of intact seeds it did greatly stimulate the unrolling of the cotyledons, a post-germination event. An investigation of endogenous hormone levels in sugar maple (*Acer saccharum*) seeds gave some ambiguous results (Webb *et al.*, 1973a). In this species the most obvious change was the large decline in ABA levels from the onset of stratification. The acidic gibberellin-like fraction was initially not detected, but rose to peak by the 40th day of treatment only to subsequently decline until, most oddly, there was again no activity at the end of stratification in the germinating seed. At the time of greatest GA content the seed was still dormant.

A number of other reports have shown GA levels to increase during stratification with no apparent correlation with germination potential at the time. These include oak (*Quercus robur*) (Vogt, 1974); European beech (*Fagus sylvatica*) (El-Antably, 1976); Norway maple (Pinfield and Davies, 1978); and large-leafed lime (*Tilia platyphyllos*) (Nagy, 1980).

Apart from hazel only one other species has been investigated extensively and that is the apple (*Malus sylvestris*). Various GAs have been identified in the seed (GA_4, GA_7 and GA_9) and it is claimed that GA_4 has been shown to increase in level after 3 weeks stratification, only to decrease again by the 7th week to levels comparable with the dormant seed when measured by bioassay (Sinska and Lewak, 1970). A confirmation of the apple GAs was obtained by GLC–MS (Sinska *et al.*, 1973), however, subsequent re-examination of the original extracts has suggested that these results may be due to contamination with GA of exogenous origin (Dennis *et al.*, 1980).

The above examples have dealt with the situation in stratification-requiring woody angiosperms, and while no definite conclusions can be reached it seems that GA remains either low during the stratification period, or undergoes a transitory increase followed by subsequent decline. As such it is doubtful whether GA plays a part in the dormancy-breaking process and it is more likely to be involved with the metabolism associated with germination and seedling growth.

A number of essentially similar investigations have been carried out on seeds of gymnosperms. The seeds of loblolly pine (*Pinus taeda*) require a 2 month stratification for complete dormancy breakage. When extractions and bioassays were carried out throughout this period it was shown that the unstratified controls and the chilled seed up to 28 days contained negligible GA activities, and a relatively high level of inhibitory substances (Paul *et al.*, 1973). After 42 days stratification an increased GA-fraction was detected, while the inhibitors had declined until they approached zero activity. It was, however, concluded that the situation may not be a simple inhibitor–promotor balance, as treatment of the unstratified seed with GA_3 only slightly enhanced germination, while the 21 day stratified seed responded markedly, once more suggestive of a change in sensitivity. These authors therefore believed other metabolic changes were a necessary part of dormancy breakage by stratification.

In Douglas fir (*Pseudotsuga menziesii*) seeds a seven-fold increase in GA activity was measured after 7 weeks stratification, but again this level subsequently declined (Taylor and Wareing, 1979a). This particular seed lot was only partially dormant showing 44% germination with unstratified seed, increasing to 61% after 7 weeks treatment. In this type of situation the results of extraction and analysis are bound to be confusing as there is no way of determining the hormonal content of individual seeds with prior knowledge of their physiological state. It is interesting that the same species was much more dormant in a fresh collection (Ross, 1969) which responded markedly to only 10 days chilling. There is obviously great variation between collections from different origins, and similarly the physiological age, collection, storage and extraction methods all have an effect. These differences should be borne in mind when comparing and interpreting reports.

Sugar pine (*Pinus lambertiana*) seeds chilled for 8 weeks had a 20-fold higher level of extractable GA than unstratified seeds (Taylor and Wareing, 1979a). The same authors also examined GA levels in seed of sitka spruce (*Picea sitchensis*) which is a light-requiring seed. In this latter case the endogenous GA levels were shown to increase 10-fold after only 30 minutes red-light (660 nm) irradiation of the 9 hour imbibed seed (Taylor and Wareing, 1979b).

The hypothesis behind these studies was that GA was limiting in the dormant seed and that stratification brought about increased GA levels thus alleviating the block to growth. It can be seen that although chilling may release the block to GA biosynthesis, a simple correlation between GA levels and germination potential is not possible.

Two simple germination experiments carried out with hazel seeds seem to confirm that GA is not a limiting factor (Younis, 1982). In the first experiment (Fig. 1), seeds were imbibed and held at either 5°C or 20°C for

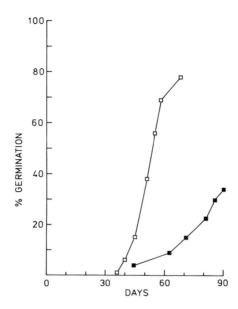

Fig. 1. Germination of hazel seeds incubated in GA$_3$ (1 mM) at 5°C. Intact fruits were imbibed and held at either 5°C (□-□-□-□) or 20°C (■-■-■-■) for 28 days; then pericarp removed and seeds incubated at 5°C with GA$_3$ in darkness (data of Younis, 1982).

28 days. After this time the pericarps were removed and exogenous GA$_3$ applied, and the seeds allowed to germinate at 5°C. Although germination took place slowly, it was more rapid and uniform in those seeds previously chilled. The unchilled seeds appear to have to make up the missing chilling time before they can respond and even then they have a slower or less uniform germination rate.

The second experiment is similar, but germination was carried out at 20°C (Fig. 2). In this case seeds chilled for 28 days were allowed to germinate in water after removal of the pericarp, while seeds imbibed at 20°C were treated with GA$_3$ after removal of the pericarp. Again there was a very rapid and uniform germination of the chilled seeds, but an appreciable lag before the unchilled seeds began to respond to the exogenous GA. If the consequence of chilling was merely the lifting of a block to GA biosynthesis, and the other metabolic events followed by GA induction, then one would have expected similar germination responses in these experiments. It is therefore possible to conclude that dormancy does not result from the lack of GA, or its activity, but from some other cause,

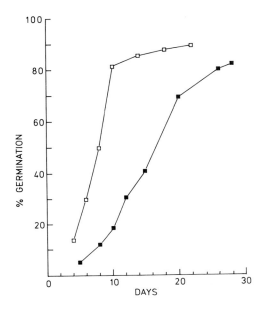

Fig. 2. Germination of hazel seeds at 20°C. Intact fruits imbibed 28 days at either 5°C (□-□-□-□), or 20°C (■-■-■-■), subsequently pericarp removed and chilled seeds incubated in water, and unchilled seeds incubated in 1 mM GA$_3$ (data of Younis, 1982).

and that sensitivity to GA may itself change in response to environmental conditions.

C. Cytokinins

Kahn (1971, 1975) has assigned to the cytokinins a permissive role in the promotor–inhibitor hypothesis of dormancy. Cytokinins have been observed to allow GA to induce germination of lettuce seed inhibited by exogenous ABA, whereas increased concentrations of applied GA alone are unable to do so. Evidence for this situation occurring endogenously has not been forthcoming. Most reports concern those species whose seeds require stratification and there is some inconsistency in the evidence. The changes in cytokinins and other hormones were measured during the cold stratification of sugar maple (Van Staden et al., 1972; Webb et al., 1973a). The reports indicate a transitory rise in cytokinin content after 20 days, at which time only a small proportion of the seeds were able to germinate. This rise is followed by a rapid decline for the remainder of the stratification period. Two points need emphasis when discussing these results. Firstly, the cytokinin only rises after the ABA levels have significantly declined. It is unlikely, therefore, that the cytokinins are carrying out any permissive role at this time, verified by the fact that the bulk of the seeds will not germinate. Secondly, the whole account of a transitory rise depends upon a single extraction and measurement, that taken at the 20th day. All of the other measurements taken are at least an order of magnitude lower, both before and after this single observation. It is to be hoped that some future study will carry out more frequent analyses during the stratification of *A. saccharum* in order to verify the authenticity of this increased cytokinin level, which is that most frequently quoted as a textbook example.

Similar transitory increases have been reported as occurring during the chilling of seeds of a number of species including *Protea compacta* and *Leucadendron daphnoides* (Brown and Van Staden, 1973); apple (Borkowska and Rudnicki, 1975); and sugar pine (Taylor and Wareing, 1979a). In contrast no such changes are detectable in walnut (*Juglans regia*) (Martin *et al.*, 1969); sycamore (Webb *et al.*, 1973b); and Douglas fir (Taylor and Wareing, 1979a).

The significance, if any, of the cytokinin levels in dormant seeds remains elusive, and while exogenous cytokinin treatments may occasionally give circumstantial support to the general hormonal hypothesis, the measurements of endogenous levels seem to lack any cohesive component.

D. Ethylene

It had been known since the 1920s that ethylene would stimulate germination of seeds in certain species, but intensive investigations were not carried out until the 1960s. By that time ethylene had come to be designated a plant hormone, and in contrast to the situation applying to all the other hormones, the rapid, simple, and accurate measurement of ethylene was made generally available with the advent of the GLC assay. Further impetus was added when it was established that many weed seeds responded to applied ethylene. This application has been tried in an attempt to eradicate witch-weed (*Striga asiatica*) in the U.S.A. (Egley and Dale, 1970; Eplee, 1975).

In non-dormant varieties and after-ripened dormant varieties of peanut (*Arachis hypogaea*) the evolution of ethylene was shown to be associated with the onset of radicle growth (Ketring and Morgan, 1969). Other peaks of production followed during early seedling growth. The dormant cultivars of the same species produced low levels, which remained below those necessary for the induction of germination (Ketring and Morgan, 1972). It is possible that the inability to produce sufficient ethylene may be a block to germination in this species, although further evidence is required to verify this.

Another species in which ethylene production may be a limiting factor is cocklebur (*Xanthium pennsylvanicum*). Esashi and coworkers have shown that dormant seeds have a far lower capacity for ethylene production than non-dormant after-ripened seeds, and that dormancy is alleviated by application of exogenous ethylene (e.g., Katoh and Esashi, 1975). Carbon dioxide is also required for both production and effect of ethylene. During long-term storage in the imbibed state, the seed entered a secondary dormancy stage but ethylene production was unaffected. If, however, the imbibition was at 8°C, ethylene did lose some of its effectiveness, whereas other germination stimulants did not (Esashi *et al.*, 1978). It is possible that this is a reflection of the site of action for ethylene being a temperature-sensitive membrane (Section VII).

In the cocklebur seed there is a 10 hour period from the beginning of imbibition to the onset of ethylene evolution, which suggests a necessity for the seed to develop the biosynthetic apparatus. In order to establish if the dormant seed was unable to produce ethylene because of a general block to protein synthesis, Satoh and Esashi (1979) examined the incorporation of labelled leucine in dormant and in after-ripened seed. Surprisingly, they were able to show that initially, protein synthesis was actually higher in dormant axes than in those that had been after-ripened, and it was not until after 12 hours that the situation was reversed. Treatment with cycloheximide inhibited the germination of the non-dormant seed and supressed ethylene evolution, however 3 mM cycloheximide slightly stimulated the germination of the dormant seed, but with no enhancement of ethylene production. They concluded that the maintenance of dormancy in this species involved some proteinaceous system.

Ethylene-stimualted germination of non-dormant upper cocklebur seed was associated with enhanced respiration, and the inhibitors of alternative respiration, benzohydroxamic acid and salicylhydroxamic acid, inhibited the germination of both non-dormant seed and the ethylene-stimulated dormant seed (Esashi *et al.*, 1979b). It was suggested that secondary dormancy in this species is a result of the inactivation of the alternative respiratory (cyanide-insensitive) system, and that ethylene may exert its stimulatory activity by the enhancement of this system (see Section V).

The possibility that relatively cheap ethylene treatments could be used for weed eradication has led to a number of investigations being carried out on such species. One recent study has been on redroot pigweed (*Amaranthus retroflexus*) seeds in which it has been demonstrated that ethylene increased germination under a number of environmental conditions (Schonbeck and Egley, 1980). It was shown that high temperatures and light increased the sensitivity to ethylene. One interesting observation was that although water stress reduced germination, the sensitivity to exogenous ethylene was heightened. The authors suggested that stress conditions reduced endogenous ethylene production, and thus endogenous levels, and that this resulted in increased sensitivity. In a further report (Schonbeck and Egley, 1981) the interactions of various factors were described. Ethylene, light and high temperature gave maximum effect after 12 hours imbibition. By 48 hours all three applied together could bring about 75% germination, but when given singly each was ineffective. They proposed that ethylene had an effect in two separate phases of dormancy in redroot pigweed, the first being during primary dormancy which could be broken early after imbibition. In those seeds that were unable to germinate at that time because of some unfavourable condition, such as an unsuitable light or temperature regime, there was an induction of secondary dormancy. In these seeds ethylene sensitivity only gradually redeveloped over a period of days.

A similar observation that sensitivity to ethylene can vary with time and in response to environmental factors has been made by Adkins and Ross (1981a) in a study of dormancy in seeds of wild oat (*Avena fatua*). In this persistent weed there appears to be a complex series of interacting mechanisms of dormancy. In the freshly shed seed dormancy is very deep, with germination prevented by several systems, the sum of which cannot be overcome by endogenous ethylene production nor by exogenous applicaton. As the imbibed seed ages in soil under warm conditions (25°C) the sensitivity to ethylene increases as some limiting system decays, such that partially dormant seed returned to cooler temperatures can now be caused to germinate by either endogenously produced or applied ethylene.

It was demonstrated that non-dormant after-ripened caryopses of wild oat produced small amounts of metabolic ethylene even when still dry, and that upon imbibition this production rapidly increased five-fold before dropping to a low level after 8 hours; this response was not seen with dormant seed of the same age. If the applied concentration were too high, or if the seeds were at an advanced stage of after-ripening, an inhibitory effect was observed. This varying response may explain why contradictory results have been reported (Chancellor *et al.*, 1971).

Inhibition of germination by exogenous ethylene has been reported in a study of light-stimulated germination of *Potentilla norvegica* (Suzuki and Taylorson, 1981). In this case the sensitivity to ethylene inhibition varied with time, being most significant during and immediately following the irradiation exposure. These authors suggested that ethylene exerts its inhibitory effect by binding loosely to the site of phytochrome action, presumably a membrane. They concluded that the association is relatively loose from the observation that high temperature or venting of the gas ended its action.

Another group who examined the role of ethylene in seed dormancy is that of Hall and coworkers. They have shown that ethylene and light had a synergistic effect in breaking the dormancy of seeds of *Spergula arvensis* (Olatoye and Hall, 1973). In common with many other species it was discovered that the presence of CO_2 during the early stages of imbibition was necessary for the ethylene action (Jones and Hall, 1979). In a study of respiration during ethylene-stimulated germination of *S. arvensis* they measured the $C_6:C_1$ ratios of evolved CO_2 and concluded that the ethylene was not acting through an enhancement of the pentose phosphate pathway (Jones and Hall, 1981), an hypothesis which will be discussed later (Section V).

This particular group have also looked for the site of ethylene binding by using cell-free preparations from the cotyledons of common bean (*Phaseolus vulgaris*). They have characterized the binding activity associated with subcellular particles, which they suggested were membranes derived from the Golgi bodies or endoplasmic reticulum (ER) (Bengochea *et al.*, 1980; Hall *et al.*, 1982).

Further studies by the same group have attempted to locate the ethylene binding site in bean cotyledons by using high resolution autoradiography (Evans *et al.*, 1982a) and binding assays on fractions separated by analytical ultracentrifugation (Evans *et al.*, 1982b). These two reports have demonstrated specific binding of ethylene to the endoplasmic reticulum and the protein body membrane. When isolated by a non-aqueous method, to prevent lysis, 38% of the total binding activity was associated with the protein bodies.

When we look for a role for ethylene in seed metabolism the evidence favours the hypothesis that ethylene affects the dormant seed by interacting with some membrane fraction. The association with phytochrome action and binding studies both support this hypothesis.

One can speculate that a possible block to ethylene biosynthesis in the dormant seed may be the unavailability of the initial substrate, methionine. This is an essential amino acid found as a constituent of certain seed storage proteins (Chapters 1 and 3 of Volume 1). The initial appearance of methionine after the onset of protein mobilization may induce this biosynthetic activity. If one binding site for ethylene is on the limiting membrane of the protein bodies, as reported by Evans *et al.* (1982a, b) and if binding accelerates the entry of proteolytic enzymes into these bodies, this could be a possible controlling site for ethylene action.

One difficulty with investigation of the mode of action of ethylene is that it often appears to be autocatalytic, in that applied ethylene induces further endogenous production. An extra complication is that ethylene is commonly evolved in response to tissue damage and the onset of axis growth may initiate such a 'wound' reponse. This effect may possibly be the explanation for the burst of ethylene production so often observed just prior to radicle emergence. As with all apparent hormonal–physiological correlations it is important to be able to distinguish any ethylene associated with the metabolism of dormancy breakage from that which may be a by-product of the renewed growth of the axis.

E. Auxins

As exogenous auxin does not appear to stimulate germination of dormant seeds it is not surprising that few investigations have been made on auxin metabolism during dormancy. On the other hand, there are many reports of the occurrence of auxins in seeds and they are often reported to increase in concentration during germination, perhaps indicative of a role in the germination processes (Tillberg, 1977).

In seeds of loblolly pine no auxin activity was detected in unstratified samples although significant inhibitory activity, which may have masked auxin activity, was detected in the oat (*Avena sativa*) coleoptile bioassay used (Paul et al., 1973). During the stratification period this inhibitory activity diminished and some positive activity was detectable by the 28th day. This positive activity remained constant throughout the stratification treatment, only rising when the seeds were allowed to germinate.

A carefully controlled study of the changes in auxin (indole-3-acetic acid, IAA) levels in Norway maple seeds during stratification and germination has been carried out recently by Tillberg and Pinfield (1981). In this species the IAA levels fell from 17 ng per g to 8 ng per g fresh weight after 53 days at 5°C. Thereafter, until 150 days, the concentration steadily increased to about 40 ng per g fresh weight, whereas in the control samples, held at 17°C, the concentration remained low. This investigation used a fluorometric determination of the indole-α-pyrone derivative as an assay, and included the addition of [1-^{14}C]IAA prior to extraction in order to monitor any losses.

However, as the excised embryos returned to 20°C were already capable of germination as the auxin levels reached a minimum, it was concluded that IAA was associated with the germination processes, rather than playing any significant role in overcoming dormancy.

F. Short Chain Fatty Acids

One of the most perplexing natural forms of dormancy loss to understand is that of temperature-dependent after-ripening in dry storage. It has been proposed that volatile fatty acids (chain length C_6-C_{10}) may act as inhibitors of germination (Berrie et al., 1975) and that their loss by evaporation, or degradation, is the essential process in dry after-ripening in *Avena* spp. (Berrie et al., 1979). It was demonstrated that the C_7, C_8 and C_9 fatty acids were particularly effective in reducing germination when applied to the non-dormant cultivated oat. It was further shown that while these fatty acids were present in the developing seed, the most effective inhibitor (C_9) had decreased by harvest to insignificant levels in this species, which does not exhibit post-harvest dormancy. However, in the wild oat the levels of the most inhibitory acids remained high well after harvest but declined during dry storage, along with the loss of dormancy. The loss of the C_7, C_8 and C_9 acids was highly significant, dropping from 2458 ng per grain to 25 ng per grain.

Although this appears to be an attractive and simple explanation for the loss of dormancy during dry storage of wild oat the hypothesis has since been disputed by Metzger and Sebesta (1982). These authors did not find such a large difference in the levels of the short chain acids when comparing dormant and after-ripened seeds. For instance, they were able to measure 450 ng C_9 per grain in the non-dormant seed (95% germination) against 603 ng C_9 per grain in the dormant seed (10% germination). In addition they were not able to show any changes in either short chain fatty acid content or in germinability of the seed after exposure to a partial vacuum for one week, thereby casting doubt upon the availability of these constituents for evaporation.

Unfortunately the wild oat is well recognized as exhibiting wide varietal differences in its dormancy characteristics, and as these two studies had used genetically heterogeneous stocks of different provenance, the explanation for the divergent observations and results may lie here.

IV. PHYTOCHROME CONTROL OF METABOLISM

In elementary terms the phytochrome system can be thought of as having two major functions in seeds. Firstly, to distinguish light from darkness, that is, to determine whether the seed is on or near the soil surface. Secondly, to interpret the light quality, that is, to assess whether the seed is under an appreciable leaf canopy, and therefore liable to experience intense competition for photosynthetically active light. The latter property may also be related to the nature of seed coats, which usually contain chloroplasts during seed development (Section 5.III,A of Volume 1). Seed coats thus impose a light regime enriched in far-red wavelengths upon the developing embryo, hence inhibiting germination.

For a comprehensive account of phytochrome action in the control of seed germination the reader is referred to Bewley and Black (1982). In the following discussion only recent work on how light may affect metabolism in the dormant seed will be considered.

Many early investigations were carried out in an attempt to discover if light, operating through the phytochrome system, controlled the balance of growth hormones in the seed. Results of these early studies were often confusing, and in some cases contradictory, and this approach appears to have become unfashionable. A recent study of this type, however, has been carried out by Taylor and Wareing (1979b) using seed of sitka spruce. In this species maximum germination takes place under continuous fluorescent light, or after 10 minutes exposure to red light. Concentrations of cytokinin increased 1.7-fold one hour after the red-light treatment, whereas under continuous light the concentration doubled within 24 hours, but by 48 hours had dropped back to less than the initial value. Gibberellin activity showed a rapid increase during a 30 minute red-light exposure and then declined during an 8 hour dark incubation, again to less than the original level. Although these results are interesting it is difficult to properly assign any significance to them without knowledge of the functions of these substances in the seed.

One recent investigation that appears to link phytochrome, growth hormones and germination has been reported by Leung and Bewley (1981a, b). These workers used the classical Grand Rapids lettuce seed system in which it has previously been shown that many enzymes increase in activity following red irradiation. In all the earlier reports the enzyme increases occurred after radicle protrusion. These authors have now shown that imbibed dormant seeds contain the same activity of α-galactosidase [EC 3.2.1.22] as the dry seeds but that red-light irradiation, for a minimum of 2 minutes, causes a subsequent increase in α-galactosidase activity prior to germination. The phytochrome control was shown to be tightly linked to enzyme induction, as a far-red irradiation given after a 5 second dark interlude failed to prevent the increase, while after only 2 seconds darkness it could do so. The far-red irradiation given after longer intervals than 5 seconds did inhibit germination, however, even though enzyme activities were not depressed (Leung and Bewley, 1981a). The red light effect could be mimicked by application of exogenous GA_3 to the seeds, again both enzyme activity and germination were stimulated. Interestingly there was a shorter lag period between GA treatment and increase in enzyme activity than that following light-induction (Leung and Bewley, 1981b). A further indication that this process may be regulated by a phytochrome–hormone link was demonstrated by some elegant experiments with excised seed parts. Firstly, it was shown that enzyme activity in the cotyledons did not increase if the axis was removed, but incubation of de-tipped seeds in the presence of isolated irradiated axes did result in enzyme induction. Secondly, they demonstrated that a mixture of GA_3 and benzyladenine could substitute for the presence of the irradiated axes.

These experiments give some of the best support for the classical hormonal concept, as the authors were able to conclude that the red-light perception took place in the axis and was followed by the appearance of a diffusible factor that was translocated into the cotyledons where enzyme induction occurred. In addition, a GA_3 and a benzyladenine mixture could substitute for the axis, whereas in whole seeds GA_3 acted more quickly than an inductive light treatment. The fact that enzyme induction could still proceed without subsequent germination, if far-red light or chemical inhibitors were applied, only confirms that the processes of dormancy breakage are separable from those of germination.

Apart from the above examples of phytochrome action through apparent hormonal changes there are many reports indicating that light may alter the metabolism of dormant seeds. Most of these reports discuss increases in respiration or in protein synthesis. However, most stimulation is evident only after axis growth has been initiated and is therefore probably a facet of the physiology of germination.

One other avenue of investigation has produced evidence that red-irradiated lettuce seeds, prevented from germinating by hypertonic solutions of mannitol, show protein and lipid digestion not evident in controls kept in the dark (Nabors et al., 1974). Normally this hydrolysis would lead to decreased water potential and thus water uptake, resulting in radicle growth (Section 3.IV). Although there is some contradictory evidence from other studies (e.g. Carpita et al., 1979) this still remains an attractive hypothesis worthy of further investigation.

V. RESPIRATION

The metabolism and growth associated with germination require an active synthesis of ATP (Section 3.III,A). The early work of Pollock and Olney (1959) demonstrated that a sharp rise in respiratory rate occurred during stratification of sour cherry (*Prunus cerasus*) seeds prior to the enhancement of the capacity for growth of the embryonic axis. Thus a simple hypothesis was proposed, that dormancy was imposed by a block in energy metabolism and was broken by increased availability of phosphate acceptors such as glucose, presumably via hydrolysis of stored forms. The observation that dormancy could be alleviated in many species by elevated oxygen concentrations seemed to support this hypothesis. Again, increased respiration was the key to dormancy breakage.

However, it became apparent that many chemical inhibitors of respiration could also promote germination and this paradox is still with us.

Many of the inhibitors used were those known to block the terminal cytochrome oxidase [EC 1.9.3.1] of the mitochondrial electron transport chain. As a consequence this led to the proposal that the alternative respiratory (cyanide-insensitive) pathway was involved in the termination of dormancy (Hendricks and Taylorson, 1972). However, other inhibitors of respiration that did not act directly on the electron transport chain could likewise bring about germination in some species, and this led to a proposal that the pentose phosphate pathway was involved in some crucial step (Roberts, 1969, 1973). Attempts have been made to examine these metabolic events *in vivo* and in some cases to draw together the two hypotheses, but so far the situation remains unresolved.

Detailed arguments have been presented elsewhere, but in essence Roberts has suggested that the pentose phosphate pathway employed a cyanide-insensitive oxidase system, and in the dormant seed this oxidase is in competition with the conventional electron transport system. An inhibition of the conventional system or oxidation of NADPH stimulates the pentose phosphate system and some intermediate is made available that allows breakage of dormancy (Roberts and Smith, 1977).

An attempt has been made by Taylorson and Hendricks (1977) to bring together the two hypotheses. They have proposed that the cyanide-type stimulation of germination is mediated by an inhibition of catalase [EC 1.11.1.6] activity, which would normally hold hydrogen peroxide (H_2O_2) at a depressed level (the H_2O_2 presumably coming from glyoxysomal β-oxidation of fatty acids, see Section 6.VI). The resultant sparing of the H_2O_2 allows peroxidase [EC 1.11.1.7] activity, through a linked quinone redox system, to oxidise any available NADPH. In this way the pool of NADP is kept at a high concentration and this would allow the pentose phosphate pathway to proceed rapidly. There is unfortunately no ascribed function for the pentose phosphate pathway in the metabolism of dormancy breakage. An obvious function would be to provide the biosynthetically useful reducing power of NADPH, but the scheme outlined above seems to preclude this. None of the various intermediates of the pentose phosphate pathway has yet been shown to stimulate germination when applied exogenously.

A number of reports have suggested that the activity of the pentose phosphate pathway is higher in non-dormant compared to dormant seeds of wild oat (e.g. Simmonds and Simpson, 1972). However, Adkins and Ross (1981b) have recently shown that while cold-soil imbibition of wild oat does indeed increase the activities of two enzymes belonging to this pathway, glucose-6-phosphate dehydrogenase [EC 1.1.1.49] and 6-phospho-gluconate dehydrogenase [EC 1.1.1.44], this treatment does not terminate dormancy. Conversely, warm (25°C) soil incubation does break dormancy but does not give significant changes in these two dehydrogenase activities.

In two species which respond positively to cold stratification, sour cherry and hazel, there is some evidence that the pathway may be of some importance. La Croix and Jaswal (1967) showed that the pentose phosphate pathway operates at increased rates in the stratified sour cherry seed. They used the $C_6:C_1$ ratio method which is claimed to distinguish between the pathway and normal glycolysis, although many criticisms can be raised over its validity. Using partially purified extracts of stratified hazel seed, Gosling and Ross (1980) have demonstrated increased specific activities of both dehydrogenases, the increases occurring concurrently with the release from dormancy.

These types of observation are not ubiquitous. Satoh and Esashi (1980) were unable to show any association between dormancy breakage of *Xanthium* seed and the pentose phosphate pathway. These authors used the $C_6:C_1$ ratio technique and also measured dehydrogenase activities. Similarly, Jones and Hall (1981) could find no evidence linking ethylene-stimulated dormancy breakage of *Spergula arvensis* with pentose phosphate pathway activity. Their results tended to suggest a decreased participation during the crucial phase. These authors did demonstrate an increased respiratory rate 60 hours after ethylene exposure, but attributed this to the heightened activity of the germinating seeds.

The extensive and thorough investigation into the dormancy of *Xanthium* seeds carried out by Esashi and his colleagues has been mentioned (Section III,D). Using inhibitors of both the conventional and the alternative respiratory systems they argue that alternative respiration is necessary for germination, and that secondary dormancy arises from its inactivation (Esashi *et al.*, 1979a, b). They further claim that ethylene exerts its effect through a stimulation of the alternative system (Esashi *et al.*, 1979b).

When genetically homogeneous lines of wild oat were investigated a similar situation was revealed. Sodium azide, which blocks conventional respiration, stimulates germination of dormant seed, whereas salicyl-hydroxamic acid, an inhibitor of alternative respiration, completely inhibits the azide-stimulated germination. However, this latter inhibitor has no effect when applied to genetically non-dormant lines of wild oat (Upadhyaya et al., 1982). These authors, therefore, conclude that although alternative respiration is necessary for the azide-stimulated germination of dormant fresh seed, it does not have a role in the germination of after-ripened seed, nor in genetically non-dormant varieties.

In two species with a stratification requirement, pear (*Pyrus communis*) and sugar pine, it has been suggested that alternative respiration is not of great importance. Pear seeds show a progressive linear increase in respiratory capacity during stratification (Alscher-Herman et al., 1981), but inhibitor studies suggest that alternative respiration remains at a constant level which is only a small part of the total. Hence it was concluded that the conventional cytochrome oxidase-mediated pathway gave the increase in respiratory capacity.

When dormant sugar pine seeds are stratified they undergo a rapid increase in respiration rate and in ATP content (Murphy and Noland, 1982a, b). Both then plateau for some time, until after 60 days at $5°C$ there is a second significant rise. Those seeds held at $25°C$ show a similar early increase during the first four days, but this is followed by a decline in both respiration and ATP concentration with no subsequent secondary increase. During the plateau phase of the stratified seed, between days 5 and 60, the conventional cytochrome oxidase (cyanide-sensitive) pathway accounts for about 80% of the respiratory activity, whereas in those seeds held at $25°C$ it constitutes only 60% of the total activity. One conclusion drawn from this work is that high temperature imbibition and incubation is actually detrimental to the metabolism of the seed. A similar suggestion has been made by Gosling and Ross (1981) concerning the $20°C$ incubation of hazel seed, where peroxidase activities show a steady increase, while not changing in the stratified seed. These 'warm controls' may be undergoing accelerated ageing.

Thus the only clear statement that can be made about respiratory metabolism during dormancy is that those treatments leading ultimately to the termination of dormancy give enhanced respiratory activity. The roles of the pentose phosphate pathway and cyanide-insensitive alternative respiration remain enigmatic. However, as this is a field in which well established techniques are available it is to be hoped that some understanding will soon be obtained.

VI. MOBILIZATION OF RESERVES

Many recent reviews and reports have uncompromisingly stated that mobilization and utilization of stored reserves does not begin until germination has been initiated. While this may be true of those seeds not exhibiting embryo dormancy, where comparison is made solely against the dry and quiescent seed, it is certainly not true in many seeds that require stratification. Bewley and Black (1982) correctly point out that one reason for this confusion is that a number of reports detailing such metabolism do not contain data from parallel studies on warm controls. The inadequacy of uncontrolled experiments is obvious, however, the question of what is an appropriate control is rarely discussed in the literature. When controls have been carried out the usual method adopted is to hold imbibed seeds at temperatures ranging from 17°C to 26°C. These temperatures are almost certainly too high; the seeds in question have evolved to fit an environment in which they would be subjected to an over-wintering regime. To expose them to an unnaturally high temperature compared to that experienced in their normal niche may in itself give misleading results. It is quite probable that these high temperatures may lead to unusual, and perhaps damaging, metabolism (see Section V). Greater thought needs to be given to what does constitute an appropriate control in stratification experiments.

The extensive studies on the complex dormancy mechanisms of ash carried out by Villiers (1971, 1972) showed that in the newly shed seed, which has to undergo a period of physiological and anatomical maturation, there is extensive metabolism involving interconversion of reserves. During this phase, which occurs at warm temperatures, cell wall development and organelle proliferation are evident. Changes in the food reserves include a large decrease in lipid content, which falls from 20% to 4% of dry weight, and the appearance of starch laid down in newly formed plastids. There is an increase in soluble protein content and a proliferation of the ER and Golgi bodies.

After the completion of this differentiation phase, ash then requires a cold-stratification for the alleviation of embryo dormancy. During this period the nucleoli swell and increase in complexity, concomitant with an increase in cellular RNA. The controls, which also show some evidence of metabolic activity despite retaining dormancy, were held at 22°C.

The early work of Pollock and Olney (1959) showed that seeds of sour cherry mobilized phosphate during the chilling treatment, which gave a good indication that reserves were mobilized during dormancy, and that dormancy breakage could be associated with changes in metabolic pattern. Subsequently it has been reported that in the same species, seeds held at 5°C showed a decrease in lipid content (La Croix and Jaswal, 1973). Although this decrease in lipid content is evident in both the cotyledons and the embryonic axis, it is proportionally much greater in the axis, as is the matching increase in free sugars. It is obviously important to separate axis metabolism from that of the cotyledons when carrying out such analyses, as it is the axis which will initially be the site of growth at the onset of germination.

A comparable situation is to be found in Douglas fir seeds where nearly 40% of lipid reserves have been reported to disappear by the third day of stratification while no changes were detected in controls held at 26°C (Ross, 1969). Although only a small increase in sugars was detected, 40% of lipid-derived carbon was respired, and some was converted to starch. It should be noted that this particular seed lot required only 10 days total cold-treatment in contrast to the much longer stratification times used by Taylor and Wareing (1979a).

A study of reserve mobilization during stratification of Norway maple with parallel controls held at 17°C, has recently been published by Davies and Pinfield (1980). Here the contents of free sugars, sucrose, and amino acids all increased, with the sugar increase occurring mainly in the cotyledons. Although these changes were significant, no apparent decrease in the lipid content could be detected; however, it was pointed out that the increase in sugars was very small in comparison to the total lipid content. The activity of the enzyme isocitrate lyase [EC 4.1.3.1] and the content of fructose-1,6-bisphosphate were also shown to increase dramatically during the cold treatment, the former not being detectable at all in the unchilled samples. The increases in free amino acids were concomitant with significant changes in proteolytic enzyme activities assayed with model substrates. This report clearly shows that reserve mobilization is a characteristic of the after-ripening process in Norway maple, prior to the onset of germination.

Perhaps the most extensive investigations on metabolism during cold after-ripening published to date have been carried out on apple seeds (Lewak and Rudnicki, 1977). In this species also there is an increase in free amino acid content accompanied by enhanced protease activities (Lewak *et al.*, 1975). The apple seed stores considerable lipid which gradually diminishes during the cold treatment, with a parallel increase in starch formation (Kawecki, 1970, and cited in Zarska-Maciejewska and Lewak, 1976; Lewak and Rudnicki, 1977). This group showed that the apple embryo contains both acidic and alkaline lipase [EC 3.1.1.3] activities. Most interestingly the acidic lipase, which increases first during the stratification, has a temperature optimum of 5°C, coincident with that of the optimal chilling temperature (Zarska-Maciejewska and Lewak, 1976). A similar lipase in extracts of hazel cotyledons also shows optimal activity at 5°C when assayed *in vitro* against a substrate of hazel oil (A. Smith and J. D. Ross, unpublished data).

Although not conclusively demonstrated it is very probable that some of the products of the lipolytic activity in apple embryos are partially conserved as starch, which has been shown to accumulate after the first 10 days stratification (Dawidowicz-Grzegorzewska and Lewak, 1978).

A recent study has been made of the ultrastructural and enzymic changes associated with the reserves stored within the hazel embryonic axis during the dormancy-breaking stratification treatment (Younis, 1982).

The hazel seed requires about 35 days at 5°C before attaining the maximum potential for germination on subsequent transfer to higher temperatures. During this period at 5°C, major reorganization takes place within the cortical cells of the radicle. After an initial period of imbibition and the associated increase in cellular volume common to seeds at both 5°C and 20°C, the cold-treated seeds rapidly begin to show evidence of heightened metabolic activity. The protein bodies become granular and internal dissolution begins by seven days; this continues until by the 28th day they appear as almost empty vacuoles (Fig. 3A, B). The lipid bodies also show some diminishment while migrating to the periphery of the cell (Fig. 3B). These events are well under way before any enhancement of the germination potential can be detected. Massive proliferation of membranes and membranous organelles reflects this increased metabolic activity. Not all the hydrolysed lipid is used for respiration or membrane biosynthesis as starch formation also occurs. Protease activity appears to increase linearly with chilling time, as does the activity of isocitrate lyase (Fig. 4A, B). The activities of both these enzymes respond rapidly in the unchilled seed following application of GA.

It is therefore possible to see a common pattern of reserve mobilization and utilization in many seeds that require stratification, in contrast to the claim, often repeated, that this is purely a consequence of the onset of germination. Indeed, this pattern appears to have been shown in all cases investigated. Whether these observations can serve as a basis for a model of dormancy breakage is debatable. There is some supporting circumstantial evidence in that application of exogenous GA, which generally brings about rapid germination in these species, also induces similar qualitative and quantitative changes in the complement of hydrolytic enzymes. However, the rapid responses to GA may be only an accelerated occurrence of the slower changes taking place at the cold stratification temperature.

VII. CONCLUSION

It is tempting to speculate that the dormant embryo is unable to resume growth as a consequence of a partial inhibition of protein synthesis. This inhibition could be the result of a number of separate, and not mutually exclusive factors, such as compartmentation of the proteases and their substrates, the presence of protease inhibitors, as shown in Scots pine, *Pinus sylvestris* (Salmia, 1981) and other species (Section 3.III of Volume 1), and an insufficiency of the protein synthesis machinery.

There are, of course, examples where this type of dormancy is unlikely,

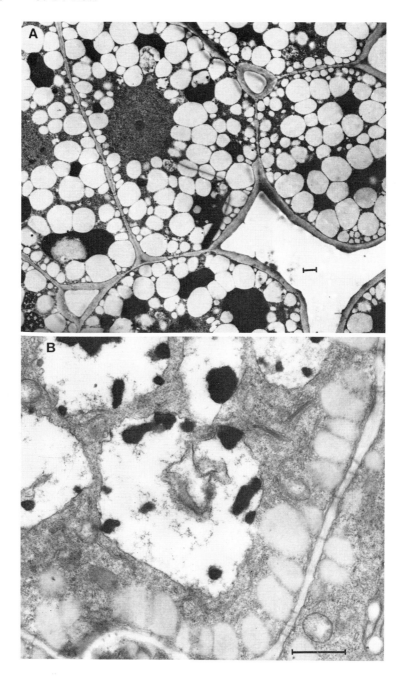

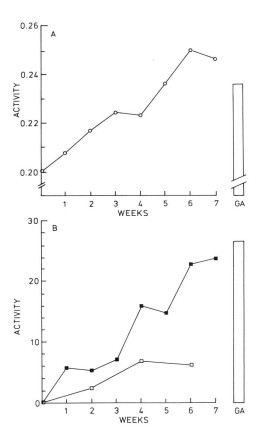

Fig. 4(A) Protease activity in extracts of hazel seeds during seven weeks stratification, assayed at 35°C against azocasein. Results expressed as A_{335}nm h^{-1} mg^{-1} protein. Bar represents activity of non-chilled seeds after incubation at 20°C in 1 mM GA_3 for 4 days (from Younis, 1982). (B) Isocitrate lyase activity in extracts of hazel seeds during 5°C stratification and 20°C warm-imbibition. Results expressed as nmoles substrate utilized min^{-1} mg^{-1} protein. Stratified seeds (■-■■-■), warm-imbibed seeds (□-□-□-□). Bar represents activity in non-chilled seeds after incubation at 20°C in 1 mM GA_3 for 4 days (from Younis, 1982).

Fig. 3(A) Electronmicrograph of cells of hazel embryonic axis. Fixation by glutaraldehyde-osmium tetroxide, after one day imbibition. Cells densely packed with reserve protein and lipid bodies. Bar represents 1 μm (from Younis, 1982). (B) Electronmicrograph of part of a cell of the hazel embryonic axis after four weeks stratification. Note dissolution of protein bodies, migration of stored lipid, and proliferation of membranes. Fixed as Fig. 3A, bar represents 1 μm (from Younis, 1982).

such as in the lower cocklebur seed, where it has been reported that the rate of protein synthesis is actually higher in the dormant than in the non-dormant seed for the first 9 hours after imbibition, and it is not until the 12th hour that the situation is reversed (Satoh and Esashi, 1979). In contrast to this, experiments with pear seeds have demonstrated that there is an increased capacity for protein synthesis resulting from changes in the constituent machinery. Increases in the activities of aminoacyl-tRNA synthetases were detectable after only 5 days stratification (Tao and Khan, 1974), whereas isolated polyribosomes showed an enhanced capacity for protein synthesis peaking after 13 days, which coincided with the period required for renewed growth of excised embryos (Alscher-Herman and Khan, 1980). In addition the embryonic axis of Norway maple has been shown to accumulate rRNA during stratification (Slater and Bryant, 1982); this would also tend to support the proposal that dormancy breakage is linked with an enhancement of the capacity for protein synthesis.

One of the earliest intracellular events visible by transmission electron microscopy in the hazel axis is the breakdown of storage proteins within their enveloping membrane. It has been tentatively suggested by Bewley and Black (1982) that some dormancy breaking phenomena may result from the characteristics of the membrane state in relation to temperature. It is possible to employ this concept to propose that low temperatures, which result in the protein body membrane assuming a quasi-crystalline state, cause the injection of extant proteases into the matrix of the protein body. At higher temperatures, with the membrane being in the liquid-crystalline phase, the enzymes may remain dissolved in, or strongly associated with the membrane. In this manner low temperatures could induce the enzymic breakdown of the stored protein, which would thereby provide the amino acids necessary for the *de novo* protein synthesis required for growth.

Bewley and Black (1982) also draw attention to the fact that many chemicals such as ethanol, ethyl ether, and chloroform are all variously reported to break dormancy in some species, and Taylorson and Hendricks (1979) have pointed out that such anaesthetics are believed to have their physiological action through their effects on membranes. Another point, made by Raven and Rubery (1982), which may be pertinent to this line of thought is that the major plant hormones are all relatively hydrophobic and lipid soluble when present as the uncharged species. Thus they too could affect membrane properties when applied exogenously to the dormant seed.

The most difficult step in the formulation of any model of seed dormancy is linking the perception of the environment to consequent changes in metabolism. It is therefore pleasing to see many different lines of inquiry converging on one area. The physicochemical properties of membranes and how they may hold the key to the block to protein synthesis and subsequent germination are likely to be the focus of much future work. Although this may be only a reflection of contemporary fashion, as has so often been the case in the past, it is to be hoped that our understanding of this fascinating subject will soon be usefully enlarged.

REFERENCES

Adkins, S. W., and Ross, J. D. (1981a). *Plant Physiol.* **67**, 358–362.
Adkins, S. W., and Ross, J. D. (1981b). *Plant Physiol.* **68**, 15–17.
Alscher-Herman, R., and Khan, A. A. (1980). *Physiol. Plant.* **48**, 285–291.
Alscher-Herman, R., Musgrove, M., Leopold, A. C., and Khan, A. A. (1981). *Physiol. Plant.* **52**, 156–160.
Amen, R. D. (1968). *Bot. Rev.* **34**, 1–31.
Arias, I., Williams, P. M., and Bradbeer, J. W. (1976). *Planta* **131**, 135–139.
Bengochea, T., Dodds, H. J., Evans, D. E., Jerie, P. H. Niepel, B., Sharri, A. R., and Hall, M. A. (1980). *Planta* **148**, 397–406.
Berrie, A. M. M., Don, R., Buller, D. C., Alam, M., and Parker, W. (1975). *Plant Sci. Lett.* **6**, 163–173.
Berrie, A. M. M., Bullet, D., Don, R., and Parker, W. (1979). *Plant Physiol.* **63**, 758–764.
Bewley, J. D., and Black, M. (1978). 'Physiology and Biochemistry of Seeds', Vol. I. Springer-Verlag, New York.
Bewley, J. D., and Black, M. (1982). 'Physiology and Biochemistry of Seeds', Vol. II. Springer-Verlag, New York.
Borkowska, B., and Rudnicki, R. (1975). *Fruit Sci. Rep.* **2**, 1–16.
Bradbeer, J. W. (1968). *Planta* **78**, 266–276.
Brown, N. A. C., and Van Staden, J. (1973). *Physiol. Plant.* **28**, 388–392.
Carpita, N. C., Nabors, M. W., Ross, C. W., and Petretric, N. L. (1979). *Planta* **144**, 225–233.
Chancellor, R. J., Parker, C., and Teferedegn, T. (1971). *Pestic. Sci.* **2**, 35.
Davies, H. V., and Pinfield, N. J. (1980). *Z. Pflanzenphysiol.* **96**, 59–65.
Dawidowicz-Grzegorzewska, A., and Lewak, St (1978). *New Phytol.* **81**, 99–103.
De Klerk, G. J., and Linskens, H. F. (1979). *Biochem. Physiol. Pflanzen.* **174**, 537–545.
Dennis, F. G., Martin, G. C., Gaskin, P., and MacMillan, J. (1980). *Planta* **147**, 376–377.
Eeuwens, C. J., and Schwabe, W. W. (1975). *J. Exp. Bot.* **26**, 1–14.
Egley, G. H., and Dale, J. E. (1970). *Weed Sci.* **18**, 586–589.
El-Antably, H. M. M. (1976). *Biochem. Physiol. Pflanzen.* **170**, 51–58.
Eplee, R. E. (1975). *Weed Sci.* **23**, 433–436.
Esashi, Y., Okazaki, M., Yanai, N., and Hishinuma, K. (1978). *Plant, Cell Physiol.* **19**, 1497–1506.
Esashi, Y., Ohhara, Y., Okazaki, M., and Hishinuma, K. (1979a). *Plant, Cell Physiol.* **20**, 349–361.
Esashi, Y., Wakabayashi, S., Tsukada, Y., and Satoh, S. (1979b). *Plant Physiol.* **63**, 1039–1043.
Evans, M., Black, M., and Chapman, J. (1975). *Nature* **258**, 144–145.
Evans, D. E., Bengochea, T., Cairns, A. J., Dodds, J. H., and Hall, M. A. (1982a). *Plant, Cell Environ.* **5**, 101–107.

Evans, D. E., Dodds, J. H., Lloyd, P. C., apGwynn, I., and Hall, M. A. (1982b). *Planta* **154**, 48-52.
Frankland, B., and Wareing, P. F. (1966). *J. Exp. Bot.* **17**, 596-611.
Frankland, B., Jarvis, B. C., and Cherry, J. H. (1971). *Planta* **97**, 39-49.
Gosling, P. G., and Ross, J. D. (1980). *Planta* **148**, 362-366.
Gosling, P. G., and Ross, J. D. (1981). *Phytochem.* **20**, 31-33.
Hall, M. A., Cairns, A. J., Evans, D. E., Smith, A. R., Smith, P. G., Taylor, J. E., and Thomas, C. J. R. (1982). *In* 'Plant Growth Substances 1982' (P. F. Wareing, ed.), pp. 375-383. Academic Press, London.
Heckler, M., and Köhler, K. (1979). *Dev. Biol.* **69**, 270-280.
Hendricks, S. B., and Taylorson, R. B. (1972). *Nature* **237**, 169-170.
Ho, D. T.-H., and Varner, J. E. (1976). *Plant Physiol.* **57**, 175-178.
Ihle, J. N., and Dure, L. S. (1972). *J. Biol. Chem.* **247**, 5048-5055.
Jarvis, B. C., and Shannon, P. R. M. (1981). *New Phytol.* **88**, 31-40.
Jones, J. F., and Hall, M. A. (1979). *Plant Sci. Lett.* **16**, 87-93.
Jones, J. F., and Hall, M. A. (1981). *Ann. Bot.* **48**, 291-300.
Kahn, A. A. (1971). *Science* **171**, 853-859.
Kahn, A. A. (1975). *Bot. Rev.* **41**, 391-420.
Karssen, C. M. (1982). *In* 'Plant Growth Substances 1982' (P. F. Wareing, ed.), pp. 623-632. Academic Press, London.
Katoh, H., and Esashi, Y. (1975). *Plant, Cell Physiol.* **16**, 687-696.
Kawecki, Z. (1970). *Roczn Nauk. Roln. ser A.* **96**, 35-51.
Ketring, D. L., and Morgan, P. W. (1969). *Plant Physiol.* **44**, 326-330.
Ketring, D. L., and Morgan, P. W. (1972). *Plant Physiol.* **50**, 382-387.
King, R. W. (1976). *Planta* **132**, 43-51.
La Croix, L. J., and Jaswal, A. S. (1967). *Plant Physiol.* **42**, 479-480.
La Croix, L. J., and Jaswal, A. S. (1973). *Can. J. Bot.* **51**, 1267-1270.
Leung, D. W. M., and Bewley, J. D. (1981a). *Nature* **289**, 587-588.
Leung, C. W. M., and Bewley, J. D. (1981b). *Planta* **152**, 436-441.
Lewak, St, and Rudnicki, R. M. (1977). *In* 'The Physiology and Biochemistry of Seed Dormancy and Germination' (A. A. Kahn, ed.), pp. 193-217. Elsevier/North Holland Biomedical Press, Amsterdam.
Lewak, St, Rychter, A., and Zarska-Maciejewska, B. (1975). *Physiol. Vég.* **13**, 13-22.
Martin, G. C., Mason, M. I. R., and Ford, H. L. (1969). *J. Amer. Soc. Hort. Sci.* **94**, 13-17.
Metzger, J. D., and Sebesta, D. K. (1982). *Plant. Physiol.* **70**, 1480-1485.
Murphy, J. B., and Noland, T. L. (1982a). *Plant Physiol.* **69**, 428-431.
Murphy, J. B., and Noland, T. L. (1982b). *Plant Physiol.* **70**, 1410-1412.
Nabors, M. W., Krugrens, P., and Ross, C. W. (1974). *Planta* **117**, 361-365.
Nagy, M. (1980). *Acta Agron. Acad. Sci. Hung.* **29**, 1-11.
Olatoye, S. T., and Hall, M. A. (1973). *In* 'Seed Ecology' (W. Heydecker, ed.), pp. 233-249. Butterworths, London.
Paul, K. B., Patel, C. S., and Biswas, P. K. (1973). *Physiol. Plant.* **28**, 530-534.
Pinfield, N. J., and Davies, H. V. (1978). *Z. Pflanzenphysiol.* **90**, 171-181.
Pinfield, N. J., and Stobart, A. K. (1969). *New Phytol.* **68**, 993-999.
Pinfield, N. J., and Stobart, A. K. (1972). *Planta* **104**, 134-145.
Pollock, B. M., and Olney, H. O. (1959). *Plant Physiol.* **34**, 131-142.
Proctor, J. T. A., and Dennis, F. G. (1968). *J. Am. Soc. Hort. Sci.* **93**, 110-114.
Raven, J. A., and Rubery, P. H. (1982). *In* 'The Molecular Biology of Plant Development' (H. Smith and D. Grierson, eds), pp. 28-48. Blackwell Scientific Publications, Oxford.

Roberts, E. H. (1969). *In* 'Society for Experimental Biology Symposia XIII', pp. 161–192. Cambridge University Press, Cambridge.
Roberts, E. H. (1973). *In* 'Seed Ecology' (W. Heydecker, ed.), pp. 189–218. Butterworths, London.
Roberts, E. H., and Smith, R. D. (1977). *In* 'The Physiology and Biochemistry of Seed Germination' (A. A. Khan, ed.), pp. 385–412. Elsevier/North Holland Biomedical Press, Amsterdam.
Ross, J. D., and Bradbeer, J. W. (1968). *Nature* **220**, 85–86.
Ross, J. D., and Bradbeer, J. W. (1971a). *Planta* **100**, 288–302.
Ross, J. D., and Bradbeer, J. W. (1971b). *Planta* **100**, 303–308.
Ross, S. D. (1969). *Bot. Gaz.* **130**, 271–275.
Salmia, M. A. (1981). *Physiol. Plant.* **53**, 39–47.
Satoh, S., and Esashi, Y. (1979). *Physiol. Plant.* **47**, 229–234.
Satoh, S., and Esashi, Y. (1980). *Physiol. Plant.* **48**, 243–246.
Schonbeck, M. W., and Egley, G. H. (1980). *Plant Physiol.* **65**, 1149–1154.
Schonbeck, M. W., and Egley, G. H. (1981). *Plant Physiol.* **68**, 175–179.
Shannon, P. R. M., Sharrard, R. M., and Jarvis, B. C. (1981). *Plant, Cell Physiol.* **22**, 1293–1304.
Simmonds, J. A., and Simpson, G. M. (1972). *Can. J. Bot.* **50**, 1041–1048.
Sinska, I., and Lewak, St (1970). *Physiol. Vég.* **8**, 661–667.
Sinska, I., Lewak, St, Gaskin, P., and MacMillan, J. (1973). *Planta* **114**, 359–364.
Slater, R. J., and Bryant, J. A (1982). *Ann. Bot.* **50**, 141–149.
Sondheimer, E., Tzou, D. S., and Galson, E. C. (1968). *Plant Physiol.* **43**, 1443–1447.
Suzuki, S., and Taylorson, R. B. (1981). *Plant Physiol.* **68**, 1385–1388.
Tao, K. L., and Khan, A. A. (1974). *Biochem. Biophys. Res. Commun.* **59**, 764–769.
Taylor, J. S., and Wareing, P. F. (1979a). *Plant, Cell Environ.* **2**, 165–171.
Taylor, J. S., and Wareing, P. F. (1979b). *Plant, Cell Environ.* **2**, 173–179.
Taylorson, R. B., and Hendricks, S. B. (1977). *Annu. Rev. Plant Physiol.* **28**, 331–354.
Taylorson, R. B., and Hendricks, S. B. (1979). *Planta* **145**, 507–510.
Tillberg, E. (1977). *Plant Physiol.* **60**, 317–319.
Tillberg, E., and Pinfield, N. J. (1981). *Physiol. Plant.* **53**, 34–38.
Trewavas, A. (1981). *Plant, Cell Environ.* **4**, 203–228.
Tuan, D. Y. H., and Bonner, J. (1964). *Plant Physiol.* **39**, 768–772.
Upadhyaya, M. K., Naylor, J. M., and Simpson, G. M. (1982). *Physiol. Plant.* **54**, 419–424.
Van Staden, J., Webb, D. P., and Wareing, P. F. (1972). *Planta* **104**, 110–114.
Villiers, T. A. (1971). *New Phytol.* **70**, 751–760.
Villiers, T. A. (1972). *New Phytol.* **71**, 153–160.
Villiers, T. A., and Wareing, P. F. (1960). *Nature* **185**, 112–114.
Vogt, A. R. (1974). *For. Sci.* **20**, 187–191.
Webb, D. P., and Wareing, P. F. (1972). *Planta* **104**, 115–125.
Webb, D. P., Van Staden, J., and Wareing, P. F. (1973a). *J. Exp. Bot.* **24**, 105–116.
Webb, D. P., Van Staden, J., and Wareing, P. F. (1973b). *J. Exp. Bot.* **24**, 741–750.
Williams, P. M., Ross, J. D., and Bradbeer, J. W. (1973). *Planta* **110**, 303–310.
Williams, P. M., Bradbeer, J. W., Gaskin, P., and MacMillan, J. (1974). *Planta* **117**, 101–108.
Wood, A., and Bradbeer, J. W. (1967). *New Phytol.* **66**, 17–26.
Younis, S. A. (1982). 'Intracellular Reorganisation During Stratification of Hazel Seeds (*Corylus avellana* L.)'. Ph.D. Thesis, University of Reading, U.K.
Zarska-Maciejewska, B., and Lewak, St (1976). *Planta* **132**, 177–181.

CHAPTER 3

Early Events in Germination

E. W. SIMON

I.	Introduction	77
II.	The Phase of Imbibition	78
	A. Imbibition	78
	B. Leakage	83
	C. Damage from Soaking	91
	D. Chilling Injury during Imbibition	92
III.	Metabolism	94
	A. Respiration	95
	B. Protein Synthesis	101
	C. RNA Synthesis	103
	D. DNA Synthesis and Cell Division	106
	E. Polyamines	108
	F. The Lag Phase and Subsequent Cell Elongation	108
IV.	Concluding Discussion	110
	References	111

I. INTRODUCTION

Seeds do not normally germinate until they have undergone a considerable period of growth and development, accumulating reserves and finally becoming air-dry. They are then quiescent and can be stored for months or years without harm, but once supplied with water they become hydrated again and embark on a second, but different, phase of activity. This phase results in outgrowth of the root and later the shoot at the expense of the reserve materials. The dry seed therefore lies at the turning point in seed life, between the phase of development and the phase of germination. For this reason the events that occur as a seed imbibes and prepares to germinate acquire a special interest. These events may reveal how it is possible for the addition of water to a dry seed to start that seed off on a new pattern of development.

The account that follows centres on the normal process of germination and omits discussion of seeds that are dormant (Chapter 2), seeds that have lost viability through prolonged storage or artificial ageing, and those seeds that remain moist when mature, the so-called recalcitrant seeds (King and Roberts, 1979). The early events considered in this chapter are those leading up to visible germination, the protrusion of the root through the testa.

Three phases of development can be recognized: (i) imbibition; (ii) the lag phase; and (iii) germination. Air-dry seeds have a low water content, around 15% or less, but given access to water they will imbibe rapidly, with tissues generally reaching a water content between 30% and 50% in a day or two (Hunter and Erickson, 1952). At this time the seed enters a lag phase: it has swollen and become heavier, and is now metabolizing actively. If the temperature is appropriate and the supplies of oxygen adequate, the seed will in due course enter the third phase, that of germination (Esashi and Leopold, 1968).

These three phases can be seen in Figure 1. Although the seed shows no obvious and outward sign of change during the lag phase, the effect of temperature on the duration of this phase can be taken as evidence that preparations are being made at the metabolic level for the eventual act of germination. Section II of this chapter considers events during the phase of imbibition; Section III discusses metabolism from the beginning of imbibition up to the time of germination.

II. THE PHASE OF IMBIBITION

A. Imbibition

The seeds of most plant species are air-dry at the time they are shed from the parent plant, with a water content below about 15% of fresh weight. The precise water content of seeds depends on the nature of the major reserves they contain; starch rich seeds have a higher water content than those that store oil (Touzard, 1975). Seed water content depends also on the relative humidity of the atmosphere in which the seeds are stored. This provides a way of determining seed water potential, the seeds being brought to equilibrium with air at a particular relative humidity and hence known water potential. Air-dry seeds of rape (*Brassica napus*), wheat (*Triticum aestivum*) and maize (*Zea mays*) have water potentials of the order of -1000 bars (-100 MPa, Shaykewich and Williams, 1973; Shaykewich, 1973), a figure in accord with the finding that *Xanthium strumarium* seeds

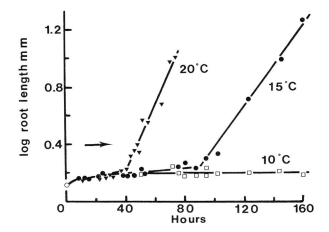

Fig. 1. Root extension in cucumber seeds. The root elongates little in the 10-hour phase of imbibition, remains constant during the lag phase at 20°C, and then begins to grow rapidly; when it is 2.5 mm long (arrow) it bursts through the testa. At 15°C the lag phase is longer while at 10°C there is no germination, the root never entering the phase of rapid extension (data of Simon *et al.*, 1976).

are unable to take up water from saturated lithium chloride solution, which has an osmotic potential of about −965 atmospheres (−98 MPa, Shull, 1913).

Exposure of air-dry seeds to free water with a water potential of zero sets up a very high potential gradient and imbibition results. Waggoner and Parlange (1976) have followed the imbibition of water by embryos of pea (*Pisum sativum*), that is, pea seeds from which the testa has been removed. There is an initial period of rapid water uptake which lasts about 30 minutes in peas (Fig. 2), and 5–10 minutes in soybean (*Glycine max*); the initial inrush of water into soybean cotyledons is so fast that Parrish and Leopold (1977) consider it to be 'uncontrolled'. There follows a slower, linear phase of water uptake lasting 5–10 hours after which imbibition slackens off and finally comes to a halt. Sectioning of pea embryos and treating the cut surface with cobalt chloride reveals a clear boundary between the outer region, which has been wetted, and the inner tissues, which are still dry. The existence of this wetting front is evidence that the dry seed tissues offer a greater resistance to the inflow of water than those that have already been wetted. As imbibition progresses the wetting front moves steadily inwards and the tissues already moistened become wetter.

It is well known that seeds of some species have a hard testa which prevents imbibition until such time as it has been damaged mechanically or

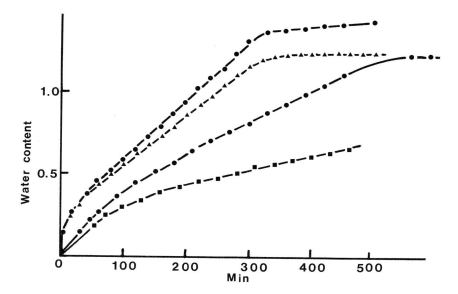

Fig. 2. The water content of imbibing pea embryos. 'Water content' is the volume of water per unit dry volume in the wetted region of the peas: the central 'dry' region of the peas contains little water and is not included in the data. Each curve records data obtained with a batch of 10 embryos (data of Waggoner and Parlange, 1976).

softened by organisms of decay (Chapter 1). Even a relatively thin seed coat like that of pea (Fig. 15 of Chapter 1, Volume 1) delays imbibition. Peas with an intact seed coat show little or no imbibition for the first 3 hours or so (Fig. 3). However, if the testa is removed, the resulting embryos imbibe rapidly from the start. Commercial seed lots of pea are likely to have some intermediate behaviour (Fig. 3), because they have been damaged during harvest, sorting or packing, and the testa has been cracked (Matthews *et al.*, 1980).

Figures 1, 2 and 3 show the time course of imbibition by large seeds. In small seeds the whole process of imbibition may be completed in a relatively short time — about 30 minutes for celery (*Apium graveolens*).

The low water potential of dry seeds or embryos can be attributed principally to matric forces which rapidly diminish on wetting. The steep gradient of water potential thus vanishes when the wetting front has reached the central tissues and imbibition then stops.

Imbibition can be made into a more gradual process by placing embryos in a solution of very low water potential instead of water. The rate of imbibition is also reduced if seeds are given only limited access to water:

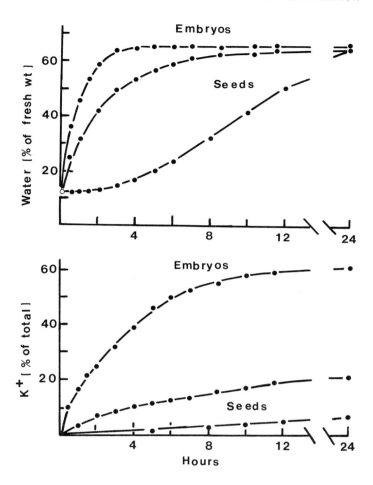

Fig. 3. Imbibition of water (above) and leakage of potassium (below) from 50 pea seeds or embryos cv. Kelvedon Wonder placed in 100 mL water. Two curves are shown for seeds in each graph: the lower curves were obtained with greenhouse-grown seed which are almost all intact, and the upper curve with commercial seed which has some degree of testa damage (from Simon and Mills, 1983).

seeds placed in a relatively dry soil, or exposed to moist air imbibe more slowly than those placed on moist filter paper, and they in turn imbibe more slowly than seeds immersed in water.

Water is not the only substance to enter a seed during imbibition. Côme (1975) has pointed out that the inflow of water occludes the pores and channels through which oxygen gas might otherwise diffuse to an imbibing embryo. Without gaseous diffusion, the small amount of oxygen carried to the embryo dissolved in the water of imbibition assumes a special significance in supporting respiration. In addition, under experimental conditions (or in the soil), seeds may be exposed to a solution rather than to pure water. Solutes then enter the seed as well as water, for the imbibed seeds weigh more than controls imbibed in water and if the imbibed seeds are washed well, their weight falls to that of the controls (Atkins, 1909). Presumably the solutes enter the apoplast from which they can be easily removed by washing.

As seeds imbibe, their volume increases. By the time air-dry seeds have imbibed sufficiently to attain a water content of 40%, their volume has increased by 30%-40% (Kollöffel, 1967; Shaykewich, 1973). This swelling is in part a reversal of the shrinkage that occurs during the final stage of development when cells decrease in size and their walls become corrugated (Lott, 1974). Imbibing seeds do not swell simultaneously and uniformly throughout their tissues. Clearly the wetting front reaches the outer regions of a seed some time before it penetrates to the inner tissues, so that in pea, for instance, the testa and axis are completely wetted at a relatively early stage of imbibition (Houben, 1966). Spurńy (1964, 1973) has used time-lapse photography to follow the volume changes that occur during the imbibition of pea seeds. During the first hour the testa swells to the extent of leaving a space between it and the embryo. This space may be water filled, or it may be occupied by a gas phase. As water permeates the embryo, it too swells, eventually occupying the whole volume inside the testa, then by 4-5 hours the embryo begins to stretch the slightly elastic testa. The root tip bursts through the testa by 25 hours under these conditions (21°C).

Isolated cotyledons may swell unevenly when placed in water, as reported for some varieties of common bean (*Phaseolus vulgaris*) (Pollack and Manalo, 1970; Retzlaff, 1970) and soybean (Duke and Kakefuda, 1981). The tensions that arise may cause the cotyledons to develop transverse cracks and to shed thin strips of tissue into the water. After 30-60 minutes the water around soybean embryos 'becomes turbid and chalky in colour, suggesting that cellular rupture has occurred' (Duke and Kakefuda, 1981). Isolated axes behave in the same way; damage to the surface layers of cells can be detected by microscopy in the first minute of imbibition, the cells becoming ruptured so that their cytoplasmic contents are free to leak out (Dunn et al., 1980). These symptoms of damage due to imbibitional swelling in soybean can be prevented almost entirely by raising the initial water content of the tissues to 17% or more. The presence of the seed coat, which presumably moderates the inflow of water, also alleviates damage.

The imbibition of water by seeds is accompanied by a release of gas, some of it air that was originally present in such places as the interstices of cell walls, starch grains and intercellular spaces. The displacement of this gas may be quite slow; some is still present in the starch of pea cotyledons 5 days after the start of imbibition and can be seen in illustrations of fresh sections (Figs 2, 3, 5 of Flinn and Smith, 1967).

In addition to this gas that is slowly displaced by water, some gas is released within minutes of wetting dry material (Haber and Brassington, 1959; Côme, 1971; Parrish and Leopold, 1977). In Warburg experiments in which water is tipped onto dry material from a side-arm, gas is released rapidly by both viable and heat-killed seeds, as well as by plant materials such as tea leaves (*Camellia sinensis*) and potato starch (*Solanum tuberosum*). If dry seeds are placed instead in a water-saturated atmosphere, both imbibition and gas release are slowed down, but the total amount of gas evolved remains unchanged. The quantity of gas released by seeds varies from 0.45 to 1.8 mL per g seed depending on the species. This gas is thought to be adsorbed on surfaces or occluded within dry seed constituents.

The gas released on imbibition is thus a mixture of that displaced from the gas phase and that released from surfaces. No chemical analysis of this gas mixture has yet been made, but the evidence available suggests that it may be air with as much as 25%–50% CO_2 (Côme, 1971) and a high content of water vapour (Parish and Leopold, 1977).

As the water front penetrates the seed during imbibition it will displace gas that was already present and release gas that was held by surface forces. This gas may accumulate within the seed as a bubble, diminishing in size as it becomes more and more confined by the advancing water front. Parrish and Leopold (1977) suggest that the bubble of gas may become so compressed as to enhance its solubility in the water of imbibition. Alternatively, if the bubble is not strictly spherical from the start, the advancing water front may cause it to balloon out in one direction so that it eventually forms a puncture through the wet outer tissues; the gas will then escape from the seed rather than becoming compressed and solubilized inside it. Gas escaping from a seed in this way may well account for the multitude of small bubbles that first adhere to the surface of seeds immersed in water, and later float up to the surface.

B. Leakage

When dry seeds are placed in water solutes leak out of them. The principal facts about leakage, established mainly through experiments with a few large-seeded legumes have been reviewed elsewhere (Simon, 1974, 1978; Simon and Mills, 1983). In many experiments leakage is followed either by monitoring the appearance of potassium in the bathing liquid or by measuring its total conductivity. The two measures show essentially the same pattern, leakage being fastest at the start of imbibition and coming to a halt after about a day (Fig. 3). By this time pea seeds have lost about 10% of their potassium. If the testas are removed from pea seeds and the resulting embryos are placed in water, leakage starts off much more rapidly and continues until about 60% of the potassium has been moved out of the tissues. As with imbibition, seeds with cracked testas occupy an intermediate position between seeds and embryos; they imbibe faster than intact seeds, and lose more solutes. Perhaps because leakage is enhanced by removal of the testa, most work on leakage has been done with isolated embryos, or seeds in which the testa has been deliberately damaged by scarification. One outcome of surgical removal of the testa is that the resulting embryos give rise to weaker seedlings than intact seeds, with shorter stems and less dry weight in root and shoot (Larson, 1968). Respiration rate is also affected, the respiration of embryos imbibed for 24 hours without testas being less than half that of embryos imbibed within seed coats (Powell and Matthews, 1978).

Leakage from pea embryos is most rapid in the first moments of imbibition, the rate then slowing down until by 10–20 minutes it reaches a steady rate that is maintained for the next half-hour or more, finally coming to a halt after about a day (Fig. 3; Simon and Raja Harun, 1972; Matthews and Rogerson, 1976; Powell and Matthews, 1981). Similar results have been obtained with soybean cotyledons (Parrish and Leopold, 1977) and scarified seeds of birdsfoot trefoil (*Lotus corniculatus*) (McKersie and Stinson, 1980).

Many different substances leak out of seeds and embryos including amino acids, organic acids, sugars, phenolics, phosphate and potassium ions, gibberellic acid and proteins. The rate of leakage is not the same for each substance. Thus McKersie and Stinson (1980) record that in the first 2 hours scarified *Lotus* seeds lose 13% of their potassium and 9% of soluble protein, but only 3.6% of sugar, 3.1% of phosphate and 1.4% of amino acids. As the process of imbibition was not completed within the 2-hour period, these differences may reflect in some measure any unevenness of distribution within the seed, for the wetting front would not have reached solutes that were located mainly towards the centre of the seed in that time. Duke and Kakefuda (1981) have followed the loss of soluble enzymes such as glucose-6-phosphate dehydrogenase [EC 1.1.1.49] and glutamate dehydrogenase [EC 1.4.1.2] from embryos of pea and peanut (*Arachis hypogaea*). After 6 hours imbibition up to 1% or 2% of the total enzyme activity appeared in the supernatant (a figure which may be compared to the loss of 50% of potassium in the same time — Fig. 3). There was also some minor loss of mitochondrial marker enzymes such as cytochrome oxidase [EC 1.9.3.1] and fumarase [EC 4.2.1.2], amounting to 5% of the total at most. Embryos of soybean and common bean leak more, losing up to 17% of both soluble and mitochrondrial enzymes, because in the absence of a testa the cotyledons of these species crack open and split off strips of tissue during imbibition.

Leakage is most rapid during the early stages of imbibition when air-dry seeds or embryos are first placed in water. As imbibition progresses, the rate of leakage declines. Likewise if air-dry seeds or embryos are allowed to take up a small amount of water from moist air or from moist filter paper, such partially imbibed material shows less leakage when immersed in water. On the other hand, if embryos are immersed in water for 30 minutes and then dried back over calcium chloride to their original weight, they can be allowed to imbibe again for 30 minutes, and so on for several cycles; such embryos, which always remain relatively dry, leak rapidly each time they are immersed in water (Simon and Raja Harun, 1972). Leakage from peas is largely suppressed once they have reached a water content of 30% or more, which corresponds to a water potential of -80 bars (-8 MPa) or higher (Simon and Wiebe, 1975). The threshold water content for leakage from root-shoot axes of soybean is 17% (Dunn et al., 1980) and from axes of Lima bean (*Phaseolus lunatus*), 20%-25% (Pollock, 1969).

In summary, it is clear that rapid leakage only occurs when dry tissues are first placed in water. Such leakage is by no means restricted to imbibition by dry seeds, for solutes also leak away from imbibing mosses (Gupta, 1976), liverworts (Dhindsa and Bewley, 1977), lichens (Farrar and Smith, 1976) and nematodes (Crowe et al., 1979).

Several lines of thought point to a cytoplasmic rather than apoplastic origin for the solutes that leak out in the early stages of imbibition by pea embryos. First, the range and variety of solutes in question seems to mirror the composition of cytoplasm. Imbibing embryos transferred to fresh water at frequent intervals continue to leak, although apoplastic solutes would presumably be washed away (Matthews and Rogerson, 1976). The same conclusion emerges from experiments in which embryos are subjected to a series of imbibition and drying cycles (Simon and Raja Harun, 1972). Finally, it is hardly credible that 60% of the potassium of a pea seed could be extracellular, since potassium is a major constituent of phytin located inside protein bodies (Chapter 4 of Volume 1).

Two major hypotheses have been put forward to account for leakage during imbibition. The first, proposed initially by Larson (1968) and Perry and Harrison (1970), lays stress on the indiscriminate force imposed upon cell membranes by the inrush of water during imbibition. The gradient of water potential is so high at the start of imbibition that it could well disrupt the organization of cell membranes, scattering the component phospholipids and proteins far from their original positions. In the absence of intact membranes, the cytoplasmic contents of these cells would be dispersed by the turbulent flow of water, hardly any of the original solutes remaining within the cells. The presence of such dead and disorganized cells can be detected with Evan's Blue, a dye which is excluded from cells with intact membranes (Gaff and Okong'O-Ogola, 1971; Kanai and Edwards, 1973). The dye penetrates the outermost cells of cotyledons of soybean and pea imbibing in 1% Evan's Blue (Duke and Kakefuda, 1981; Simon and Mills, 1983). In pea the cells that stain occupy less than 5% of the volume of the embryo. When dry soybean axes are placed in water some of the surface cells can be seen to rupture allowing cytoplasmic contents to leak out (Dunn et al., 1980). A more or less complete loss of enzymes from these peripheral cells could well account for the loss of up to 5% of the enzyme present in whole embryos of pea or peanut (Duke and Kakefuda, 1981). The inner cells of imbibed pea embryos, occupying 95% of the volume, fail to stain with Evan's Blue. This indicates that they have retained intact membranes. These cells are presumably not wetted until the inflow of water has slackened to such an extent that water can pass through membranes rather than brushing them aside. The testa of pea may act in the same way, for it moderates the inflow of water to the embryo and less than 0.7% of the embryo then stains with Evan's Blue (Mills, 1983).

The tetrazolium test has also been used to assess the state of cell membranes. Pea embryos imbibed in water show a poor and patchy stain, up to half the volume of the embryo remaining unstained. At first this was taken as evidence that the outer cells were 'dead', but it has subsequently become clear that, as in the tetrazolium test for freezing injury (Steponkus, 1970), failure to stain results from lack of dehydrogenase substrate (Powell and Matthews, 1981; Mills, 1983). The tetrazolium test thus provides evidence of substrate loss from embryo cells, but leaves open the question of mechanism.

If membrane rupture were the only mechanism of leakage it would be difficult to account for: (i) the loss of 60% of the potassium from pea embryo (Fig. 3); (ii) the loss of tetrazolium substrate from as much as 50% of the volume of pea embryos; (iii) the continuation of leakage for a period of hours (Fig. 3), long after the outermost cells had become wetted; and (iv) leakage occurring even when imbibition is slow, as when pea embryos are placed in saturated lithium chloride or 55% sucrose (Simon and Mills, 1983).

It seems that some additional mechanisms must be involved in leakage. One possibility is that membranes are not ruptured by the inflow of water but on the contrary are reorganized or repaired (Simon, 1974, 1978). The phospholipids and proteins of membranes normally adopt the bilayer configuration by virtue of their amphipathic nature, forming a central hydrophobic domain coated by the polar and charged regions which face the aqueous environment on either side of the membrane. The water has a role in stabilizing the bilayer configuration, both in living cells and under *in vitro* conditions.

Many of the observations on leakage can be understood if it is supposed that the membrane constituents come to adopt some other configuration under the dry conditions that prevail in maturing seeds. During imbibition the reverse changes would happen, cell cytoplasm becoming hydrated and lamellar membranes being reformed from their constituent molecules. Solutes cannot leak out of dry cells because of the lack of solvent, but leakage becomes a possibility once they are hydrated. It is suggested that each cell would experience a short period of leakiness as its water content rose and before the membranes had reverted fully to their semi-permeable lamellar condition. The reassembly of phospholipid and protein molecules to form lamellar membrane could not happen in the face of the violent inrush of water that the outer cells of pea embryos experience. Orderly reassembly could only occur in cells that are hydrated relatively slowly, such as those towards the interior of a large seed.

Each of the inner cells of a large seed would therefore contribute to the overall leakage from a seed or embryo as the wetting front reached it, so that leakage would continue at least until imbibition had come to a halt, and a relatively high proportion (up to 60% according to Fig. 2) of the solutes present could have been released. The chief characteristics of the two hypotheses about leakage are summarized in Table I along with an indication of the situations to which each is thought to apply.

When the second hypothesis was first proposed (Simon, 1974) it seemed to be in accord with biophysical work on animal membrane systems and

Table I. Hypotheses on the leakage of solutes from seeds

	Membrane rupture	Membrane repair
Rate of imbibition	Must be rapid	Moderate or slow
Cells affected	Only the outermost cells	Inner cells
Permeability to Evan's blue	Can enter cells	Excluded from cells
Solute release	Nearly 100% release of solutes from cells with membrane rupture	Up to half the potassium present is released from pea embryos
Time scale	Could account for initial rapid leakage from pea embryos	Could account for slow, long-continued leakage from pea embryos

model phospholipid–water systems. The evidence encouraged the view that lamellar membrane systems required a minimum of 20% water for stability. Attention was focused on the work of Luzzati and Husson (1962) with a brain phospholipid preparation, which showed a shift at water contents below 20% from a stack of flat lamellar bilayers alternating with water layers, to a hexagonal structure in which the water was limited to a series of channels each lined by the polar heads of the phospholipids. Extrapolating from this structure, a model was derived showing what might happen when a single bilayer such as the plasma membrane suffered dehydration (Fig. 4). In section it was thought to become a series of watery channels lined by the hydrophilic head-groups of the phospholipid molecules. The upheaval consequent on imbibition, when the phospholipid molecules change from the conformation on the right of Figure 4 to that on the left, is at the time when leakage would occur, and comes to a halt as soon as the lamellar membrane has enveloped the cell completely. This hypothesis taken in conjunction with that of membrane rupture seems to account for many of the observations on leakage (Table I), and it could account also for some of the changes observed when pollen imbibes water (Heslop-Harrison, 1979; Shivanna and Heslop-Harrison, 1981). Nevertheless, these ideas have proved controversial, largely because experiments with seed phospholipids have not revealed the presence of a hexagonal phase under dry conditions (McKersie and Stinson, 1980; McKersie and Senaratna, 1983; Seewaldt *et al.*, 1981; see also Priestley and de Kruijff, 1982 for nuclear magnetic resonance (NMR) experiments on hydrating pollen). It may well be significant that plant membranes have a higher overall ratio of phosphatidylcholine to phosphatidylethanolamine than the brain phospholipid preparations used by Luzzati and Husson (1962), for there is evidence that the hexagonal phase only appears when this ratio is relatively low (Hui *et al.*, 1981). At the same time, however, it is open to question whether this overall ratio of types of phospholipid in plant membranes represents the true situation, or whether there is some lateral phase separation within the bilayer (Wunderlich *et al.*, 1978) that might result in the appearance of a hexagonal phase in some regions under dry conditions, but not in others.

The hypothesis set out in Figure 4 can also be criticized on the grounds that in view of their low water potential, dry seeds are unlikely to have any water-filled channels; there may instead be channels filled with a strong solution of cell solutes. In this sense the proposition of Figure 4 is over simple and perhaps naive. The same comment could be applied to the degree of order and regularity implied by the drawing. It must be remembered that the plasma membrane, for instance, is exposed on one side

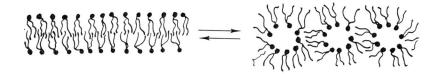

Fig. 4. Orientation of membrane phospholipids in relation to degree of hydration. Left, the lamellar configuration with the polar head-groups facing the aqueous phase on either side of the bilayer. Right, the hexagonal configuration which may be adopted at water levels below about 20%. The polar head-groups now line a series of narrow water-filled cylinders running at right angles to the page (from Simon, 1978).

to the cell cytoplasm — a most heterogeneous region containing a great variety of organelles and particles, some like protein bodies and starch grains with a higher water content than, say, the lipid bodies. This difference in water content may become significant in the later stages of seed development when the overall water content has fallen to a low level. In some localities there might be relatively moist regions adjacent to a membrane that would attract the head-groups of the phospholipids. In this way the normal bilayer arrangement would be disrupted wherever the membrane was close to a moist spot.

Another possibility might be the development of non-bilayer lipid structures within the membranes. These are in effect small micellar regions of hexagonal phase sandwiched in between the monolayers of the membrane (de Kruijff *et al.*, 1980). Such regions might arise in bilayers as they become desiccated and disappear again on imbition.

A rather different suggestion about membrane structure during imbibition stems from the knowledge that the plasma membrane must increase in surface area in concert with the swelling of individual cells, and the observation that vesicles and lipid bodies are closely associated with the plasma membrane in dry seeds (Chabot and Leopold, 1982; Vigil *et al.*, 1982). The former authors propose that 'the progress of expansion of cells requires the blebbing in of large amounts of supplementary membrane lipids, and that this blebbing process may be associated with extensive solute leakage'.

It is possible that electron microscopy might reveal the true state of membranes in dry tissues, but the results to date are equivocal. Aqueous fixatives are not reliable because the tissue may become hydrated before it is fixed. Freeze-fracture seems to be the most acceptable technique as it provides no opportunity for dry tissues to become hydrated as they are processed. Buttrose (1973) has shown that a preliminary 10 second exposure to water exercises a profound effect on the image obtained with preparations of dry barley (*Hordeum vulgare*) scutellum; the results can be read as evidence that the normal bilayer structure is absent from dry seeds, but reappears after a few seconds imbibition. In similar experiments with radicles of cowpea (*Vigna unguiculata*) seeds, Thomson and Platt-Aloia (1982) conclude that a normal bilayer is present in dry tissues, with 8.4% water. The same conclusion arises from the work of Vigil *et al.* (1982 and pers. comm. 1982) on the radicles of dry cotton seed (*Gossypium hirsutum*).

On the other hand, the appearance of lettuce (*Lactuca sativa*) seed in freeze-fracture preparations depends on water content in such a way as to suggest a membrane phase change at water contents above 20%-25% (Toivio-Kinnucan and Stushnoff, 1981). Chabot and Leopold (1982) report that as soybean tissue imbibes water progressively larger expanses of plasma membrane are formed, with an extensive increase in the number of particles embedded in the membrane, possibly resulting from the reassociation of proteins into membranes during rehydration, a view that would be consistent with the maturation of mitochondria in pea seeds during imbibition (see Section III,A).

Like seeds, some nematodes such as *Aphelenchus avenae* can survive a drying and rehydration cycle, but only if the drying process is a gradual one. Quick drying results in a loss of particles, presumed to be protein, from cell membranes (Crowe and Crowe, 1982a). Furthermore, these authors report evidence (both freeze-fracture and NMR) for the presence of hexagonal phase lipids in lobster (*Homarus americanus*) muscle microsomes with less than 20% water (Crowe and Crowe, 1982b; Crowe *et al.*, 1983).

The two hypotheses, that of membrane rupture and that of membrane repair are contrasted in Table I. It should now be clear that although there is substantial evidence for the first, the second remains a hypothesis; it can account for many of the observations, and has stimulated research on several fronts — but it remains open to doubt and equivocation. In a temporal sense, the two hypotheses are not mutually exclusive alternatives for a large seed like that of pea. Membrane rupture is limited to circumstances in which the inflow of water is very rapid, which means that it will only occur in the early stages of imbibition. Membrane repair characterizes situations in which the inflow of water is slower. Powell and Matthews (1981) have shown that the time course of leakage from pea seeds and embryos over the first 30 minutes is much the same even if the seeds or embryos are first heated to 105°C for 24 hours. This is hardly surprising, for much of the early leakage in either case will be from cells whose membrane constituents have been swept aside. The dead cells continue to leak more rapidly than the living ones even towards the end of the 30-minute period, perhaps because their proteins have become denatured, and restoration of membrane integrity is no longer possible (cf. mitochondria, Section III,A).

Some reference has already been made to the role of the testa as regards imbibition. When it is removed, imbibition becomes more rapid and leakage is enhanced (Fig. 3). A major part of the difference in leakiness between different seed lots of peas can be attributed to the condition of the testa, leakage being more rapid in seeds with damaged testas, but some of the difference is still evident when the testa has been completely removed, and so must be inherent in the embryo itself (Matthews and Rogerson, 1976; Powell and Matthews, 1978, 1979).

The testa may thus influence the extent of leakage by regulating the rate of imbibition, but it seems that it may act in other ways as well.

In peanut the testa is relatively thin and does not delay the start of imbibition or reduce the amount of water taken up after 24 hours. Nevertheless, removal of the testa enhances the leakage of potassium, phosphate, sugars and total conducting materials after 24 hours (Abdel Samad and Pearce, 1978). One way in which the testa may be acting here is by trapping a layer of gas between itself and the embryo inside; solutes might then leak out of the cells of the embryo during imbibition, but they could not move to the testa and so to the outside solution if there were an intervening gas space (Mills, 1983).

C. Damage from Soaking

The imbibition of seeds in soil may be slow and in certain weather conditions, unreliable. To ensure rapid and uniform germination it might be thought desirable to soak seeds in water before sowing them, but for many species the opposite is true. In their pioneer studies on soaking, Kidd and West placed seeds under 4 cm of water for 8–72 hours at 17°C and then set them out in sand or soil (Bailey, 1933). Broad beans (*Vicia faba*) germinated more rapidly and produced taller plants if they were first soaked for up to 72 hours. On the other hand, seeds of pea, sunflower (*Helianthus annuus*) and common bean germinated less rapidly if soaked for longer than 24 hours and produced less vigorous plants. *Phaseolus vulgaris* proved to be particularly susceptible, 6 hours soaking being enough to retard growth while 3 days soaking killed all the seeds.

The lesion underlying soaking injury is still under investigation. It is not due to microorganisms, for injury arises even under sterile conditions

(Eyster, 1940). Solutes leak out of seeds subjected to prolonged soaking just as they do from imbibing seeds, but this leakage is not thought to account for soaking injury, as pea seedlings, albeit rather weakly ones, can be raised in continuously flowing tap water (Larson and Lwanga, 1969). Also, it has not proved possible to alleviate soaking injury by supplying solutes to soaked seeds (Orphanos and Heydecker, 1968). The seeds of *Phaseolus vulgaris* represent a special case, for here morphological changes that occur as the seed is soaked cause the cavity between the cotyledons to become flooded with excess water, so reducing the supply of oxygen to the tissues of the axis (Orphanos and Heydecker, 1968). Anaerobiosis may indeed be the key to soaking injury. Crawford (1977) attributes it to self-poisoning by ethanol; seeds that are little affected by soaking such as those of rice (*Oryza sativa*) or lettuce, produce less ethanol than peas, for instance. Another consequence of soaking in water is that the syntheses of protein and RNA are slowed down (Hecker *et al.*, 1977).

If soaking injury were simply the result of subjecting seeds to anaerobic conditions one would expect that bubbling air or oxygen through the water would prevent injury. A stream of air does improve the subsequent germination of *P. vulgaris* or *V. faba*, but replacing the air with oxygen proves to be even more damaging than just soaking in water alone (Barton, 1950). Furthermore, treatment with oxygen lowered the percentage germination of pea and oat (*Avena sativa*) even though soaking in water alone had little effect on germination in these experiments. Treatment with oxygen causes the intercellular spaces of bean (*P. vulgaris*) cotyledons to become flooded with water so that they imbibe more from oxygenated than from unbubbled water (M. King, pers. comm., 1977). These findings suggest the possibility that oxygenation causes a breakdown of membrane integrity, allowing cell contents to invade intercellular spaces and so displace their gas content (Simon, 1974).

D. Chilling Injury during Imbibition

The seeds of a number of species of tropical or subtropical origin are damaged if they imbibe cold water; among the species in question are cotton, soybean, Lima bean and maize. These species are also sensitive to chilling injury in the adult stage, but the symptoms they develop in leaves, stems, fruits and so on differ from those exhibited at the time of germination.

Cotton seed will not germinate at 5°C, but if exposed to 5°C for a short period and then transferred to 31°C, the seeds will germinate, albeit tardily, and the primary root is aborted, being replaced by lateral roots which grow out from the base of the hypocotyl (Christiansen, 1963). Exposure to water at 5°C for the first few minutes of imbibition is enough to injure cotton, Lima bean and soybean; prolonged exposure causes increasing injury as seen by slower development and increased root abnormality (Pollock and Toole, 1966; Christiansen, 1968; Bramlage et al., 1978). If seeds of cotton or Lima bean are first imbibed warm (31°C) and then transferred to 5°C they are less severely damaged; 4 hours at 31°C is enough to render cotton seed immune to the cold (Christiansen, 1968; Pollock and Toole, 1966). Experiments with seeds at different water contents have shown that there is no injury in cotton and soybean seeds at water contents above 13% (Christiansen, 1969; Hobbs and Obendorf, 1972). The corresponding figure for maize is 16% and for axes and scarified seeds of Lima bean, 20% (Cal and Obendorf, 1972; Pollock, 1969). Axes of Lima bean can be repeatedly hydrated up to 30% water and dried down again to 10% with corresponding effects on growth. Whatever the mechanism of response to the cold, it is evidently a reversible one and depends directly on the moisture level at the start of imbibition.

Like other tissues, axes of Lima bean and soybean lose solutes during imbibition, but in contrast, for instance, to peas (Perry and Harrison, 1970), this leakage is notably temperature sensitive, being doubled if the temperature is reduced from 25°C to 5°C. Treatment with cold water enhances leakage from soybean cotyledons so long as it is applied at the start of imbibition; if the cotyledons are first exposed to warm water for even 1 minute, subsequent chilling causes little or no increase in leakage (Leopold and Musgrave, 1979). At both 25°C and 5°C leakage from soybean is largely suppressed if the tissues are first hydrated to 13%-20% (Hobbs and Obendorf, 1972; Pollock, 1969).

These experiments indicate that imbibitional chilling injury is similar in many respects to leakage from seeds that are not cold-sensitive. Chilling injury and leakage are most intense when *dry* seeds are placed in water, both can be induced again if the tissues are dried down after a short period of imbibition, and both are most marked in the very early stages of imbibition. Another similarity is that both processes have long-lasting effects. Cotton and soybean plants that were chilled during imbibition could be recognized weeks later by their small size and low dry weight (Christiansen and Thomas, 1969; Obendorf and Hobbs, 1970).

It is tempting to suggest that leakiness is the key factor in imbibitional chilling damage, either because it leads to a loss of cytosol materials out of individual cells, or because, operating at the level of cell organelles, it destroys the normal degree of compartmentation within cells. This in turn suggests membrane dysfunction. As chilling damage occurs during the first minutes of imbibition it bears some degree of resemblance to the membrane rupture caused by the initial violent inrush of water. This is consistent with evidence that the seed coat has a controlling influence on chilling injury. Thus intact pea seeds are slow to imbibe water (Fig. 3) and it seems this is what protects them from damage if they are placed in water at 2°C for 3 hours and then transferred to the warm. If, on the other hand, the seeds are nicked with a razor blade, water uptake is more rapid and germination of seeds chilled at 2°C is much poorer than in controls imbibed at 25°C. The seed coat thus protects pea embryos from rapid imbibition, prevents extensive leakage and stops chilling damage. By contrast, the seed coat in soybean does not reduce the rate of imbibition much and does not prevent chilling injury. Soybean seeds can, however, be protected from chilling injury, if the rate of imbibition is slowed down by placing the seeds in polyethylene glycol solution rather than water (Tully *et al.*, 1981; Musgrave and Leopold, 1981).

The evidence thus indicates that imbibitional chilling injury is related to the initial rapid inrush of water; however, it still remains difficult to understand why the lesion should be enhanced in the cold. In theory some special feature of membrane lipid composition might distinguish the particularly cold-sensitive seeds of soybean from those of pea, but no significant difference in the overall composition of pea and soybean membrane lipids has been detected (Priestley and Leopold, 1980). Nor is there any evidence of mitochondrial malfunction or of failure of protein synthesis under chilling conditions (Cohn and Obendorf, 1976; Stewart and Bewley, 1981). O'Neill and Leopold (1982) were unable to detect any bulk phase transition in soybean lipids by thermal or optical methods.

III. METABOLISM

Dry seeds contain many enzymes including those responsible for the metabolic events that were still being conducted at an appreciable rate as the seeds finally matured and dried out. Most of these enzymes are resistant to desiccation for they become active once more as soon as imbibition starts. A few enzymes, already present in the seed, undergo an activation process at some stage during germination, becoming fully active as a result of proteolysis or some conformational change (Mayer and Marbach, 1981).

No attempt is made to catalogue here all the changes in enzyme activity and metabolic pathways that occur as seeds prepare for germination; attention is focused on some of the major pathways of metabolism.

A. Respiration

The respiration of dry seeds is so slow as to be near the limit of detection. Harrington (1973) gives a figure of 0.005 mg CO_2 per g dry weight per year for the respiration of onion (*Allium cepa*) seeds with 4.2% moisture at 25°C. Clegg (1978) has argued from a consideration of the amount of water that would be bound to cell components that unless cells have at least 10% water there is unlikely to be any true metabolism, that is, enzyme-catalysed reactions that are integrated into metabolic pathways, regulated in rate and direction, and contributing to cellular maintenance. Be that as it may, there is no doubt that the rate of respiration, and of other metabolic processes, rises rapidly once imbibition commences. This activation of metabolism is clearly the outcome of two processes: (i) the enhanced respiration of individual cells as they imbibe; and (ii) the increasing number of cells that are wetted as the water front penetrates the seed. To elucidate the relationship between water content and metabolic activity it is clearly best to work with seed material that is uniformly wetted to a particular moisture content.

Edwards (1976) supplied ^{14}C-labelled acetate or leucine to seeds of charlock (*Sinapis arvensis*) at 0°C, redried them rapidly at 0°C and then maintained the seeds at low water potentials for 6 months at 25°C. In other experiments, $^{14}CO_2$ was supplied to seeds at low water potential. The rates of CO_2 fixation and utilization of acetate increased exponentially, and the rate of incorporation of leucine into protein increased linearly as the water content of the seeds increased. Only at the lowest water potentials, -171.6 to -76.2 MPa, corresponding to water contents of 4% and 6%, was metabolism reduced to 'a low rate'. Wilson (1970) applied labelled phosphate in ethanol to seeds of *Agropyron desertorum* and then placed them in atmospheres at a range of humidities for 6 days, by which time their water content had become constant. No phosphate was incorporated into ATP at water potentials lower than about -130 atmospheres (-13.2 MPa; water content not stated).

We may conclude that active metabolism commences in the cells of air-dry seeds as their water potential rises. It is possible that some facets of metabolism may become active before others, that is, at lower water potential. If this happens it could give rise to some imbalance of metabolic pathways.

The progress of respiration in imbibing seeds is generally triphasic. The rate rises rapidly in the first hours of imbibition, reaches a plateau value of variable duration and then increases further. This further rise is associated with cell division in the tissues of the root–shoot axis, but is present also on a more modest scale in the cotyledons of pea and bean in which there is no cell division. The time course of imbibition is similar. In bean seeds, water content and respiration rate both rise rapidly for the first 8 hours until the water content has reached about 50% (Fig. 5). There is then a 10 hour lag phase followed by a gradual increase in respiratory activity. The initial 8 hour period is thought to be one of physical activation of pre-existing systems, for after 6 hours imbibition, the whole process can be reversed by desiccating the beans. Neither imbibition nor the activation of respiration is much influenced by reducing the temperature from 25°C to 16°C (Öpik and Simon, 1963).

The substrates consumed in respiratory metabolism during the early period of germination are sugars or oligosaccharides (Hsu et al., 1973; Mayer and Shain, 1974). Several types of evidence suggest that the pathway of sugar breakdown in the first few hours of imbibition differs from that which is adopted in the later phases of germination and seedling growth. This conclusion emerges from inhibitor studies, work with labelled substrates and what is known about the activity of mitochondria. Thus Morohashi and Shimokoriyama (1975) have supplied labelled glucose to mung bean (*Vigna radiata*) seeds. During the first 5.5 hours of imbibition the glucose was converted predominantly to ethanol (and lactate ?) rather than Krebs cycle acids, while the opposite was true in the following 3 hours suggesting a shift from fermentation to respiration. The total oxidative activity of mitochondria extracted from seeds increased as imbibition progressed, but the glycolytic enzymes were fully active right from the start of imbibition. The authors conclude that mitochondrial activity is rate-limiting in the early hours of imbibition and judging from inhibitor experiments they conclude that the limiting step is in the electron transport chain.

Wilson and his colleagues (Spedding and Wilson, 1968; Collins and Wilson, 1972, 1975) have used tritiated water (^3H_2O) in the imbibing medium to follow metabolic pathways in the very early period of water uptake. To facilitate the introduction of the ^3H_2O to the seed tissues the testa was partially removed from mustard (*Sinapis alba*) seeds, beans (*P. vulgaris*) were dissected to yield axes and small pieces of cotyledon and barley embryos were simply removed from the grains. From the time at which Krebs cycle acids became labelled it was concluded that the cycle was not functioning in the first 15, 30 or 60 minutes, in barley, mustard and bean (axes) respectively.

There are a number of records of change in the response of seed tissues to inhibitors during the early hours of germination which suggest some change in pathway; examples are the response of lettuce to arsenate (Speer, 1973) and the response of soybean to cyanide and salicylhydroxamic acid (Yentur and Leopold, 1976; Siedow and Girvin, 1980).

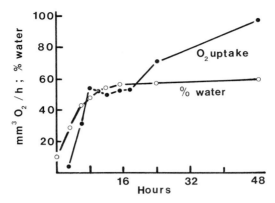

Fig. 5. Respiration and water uptake by cotyledons of germinating beans, *Phaseolus vulgaris*, at 25°C (from Öpik, 1980).

Mitochondria can be prepared from dry seeds if these are first pulverized to yield a powder which can then be ground up in a buffered medium. The application of this procedure to peas gives a pellet at 25 000 g which has little succinoxidase activity and only low malate dehydrogenase [EC 1.1.1.37] and cytochrome oxidase activities. Substantial amounts of these latter enzymes are present in the supernatant fraction. Preparations made after 3 hours or better, 6 hours imbibition have much more oxidase activity in the mitochondrial fraction and less in the supernatant (Nawa and Asahi, 1971; Sato and Asahi, 1975). The mitochondria in these preparations were said to be more 'mature'. The process of maturation is not inhibited when peas are set to germinate in a solution of cycloheximide strong enough to prevent synthesis of mitochondrial proteins in the cytoplasm (Nawa and Asahi, 1973a; Morohashi, 1980; Morohashi and Bewley, 1980a).

It seems therefore that maturation involves the transfer of preexisting enzyme protein from the supernatant to the mitochondrial fraction. Strong support for this view comes from more recent experiments in which it has been shown that a soluble succinate dehydrogenase [EC 1.3.99.1] is present in dry pea cotyledons, but not in fully imbibed ones. The disappearance of the soluble enzyme during imbibition is linked with the appearance of a membrane-bound, mitochondrial succinate dehydrogenase with very similar properties (Nakayama *et al.*, 1980). In peas, mitochondrial maturation evidently involves the association of the soluble enzyme with the inert mitochondrial membranes found in dry cotyledons. The same conclusion emerges from experiments with castor bean (*Ricinus communis*) (Albergoni *et al.*, 1964).

In the early stages of germination the respiratory activity of bean cotyledons depends on their water content (Fig. 5), and is reduced again if imbibed cotyledons are dried down. The assembly of cytoplasmic proteins into the mitochondrial membranes of pea cotyledons during imbibition can also be reversed with loss of malate dehydrogenase and cytochrome oxidase if 6 hour or 18 hour imbibed cotyledons are dehydrated (Nawa and Asahi, 1973b). A similar reversible increase in the activity of NADH-cytochrome *c* reductase [EC 1.6.99.3] in lettuce has been reported by Eldan and Mayer (1972). Some of the membrane-bound mitochondrial enzyme systems in mitochondria from *Vicia faba* are damaged by desiccation *in vitro* (Priestley and Bruinsma, 1982).

Enzyme protein may be lost in the same way from mitochondrial membranes under natural conditions, as seed water content falls during the final stages of development on the parent plant. Thus in ripening castor bean seeds there is a fall in the activity of three mitochondrial enzymes: cytochrome oxidase, malate dehydrogenase and succinate cytochrome *c* reductase (Lado, 1965). The absence of such enzymes from the mitochondria is the factor that makes it difficult to prepare active mitochondria from dry seeds. This difficulty is not insuperable. When dry seeds are ground up in buffer the oxidative enzymes present in the soluble phase become diluted and dispersed in the grinding medium so that reassembly into the mitochondria is inefficient, and the mitochondria obtained on centrifugation have low activity. However, if the dry seeds are powdered, allowed to stand for 1.5 hours with an equal volume of water, and then ground up with buffer, more active mitochondrial preparations are obtained (Kollöffel, 1970). It seems that the initial period of imbibition in a small volume of water provides the opportunity for mitochondrial reassembly.

The effect of repeated imbibition and drying cycles on mitochondrial activity has been investigated by Morohashi and Bewley (1980b). Pea cotyledons imbibed for 4 hours were subsequently dried back to their original weight over silica gel, and then set to imbibe for a further 4 hours. Mitochondria from such cotyledons were a little less active than those just imbibed once for 4 hours. However, a series of three consecutive imbibition and drying cycles was more damaging, yielding at the end mitochondria that had only 59% of the oxidase activity of those imbibed once only.

These observations can be related to the changes in membrane organization that were proposed above (Section II,B) to account for the leakage that occurs during imbibition (Fig. 4). The possible fate of membrane-bound proteins is indicated in Figure 6, which must be regarded as simply a model of what may happen. According to this diagram, membranes would undergo a major change in molecular architecture in

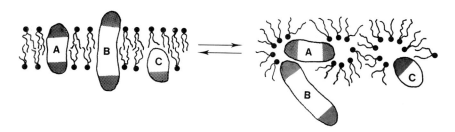

Fig. 6. Orientation of membrane components in relation to degree of hydration: a hypothetical model system based on Figure 4. On the left is shown the fluid mosaic model of a membrane under aqueous conditions showing 3 integral proteins with their polar and ionic moieties darkened. On the right is the sort of configuration which may be adopted under dry conditions (from Simon, 1978).

seeds that dry out on reaching the final stages of development. When the dry seed is hydrated once more physical forces will reestablish a lamellar membrane, the proteins ideally reappearing in the same positions and with the same orientation as they originally had during development. However, errors could easily creep in during this reassembly process, the insertion of particular proteins into the membrane being faulty (proteins appearing, for example, on the wrong side of the membrane), or even failing altogether. If this were to be repeated through several cycles, one could well expect a cumulative rundown of mitochondrial efficiency. Priestley and Bruinsma (1982) discuss an alternative model in which desiccation leads to protein denaturation.

The pattern of mitochondrial development in peanut cotyledons differs from that described for peas. In peanuts electron microscopy reveals an increase in numbers of mitochondria, and the increase in mitochondrial oxidase activity over the first 3 days can be substantially inhibited by cycloheximide (Morohashi et al., 1981). It seems that the increasing activity of mitochondria in pea cotyledons must be attributed to reassembly of preexisting units, while in peanuts there is evidence of mitochondrial biogenesis. Little is known as yet about mitochondrial development in other legume species.

As discussed above sugars may be degraded by anaerobic pathways in the early stages of imbibition if mitochondria are not yet fully functional. The same may happen at later stages of development if the growing embryo experiences anaerobic conditions resulting from the inefficient transport of oxygen in solution through the testa. Under such conditions ethanol, ethyl β-D-glucose and lactate may accumulate until the root pierces the testa, allowing oxygen to diffuse to the embryo in the gas phase. If seeds are water-logged such products of anaerobiosis accumulate in greater quantity (Sherwin and Simon, 1969; Dioreau, 1976).

The central importance of respiratory metabolism lies in the production of ATP. The quantity of ATP present in seed tissues increases by as much as 10-fold during the first hour of imbibition (Obendorf and Marcus, 1974), a rise which is thought to result partly from new synthesis (Anderson, 1979; Perl, 1982 and references therein), and partly from enzymatic formation from existing AMP (Moreland et al., 1974).

Atkinson (1977) has developed the concept of energy charge, which is a measure of the metabolic energy stored in the adenine nucleotide system:

$$\text{Energy charge} = \frac{[\text{ATP}] + 0.5\,[\text{ADP}]}{[\text{ATP}] + [\text{ADP}] + [\text{AMP}]}$$

The activity of enzymes involved in ATP generation, and those involved in ATP utilization influence the prevailing energy charge. A charge of less than 0.5 is characteristic of metabolism directly primarily towards synthesis of ATP, whereas a charge between 0.5 and 1.0 indicates the consumption of ATP in synthetic reactions.

In dry seeds the energy charge is generally low, but it rises rapidly on imbibition, consistent with the sudden upsurge of metabolic activity during this phase (Obendorf and Marcus, 1974; Rodaway et al., 1979). In lettuce the energy charge rises from 0.2 to 0.8 after 30 minutes in aerated water. In the presence of cyanide, or under nitrogen the energy charge remains at 0.3, while seeds imbibed first in aerated water and then transferred to cyanide or nitrogen show a fall in energy charge to a low value within a few minutes. These results, together with the fact that oxygen uptake during this period was cyanide-sensitive, are presumptive evidence for the operation of a normal cytochrome pathway coupled with the production of ATP in the first minutes and hours of imbibition (Hourmant and Pradet, 1981). If this interpretation is correct, it should be possible to extract fully active mitochondria from lettuce seeds in the early phases of germination.

Lettuce seed have a low anaerobic metabolism producing little ethanol or lactate even under nitrogen (Hourmant and Pradet, 1981). However, a different pattern emerges from experiments with radish (*Raphanus sativus*), for here ATP production continues under conditions that would prevent oxidative phosphorylation. During the first hour of imbibition the rise in ATP level is scarcely inhibited by anaerobiosis or the presence of 5 mM fluoroacetate or iodoacetate; by contrast though, ATP levels continue to rise slowly over the next 2 days in controls, but not under nitrogen or in the presence of the inhibitors. It seems that glycolytic phosphorylation is 'sufficient to meet the energy demands required to sustain the tissue' (Moreland et al., 1974), but cannot support germination. Although control seeds germinated in 16 hours those treated with fluoroacetate produced thin, short radicles emerging between 24 and 36 hours and those under nitrogen or treated with iodoacetate showed no sign of radicle emergence even by 48 hours.

B. Protein Synthesis

The developing seed has a major commitment to protein synthesis. The new cells produced in the developing embryo are each provided with a complement of organelles and enzymes, and in addition, massive reserves of protein may be accumulated in the storage organs (Chapter 3 of Volume 1). As seeds reach maturity, the pathways of metabolism involved in protein synthesis slow down (Müntz, 1982). There is virtually no metabolic activity in the dry seed. When the seed germinates, however, protein synthesis starts up again. Some protein synthesis, like the production of ribosomes or the enzymes of glycolysis, is presumably a repeat run along the same pathways as were operating before, but some synthesis results in the production of new proteins like α-amylase [EC 3.2.1.1], not synthesized at any earlier stage in development (Section 4.II). The mature dry seed is at a watershed between the phases of maturation and germination.

The advent of techniques for the two-dimensional separation of proteins (O'Farrell, 1975) has made it possible to monitor the level of hundreds of individual proteins in plant cells, and it has now been shown that certain proteins are uniquely associated with development. These proteins are no longer synthesized in bean (*Phaseolus vulgaris*) axes taken from seeds imbibed for 12 hours. It seems that the desiccation seeds experience as they mature and the rehydration that occurs during imbibition not only suppress the formation of such 'developmental' proteins, but also induce the synthesis of entirely new and specific 'germination' proteins in a matter of hours (Dasgupta and Bewley, 1982). It is not known how desiccation and rehydration bring about this change in the pattern of protein synthesis, but considerable progress has been made on an associated question — the fate of the apparatus required for protein synthesis during seed desiccation and the subsequent period of imbibition. The whole complex of enzymes, cofactors and structures must obviously be present in developing seeds, but how much of this is preserved in the mature dry seed, ready for use once germination starts?

Some insight into the state of the protein-synthesizing system in dry seeds can be gained by following the development of a seed. During periods of rapid synthesis of reserve proteins in pea cotyledons, the ratio of polysomes to monosomes remains high at about 0.8 (Beevers and Poulson, 1972; Poulson and Beevers, 1973). However, as the rate of protein synthesis declines while the seed is drying out, the ratio of polysomes to monosomes declines to about 0.3 or less. This transition occurs without any major decline in the total amount of RNA. The polysomes are evidently converted to monosomes as individual ribosomes become detached from mRNA molecules. When assayed, preparations of ribosomes show a corresponding loss of the ability to promote amino acid incorporation into protein in a cell-free system prepared from pea cotyledons. However, the ability of ribosomal preparations from mature pea cotyledons to incorporate phenylalanine into protein can be stimulated by the addition of the artificial message polyuridylate (poly U). These observations suggest that protein synthesis becomes restricted in dehydrating seeds through lack of mRNA, a conclusion that appears to be generally true (Section 3.IV of Volume 1).

The large body of literature on protein synthesis during imbibition and germination has been reviewed by Bray (1979), Mayer and Marbach (1981), Bewley and Black (1978) and Bewley (1982). The ability of imbibing tissues to synthesize protein is assessed by supplying a labelled amino acid and then determining how much has been converted to protein. To ensure ready access of the labelled substrate to the living tissues, it is usual to work with isolated tissues such as cereal embryos or the axes of beans. These isolated tissues take up water and substrate much more readily than intact seeds and quickly become fully imbibed. The incorporation of leucine into protein by oat embryos proceeds at increasing rates over the 2 hour period during which they imbibe water (Fig. 7). This is probably in part the result of the growing number of cells that become wetted as imbibition progresses, and in part the increasing synthetic activity of individual cells. Protein synthesis also begins within an hour of the start of imbibition in embryos of wheat (Marcus et al., 1966), barley (Stoddart and Thomas, 1973) and rye (*Secale cereale*) (Sen et al., 1975), and in axes of Lima bean (Klein et al., 1971).

Protein synthesis is resumed so promptly at the start of imbibition as to suggest that most of the biochemical and structural components necessary must already be present in the dry seed. Cell-free systems capable of protein synthesis can indeed be prepared from dry wheat embryos and pea axes (Peumans et al., 1980). The wheat germ system has a complete range of soluble components including transfer RNA (tRNA) and aminoacyl-tRNA synthetases (Weill and Parthier, 1982), together with the factors required for chain initiation and elongation (Marcus, 1982). Ribosomes prepared from dry wheat embryos are unable to synthesize protein. However, the activity of these ribosomal preparations rises rapidly on imbibition and is parallelled by an increase in the proportion of polysomes present (Marcus et al., 1966; Weeks and Marcus, 1971).

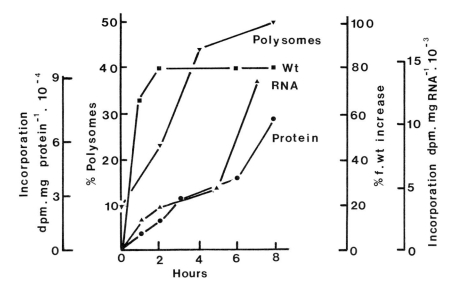

Fig. 7. Imbibition, polysome content and incorporation of leucine and uridine into protein and RNA respectively by oat embryos (data of Bewley, 1982).

In whole cereal embryos the rate of protein synthesis continues to increase even when imbibition has come to a halt. One factor that might account for this increase in oat embryos is the increasing proportion of polysomes present (Fig. 7). In wheat embryos, on the other hand, there appears to be no marked rise in polysome content during the lag period which extends from 40 minutes to 5 hours from the start of imbibition (Spiegel and Marcus, 1975). The increasing efficiency of protein synthesis during this lag phase may be attributed in part to the higher levels of ATP and GTP (Brooker *et al.*, 1977) and perhaps also to increased RNA turnover.

C. RNA Synthesis

Cell-free preparations of wheat embryos contain a type of RNA that converts ribosomes to polysomes and causes leucine to become incorporated into trichloroacetic acid (TCA)-insoluble material. The early work on this subject is reviewed by Payne (1976). The discovery that such mRNA molecules have long terminal sequences enriched in adenosine (A) residues has made it possible to isolate mRNA from cell-free preparations by affinity chromatography on oligo deoxythymidine (dT)-cellulose, which separates the polyadenylated or poly(A)-rich RNA. Messenger RNA preparations isolated in this way from dry seeds of pea, bean and rape, and from rye embryos act as templates in a wheat-germ system for the synthesis of a variety of proteins ranging in molecular weight (MW) from 10 000 to 70 000 daltons. The synthesis of such high MW proteins is evidence for the presence of intact mRNA molecules in the dry seed (Gordon and Payne, 1976; Caers et al., 1979). This mRNA was evidently transcribed at some stage prior to dehydration and then conserved in the dry seed. It is often known as long-lived mRNA, but the terms conserved mRNA or stored mRNA are perhaps more appropriate as they direct attention to the chief characteristic of this class of mRNA.

Further evidence for the presence of stored mRNA in seeds comes from experiments in which seeds are set to imbibe under conditions that block the formation of new mRNA; if protein synthesis then continues it must be utilizing preformed mRNA. Spiegel and Marcus (1975) treated wheat embryos with concentrations of α-amanitin or cordycepin (3'-deoxyadenosine) strong enough to inhibit mRNA synthesis by over 80%. Embryos were placed in the inhibitor solutions for a preliminary 30 minutes at 0°C to allow the inhibitor to enter the tissues. During the next 40 minutes at 25°C polysomes were formed as rapidly as in the controls, indicating that preformed mRNA was largely responsible for protein synthesis at this time.

The mechanism that protects such molecules of conserved mRNA from degradation during the final stages of seed maturation remains to be ascertained. It may be that its survival is ensued by its subcellular location, possibly in the nucleus or in ribonucleoprotein particles (e.g., Hammett and Katterman, 1975; Peumans et al., 1979).

Since there is a change in the spectrum of proteins synthesized during the process of germination some 'development proteins' no longer being formed, and new 'germination proteins' appearing for the first time, we may expect some change in the population of mRNA molecules at this time. Cordycepin has been used to determine the life-span of mRNA molecules during the lag phase (Caers et al., 1979). When wheat embryos are imbibed for 2 hours in a solution containing labelled adenosine and are then transferred to a solution of unlabelled adenosine together with cordycepin, the labelled poly(A)-rich RNA formed during the pulse disappears with a half-life of about 2 hours; much the same is true in radish axes (Delseny et al., 1977).

Despite this rapid loss of existing mRNA, the synthesis of new messenger molecules can be detected within the first hour of imbibition (Payne, 1977; Bray, 1979). In both radish and wheat there is a fall in the overall level of mRNA during the first hour or two of imbibition, followed by a rise to about the original level over the next few hours (Delseny et al.,

1977; Smith and Bray, 1982). It appears that degradation is at first more rapid than synthesis, but that later the situation is reversed. The initial loss of poly(A)-rich RNA is thought to represent the degradation of a 'major portion' of the stored messenger (Smith and Bray, 1982).

The conclusion from these experiments is that embryos and axes can synthesize proteins coded by conserved mRNA during early imbibition (the first 40 minutes for wheat embryos) but thereafter newly synthesised mRNA molecules must play an increasingly important role. They may account at least in part for the enhanced rate of protein synthesis that is observed during the lag phase. The fact that α-amanitin and cordycepin applied during the lag phase prevent the synthesis of protein as well as mRNA is in accord with this view (Caers *et al.*, 1979; Cheung *et al.*, 1979).

A number of workers have sought to determine how far the changing population of mRNA molecules can be related to the changing spectrum of proteins present in seeds during early germination. Does the poly(A)-rich RNA extracted from seed tissues at different times code for different proteins? Matilla *et al.* (1980) report that the mRNA from dry and 18 hour imbibed cotyledons of chick pea (*Cicer arietinum*) code for similar (but not identical) populations of polypeptides, as revealed by gel electrophoresis. The more sensitive two-dimensional separation of proteins provides a fuller picture. The pattern of products formed by cell-free translation of the bulk poly(A)-rich RNA isolated from dry wheat embryos is broadly similar to that formed *in vivo* in embryos during the first 40 minutes of imbibition. However, by 17 to 18 hours from the start of imbibition there is a marked quantitative difference in the pattern of polypeptides synthesized in the embryos, with a lower proportion of ribosomal proteins than at first (Cuming and Lane, 1979). The change in the population of polypeptides synthesized in the embryos over the first 5 to 6 hours can be suppressed by adding α-amanitin to the germination medium. This finding is consistent with the hypothesis that the transcription of new mRNA during the lag period is responsible for directing the changing pattern of protein synthesis (Thompson and Lane, 1980).

If we could attribute distinctive roles to the sequence of individual proteins that are formed during imbibition, we might well gain some understanding of the key events that decide whether a seed will germinate or not, and we should be able to evaluate more clearly the respective functions of mRNA that is stored and that which is formed afresh as imbibition progresses. Some first steps in this direction have now been taken. Mung bean axes have a species of mRNA that codes for a protein of MW 12 000 daltons. This protein is a storage albumin (Section 7.I,B,*2*). Both the messenger and the protein are present in dry mung bean axes but disappear during imbibition. The activity of the messenger fraction falls by 70% after

3 hours imbibition (Carlier *et al.*, 1980). In the same way, dry wheat embryos have a messenger for what is probably the most abundant protein present in the embryos. The protein and its messenger are present in quantity at the start of imbibition (0 to 1 hour), can be readily detected up to 5 hours, but are both degraded soon after the embryo starts growth at about 10 hours (Grzelczak *et al.*, 1982).

The predominant view emerging from this discussion is that stored mRNA may have little role once imbibition commences, and that it is liable to be degraded within a few hours. The alternative view is that some of the stored mRNA has a role at a much later stage, when germinating seeds mobilize most of their reserves. Dure and his colleagues have argued that the mRNA necessary for carboxypeptidase synthesis in the cotyledons of cotton is stored in the dry seed, ready (perhaps after polyadenylation) to function when the seed eventually germinates (Dure, 1979). Carboxypeptidase cannot be detected until 24 hours after the start of imbibition and only reaches its full activity by 3 or 4 days. This timetable is consistent with its probable role in hydrolysing stored protein, since the main period of mobilization of stored reserves occurs after radicle emergence, in the phase of seedling growth (Waters and Dure, 1966). The control of carboxypeptidase synthesis is discussed by Bewley (1982).

It is now well established that other forms of RNA as well as mRNA are synthesized from the start of imbibition (Fig. 7; Huang *et al.*, 1980 and references therein). The regions of the seed most active in RNA synthesis have been detected by autoradiography of rape seeds after imbibition in a solution of labelled uridine. After 2 hours imbibition, RNA was being synthesized only in the outermost cells of the embryo, but later the onset of synthesis spread to other cells (Payne *et al.*, 1978).

D. DNA Synthesis and Cell Division

Cell division comes to a halt at a relatively early stage of development in pea and bean seeds. The embryo then grows by cell expansion alone, although the synthesis of DNA may continue so that cotyledon cells become polyploid with 8–64C levels of DNA (Müntz, 1982; Section 3.IV of Volume 1).

The synthesis of protein and RNA is resumed as soon as seed tissues begin to imbibe, but DNA synthesis follows later. In wheat embryos the synthesis of the enzyme DNA polymerase [EC 2.7.7.7] is evidently a prerequisite for DNA synthesis (Mory *et al.*, 1975). In embryos of wheat and rye the incorporation of labelled thymidine commences after about 5 hours imbibition (Bray, 1979) but in cotton and broad bean it is delayed until later (Table II).

Table II. Timing of events following imbibition

Species	Hours since start of imbibition to the period of:				References
	DNA synthesis	Mitosis	Rapid cell expansion	Germination	
Vicia faba (seed)	28–40	56–66			Davidson (1966)
	25–40	40–50			Jakob and Bovey (1969)
		55	46	48–60	Rogan and Simon (1975)
Zea mays (grain)	45	74			Deltour and Jacmard (1974)
Gossypium hirsutum (seed)	12–18			18	Clay et al. (1975)
Phaseolus lunatus (axis)		11–12	7	12–13	Walton (1966)

The fragmentary data in Table II indicate that cell division follows only some hours after the start of DNA synthesis, at about the time of germination. The lengthening of the radicle that is entailed in germination is normally accomplished by cell elongation alone, mitosis contributing little or nothing. Germination can indeed occur in the complete absence of cell division, as when lettuce seeds are treated with γ-rays or germinated at low temperature (Haber and Luippold, 1960).

E. Polyamines

Polyamines such as spermine, spermidine and putrescine are to be found in seeds and the quantity present tends to increase in the first hours of growth leading to germination. Attempts have been made to relate this rise and the subsequent fall to other changes that occur early in germination such as the onset of mitosis (Anguillesi *et al.*, 1980), or the level of protein and RNA (Villanueva *et al.*, 1978). However, so many pathways of metabolism are starting up in the early hours of germination that a similar timing or a similar pattern of change in content is not sufficient to establish whether polyamines actually control metabolic events during germination. In animal cells, where there is also a close association between polyamine accumulation and cell growth, it has been suggested on the one hand that polyamine accumulation is the trigger for cell growth, and on the other, that polyamine biosynthesis is itself largely dependent on the accumulation of RNA (Goyns, 1982).

F. The Lag Phase and Subsequent Cell Elongation

Exposing a dry seed to water triggers off a variety of events including the uptake of water and the concomitant leakage of solutes, the activation of respiration and the maturation of mitochondria, the formation of polysomes and the synthesis of protein, RNA and nucleotide sugars. Inhibitor experiments show that there is no germination if the synthesis of protein or RNA is blocked (Brooker *et al.*, 1977; Tao and Kahn, 1976). Although the operation of these various pathways may be a necessary precondition for germination, they are not in themselves sufficient to ensure that seeds will germinate. The seeds may for instance be dormant, as in Grand Rapids lettuce which form polysomes and begin protein synthesis in the dark, but will not germinate unless stimulated by light or gibberellin (Fountain and Bewley, 1973; Bewley, 1982; Section 2.IV).

Alongside this battery of events triggered in each cell layer as it becomes hydrated must be set others that are delayed, only coming to the fore some hours after the start of imbibition. Chief among these is the start of cell extension (Section III,E). Although the oat embryos used in the experiments of Figure 7 were not dormant (J. D. Bewley, pers. comm. 1983) there was no extension, and hence no germination, in the 6 hours following the completion of imbibition. The lag period lasts at least 6 hours in oats, and is 50-90 hours in cucumber (*Cucumis sativus*), depending on the temperature (Fig. 1). Why should there be such a lag? What is it that prevents outgrowth of the root at the start of imbibition? It may be that some essential metabolic pathway only comes into operation towards the end of the lag period. DNA synthesis could be taken as an example (Table II), but this is unlikely to be the limiting factor, as seeds will germinate even in the absence of mitosis (Section III,D).

An alternative hypothesis would be that root outgrowth is delayed until the product of some particular pathway has accumulated to a sufficient extent. A further possibility is suggested by electron microscopy, for during the first 9 hours of germination a whole series of changes occur in the fine structure of root cells in rye (Sargent and Osborne, 1980). Some of these changes may be necessary prerequisites for cell outgrowth, so interposing a delay before germination can occur.

One approach that might provide insight into the events of the lag period is suggested by the finding that seeds first hydrated for a few hours and then dried back to their original weight will germinate more quickly than usual when they are finally sown. Evidently preparations completed during the first period of imbibition are stable to desiccation and need not be repeated. When pretreated oat grains are sown they begin to synthesize protein and RNA more quickly than controls (Akalehiyot and Bewley, 1980). It has also been established in wheat and annual rye grass (*Lolium rigidum*) that the pretreatment provides an opportunity for the embryo to make a small initial enlargement (Lush *et al.*, 1981).

Current thoughts on cell growth in seedlings and mature plants focus on the need for some loosening of the cell wall so that it becomes able to yield in response to the turgor pressure exerted from within. This loosening of the cell wall is linked to acidification of the wall compartment through the operation of the ATPase located in the plasma membrane (for reviews, see Marrè, 1979; Labavitch, 1981). This system is responsible for extrusion of hydrogen ions (protons, H^+) and the uptake of potassium ions (K^+) or other available monovalent cations. These energy dependent ion movements simultaneously enhance wall-loosening and provide an osmotic gradient that will in turn stimulate water inflow.

The initiation of root outgrowth by cell expansion in germinating embryos may depend on a similar mechanism. Experiments using the fungal toxin fusicoccin (a compound produced by *Fusicoccum amygdali*) support this conclusion. Fusicoccin is known to stimulate H^+ efflux, K^+ uptake and growth by cell expansion in many kinds of plant tissue (Marrè, 1979), including excised tissues already 'predisposed' to grow (Hanson and Trewavas, 1982). Initial cell expansion in the radicle of *Haplopappus gracilis* has been distinguished as a process that occurs independently of protein synthesis (Galli *et al.*, 1981). Cycloheximide provided to imbibing achenes in the absence of fusicoccin prevented germination, but fusicoccin in the presence of the same concentration of cycloheximide promoted germination (Galli *et al.*, 1981).

Fusicoccin can not only relieve dormancy (Lado *et al.*, 1974), but also hasten germination in non-dormant seed. Embryos of *Haplopappus gracilis* begin elongation as soon as they start to imbibe in 10 μM fusicoccin, whereas in water there is a 5 hour lag before germination starts (Fig. 5 in Galli *et al.*, 1975). In maize visible germination occurs after 27 hours with fusicoccin, but only after 31 hours in water (Lado *et al.*, 1975). More experiments of this type are needed to explore the effects of fusicoccin on the lag phase. Does it abolish the lag phase in other species besides *Haplopappus gracilis*, or does it perhaps just shorten the lag phase by 4 or 5 hours in all species?

IV. CONCLUDING DISCUSSION

The contrasting features of the metabolism of developing and germinating seeds were mentioned in Section I. It is perhaps appropriate in closing this chapter to draw attention to the several lines of evidence suggesting that seed dehydration followed by imbibition may be the trigger that brings about this switch in the pattern of seed life.

(i) Seeds of some species can be induced to germinate only if they have first become air dry and then rehydrated during imbibition. This applies also to immature seeds of certain species. The seeds of soybean (Adams and Rinne, 1981) and common bean (Dasgupta and Bewley, 1982) can be induced to germinate at unusually early stages of development by first air-drying.

(ii) Desiccation and rehydration terminate the synthesis of proteins normally accumulated during development and initiate the synthesis of different proteins: a 'germination pattern' (Dasgupta and Bewley, 1982).

(iii) Gibberellic acid (GA_3) will elicit the production of α-amylase in wheat aleurone cells only if the grains have been dried down to a water content of less than about 25% (Section 4.II,G).

(iv) The data on seed leakage (Section II,B) and on maturation of mitochondria in pea (Section III,A) suggest that there is a change in membrane architecture at water contents below 20%–30%. Such a change might possibly mediate the observed effects of desiccation and imbibition on seed development.

REFERENCES

Abdel Samad, I. M., and Pearce, R. S. (1978). *J. Exp. Bot.* **29**, 1471-1478.
Adams, C. A., and Rinne, R. W. (1981). *J. Exp. Bot.* **32**, 615-620.
Akalehiywot, T., and Bewley, J. D. (1980). *Can. J. Bot.* **58**, 2349-2355.
Albergoni, F., Lado, P., Marziani, G., and Marré, E. (1964). *Giorn. Bot. Ital.* **71**, 469-488.
Anderson, J. D. (1979). *Plant Physiol.* **63**, 100-104.
Anguillesi, M. C., Grilli, I., and Floris, C. (1980). *Planta* **148**, 24-27.
Armstrong, C., Black, M., Chapman, J. M., Norman, H. A., and Angold, R. (1982). *Planta* **154**, 573-577.
Atkins, W. R. G. (1909). *Sci. Proc. Royal Dublin Soc.* **12**, 35-46.
Atkinson, D. E. (1977). 'Cellular Energy Metabolism and its Regulation'. Academic Press, New York.
Bailey, W. M. (1933). *Bot. Gaz.* **94**, 689-713.
Barton, L. V. (1950). *Contr. Boyce Thompson Inst.* **16**, 55-70.
Beevers, L., and Poulson, R. (1972). *Plant Physiol.* **49**, 476-481.
Bewley, J. D. (1982). *In* 'Nucleic Acids and Proteins in Plants I. Structure, Biochemistry and Physiology of Proteins' (D. Boulter and B. Parthier, eds), Encyclopedia of Plant Physiology, New Series, Vol. 14A, pp. 559-591. Springer-Verlag, Berlin.
Bewley, J. D., and Black, M. (1978). 'Physiology and Biochemistry of Seeds in Relation to Germination', Vol. 1. Springer-Berlin.
Bewley, J. D., and Larsen, K. M. (1979). *Phytochemistry* **18**, 1617-1619.
Bramlage, W. J., Leopold, A. C., and Parrish, D. J. (1978). *Plant Physiol.* **61**, 525-529.
Bray, C. M. (1979). *In* 'Recent Advances in the Biochemistry of Cereals' (D. L. Laidman and R. G. Wyn Jones, eds), pp. 147-173. Academic Press, London.
Brooker, J. D., Cheung, C. P., and Marcus, A. (1977). *In* 'The Physiology and Biochemistry of Seed Dormancy and Germination' (A. A. Khan, ed.), pp. 347-356. North-Holland Publishing Co., Amsterdam.
Buttrose, M. S. (1973). *Protoplasma* **77**, 111-122.
Caers, L. L., Peumans, W. J., and Carlier, A. R. (1979). *Planta* **144**, 491-496.
Cal, J. P., and Obendorf, R. L. (1972). *Crop Sci.* **12**, 369-373.
Carlier, A. R., Manickam, A., and Peumans, W. J. (1980). *Planta* **149**, 227-233.
Chabot, J. F., and Leopold, A. C. (1982). *Amer. J. Bot.* **69**, 623-633.
Cheung, C. P., Wu, J., and Suhadolnik, R. J. (1979). *Nature* **277**, 66-67.
Christiansen, M. N. (1963). *Plant Physiol.* **38**, 520-522.
Christiansen, M. N. (1968). *Plant Physiol.* **43**, 743-746.
Christiansen, M. N. (1969). 'Proceedings of the 23rd Annual Cotton Physiology Conference' (J. Brown, ed.). National Cotton Council, Memphis, Tennessee.
Christiansen, M. N., and Thomas, R. O. (1969). *Crop Sci.* **9**, 672-673.
Clay, W. F., Katterman, F. R. H., and Hammett, J. R. (1975). *Plant Physiol.* **55**, 231-236.
Clegg, J. S. (1978). *In* 'Dry Biological Systems' (J. H. Crowe and J. S. Clegg, eds), pp. 117-153. Academic Press, New York.
Cohn, M. A., and Obendorf, R. L. (1976). *Crop Sci.* **16**, 449-452.

Collins, D. M., and Wilson, A. T. (1972). *Phytochemistry* **11**, 1931-1935.
Collins, D. M., and Wilson, A. T. (1975). *J. Exp. Bot.* **26**, 737-740.
Côme, D. (1971). *Physiol. Vég.* **9**, 439-446.
Côme, D. (1975). *In* 'La Germination des Semences' (R. Chaussat and Y. De Deunff, eds), pp. 27-44. Gauthier-Villars, Paris.
Crawford, R. M. M. (1977). *New Phytol.* **79**, 511-517.
Crowe, J. H., and Crowe, L. M. (1982a). *Cryobiology* **19**, 317-328.
Crowe, L. M., and Crowe, J. H. (1982b). *Arch. Biochem. Biophys.* **217**, 582-587.
Crowe, J. H., O'Dell, S. J., and Armstrong, D. A. (1979). *J. Exptl. Zool.* **207**, 431-437.
Crowe, J. H., Crowe, L. M., and Mouradian, R. (1983). *Cryobiology* **20**, 346-356.
Cuming, A. C., and Lane, B. G. (1979). *Eur. J. Biochem.* **99**, 217-224.
Dasgupta, J., and Bewley, D. J. (1982). *Plant Physiol.* **70**, 1224-1227.
Davidson, D. (1966). *Am. J. Bot.* **53**, 491-495.
Delseny, M., Aspart, L., and Guitton, Y. (1977). *Planta* **135**, 125-128.
Deltour, R., and Jacqmard, A. (1974). *Ann. Bot.* **38**, 529-534.
Dhindsa, R. S., and Bewley, J. D. (1977). *Plant Physiol.* **59**, 295-300.
Doireau, P. (1976). *Phys. Vég.* **14**, 467-486.
Duke, S. H., and Kakefuda, G. (1981). *Plant Physiol.* **67**, 449-456.
Dunn, B. L., Obendorf, R. L., and Paolillo, D. J. (1980). *Plant Physiol.* **65**, Supp. 139.
Dure, L. S. (1979). *In* 'The Plant Seed: Development, Preservation and Germination' (I. Rubenstein, R. L. Phillips, C. E. Green and B. G. Gengenbach, eds), pp. 113-127. Academic Press, New York.
Edwards, M. (1976). *Plant Physiol.* **58**, 237-239.
Eldan, M., and Mayer, A. M. (1972). *Physiol. Plant.* **26**, 67-72.
Esashi, Y., and Leopold, A. C. (1968). *Plant Physiol.* **43**, 871-876.
Eyster, H. C. (1940). *Amer. J. Bot.* **27**, 652-659.
Farrar, J. F., and Smith, D. C. (1976). *Plant Physiol.* **77**, 115-125.
Flinn, A. M., and Smith, D. L. (1967). *Planta* **75**, 10-22.
Fountain, D. W., and Bewley, J. D. (1973). *Plant Physiol.* **52**, 604-607.
Gaff, D. F., and Okong-O-Ogola, O. (197). *J. Exp. Bot.* **22**, 756-758.
Galli, M. G., Sparvoli, E., and Caroi, M. (1975). *Plant Sci. Lett.* **5**, 351-357.
Galli, M. G., Levi, M., and Sparvoli, E. (1981). *Physiol. Plant.* **51**, 321-325.
Gordon, M. E., and Payne, P. I. (1976). *Planta* **130**, 269-273.
Goyns, M. H. (1982). *J. Theor. Biol.* **97**, 577-589.
Grzelczak, A., Sattolo, M. H., Hanley-Bowdoin, L. K., Kennedy, T. D., and Lane, B. G. (1982). *Can. J. Biochem.* **60**, 389-397.
Gupta, R. K. (1976). *Biochem. Physiol. Pflanzen.* **170**, 389-395.
Haber, A. H., and Brassington, N. (1959). *Nature* **183**, 619-620.
Haber, A. H., and Luippold, H. J. (1960). *Plant Physiol.* **35**, 168-173.
Hammett, J. R., and Katterman, F. R. (1975). *Biochem.* **14**, 4375-5379.
Hanson, J. B., and Trewavas, A. J. (1982). *New Phytol.* **90**, 1-18.
Harrington, J. F. (1973). *Seed. Sci. and Technol.* **1**, 453-461.
Hecker, M., Köhler, K.-H., and Wiedmann, M. (1977). *Biochem. Physiol. Pflanzen* **171**, 401-408.
Heslop-Harrison, J. (1979). *Am. J. Bot.* **66**, 737-743.
Hobbs, P. R., and Obendorf, R. L. (1972). *Crop Sci.* **12**, 664-667.
Houben, J. (1966). *Planta* **71**, 87-97.
Hourmant, A., and Pradet, A. (1981). *Plant Physiol.* **68**, 631-635.
Hsu, S. H., Hadley, H. H., and Hymowitz, T. (1973). *Crop Sci.* **13**, 407-410.
Huang, B. F., Rodaway, S. J., Wood, A., and Marcus, A. (1980). *Plant Physiol.* **65**, 1155-1159.

Hui, S. W., Stewart, T. P., Yeagle, P. L., and Albert, A. D. (1981). *Arch. Biochem. Biophys.* **207**, 227–240.
Hunter, J. R., and Erickson, A. E. (1952). *Agronomy J.* **44**, 107–109.
Jacob, K. M., and Bovey, F. (1969). *Exptl. Cell Res.* **54**, 118–126.
Kanai, R., and Edwards, G. E. (1973). *Plant Physiol.* **52**, 484–490.
King, M. W., and Roberts, E. H. (1979). 'The Storage of Recalcitrant Seeds — Achievements and Possible Approaches'. International Board for Plant Genetic Resources, Rome.
Klein, S., Barenholz, H., and Budnik, A. (1971). *Plant and Cell Physiol.* **12**, 41–60.
Kollöffel, C. (1967). *Acta Bot. Neerl.* **16**, 111–122.
Kollöffel, C. (1970). *Planta* **91**, 321–328.
de Kruijff, B., Cullis, P. R., and Verkleij, A. J. (1980). *Trends in Biochem. Sciences* **5**, 79–81.
Labavitch, J. M. (1981). *Annu. Rev. Plant Physiol.* **32**, 385–406.
Lado, P. (1965). *Giorn. Bot. Ital.* **72**, 359–369.
Lado, P., Rasi-Caldogno, F., and Colombo, R. (1974). *Physiol. Plant.* **31**, 149–152.
Lado, P., Rasi-Caldongo, F., and Colombo, R. (1975). *Physiol. Plant.* **34**, 359–364.
Larson, L. A. (1968). *Plant Physiol.* **43**, 255–259.
Larson, L. A., and Lwanga, K. (1969). *Can. J. Bot.* **47**, 707–709.
Leopold, A. C., and Musgrave, M. E. (1979). *Plant Physiol.* **64**, 702–705.
Lott, J. N. A. (1974). *Can. J. Bot.* **52**, 1465–1468.
Lush, W. M., Groves, R. H., and Kaye, P. E. (1981). *Aust. J. Plant Physiol.* **8**, 409–425.
Luzzati, V., and Husson, F. (1962). *J. Cell Biol.* **12**, 207–219.
McKersie, B. D., and Senaratna, T. (1983). *Recent Adv. Phytochem.* **17**, 29–52.
McKersie, B. D., and Stinson, R. H. (1980). *Plant Physiol.* **66**, 316–320.
Marcus, A. (1982). *In* 'Nucleic Acids and Proteins in Plants I. Structure, Biochemistry and Physiology of Proteins' (D. Boulter and B. Parthier, eds), Encyclopedia of Plant Physiology, New Series, Vol. 14A, pp. 113–135. Springer-Verlag, Berlin.
Marcus, A., Feeley, J., and Volcani, T. (1966). *Plant Physiol.* **41**, 1167–1172.
Marrè, E. (1979). *Annu. Rev. Plant Physiol.* **30**, 273–278.
Matilla, A., Nicolas, G., Vicente, O., and Sierra, J. M. (1980). *Plant Physiol.* **65**, 1128–1132.
Matthews, S., and Rogerson, N. E. (1976). *J. Exp. Bot.* **27**, 961–968.
Matthews, S., Powell, A. A., and Rogerson, N. E. (1980). *In* 'Seed Production' (P. D. Hebblethwaite, ed.), pp. 513–535. Butterworths, London.
Mayer, A. M., and Marbach, I. (1981). *Progr. Phytochem.* **7**, 95–136.
Mayer, A. M., and Shain, Y. (1974). *Annu. Rev. Plant Physiol.* **25**, 167–193.
Mills, L. K. (1983). 'The Leakage of Solutes during Imbibition'. Ph.D. Thesis, The Queen's University of Belfast, Northern Ireland.
Moreland, D. E., Hussey, G. G., Shriner, C. R., and Farmer, F. S. (1974). *Plant Physiol.* **54**, 560–563.
Morohashi, Y. (1980). *J. Exp. Bot.* **31**, 805–812.
Morohashi, Y., and Bewley, J. D. (1980a). *Plant Physiol.* **66**, 70–73.
Morohashi, Y., and Bewley, J. D. (1980b). *Plant Physiol.* **66**, 637–640.
Morohashi, Y., and Shimokoriyama, M. (1975). *J. Exp. Bot.* **26**, 932–938.
Morohashi, Y., Bewley, J. D., and Yeung, E. C. (1981). *J. Exp. Bot.* **32**, 605–613.
Mory, Y. Y., Chen, D., and Sarid, S. (1975). *Plant Physiol.* **55**, 437–442.
Müntz, K. (1982). *In* 'Nucleic Acids and Proteins in Plants I. Structure, Biochemistry and Physiology of Proteins' (D. Boulter and B. Parthier, eds), Encyclopedia of Plant Physiology, New Series, Vol. 14A, pp. 505–558. Springer-Verlag, Berlin.
Nakayama, N., Sugimoto, I., and Asahi, T. (1980). *Plant Physiol.* **65**, 229–233.

Nawa, Y., and Asahi, T. (1971). *Plant Physiol.* **48**, 671-674.
Nawa, Y., and Asahi, T. (1973a). *Plant Physiol.* **51**, 833-838.
Nawa, Y., and Asahi, T. (1973b). *Plant, Cell Physiol.* **14**, 607-610.
Obendorf, R. L., and Hobbs, P. R. (1970). *Crop Sci.* **10**, 563-566.
Obendorf, R. L., and Marcus, A. (1974). *Plant Physiol.* **53**, 779-781.
O'Farrell, P. H. (1975). *J. Biol. Chem.* **250**, 4007-4021.
O'Neill, S., and Leopold, A. C. (1982). *Plant Physiol.* **70**, 1405-1409.
Öpik, H. (1980). 'The Respiration of Higher Plants'. Edward Arnold, London.
Öpik, H., and Simon, E. W. (1963). *J. Exp. Bot.* **14**, 299-310.
Orphanos, P. I., and Heydecker, W. (1968). *J. Exp. Bot.* **19**, 770-784.
Parrish, D. J., and Leopold, A. C. (1977). *Plant Physiol.* **59**, 1111-1115.
Payne, P. I. (1976). *Biol. Rev.* **51**, 329-363.
Payne, P. I. (1977). *Phytochem.* **16**, 431-434.
Payne, P. I., Dobrzanska, M., Barlow, P. W., and Gordon, M. E. (1978). *J. Exp. Bot.* **29**, 77-88.
Perl, M. (1982). *J. Exp. Bot.* **44**, 463-470.
Perry, D. A., and Harrison, J. G. (1970). *J. Exp. Bot.* **21**, 504-512.
Peumans, W. J., Caers, L. I., and Carlier, A. R. (1979). *Planta* **144**, 485-490.
Peumans, W. J., Carlier, A. R., and Delaey, B. M. (1980). *Plant Physiol.* **66**, 584-587.
Pollock, B. M. (1969). *Plant Physiol.* **44**, 907-911.
Pollock, B. M., and Manalo, J. R. (1970). *Proc. Amer. Soc. Hort. Sci.* **95**, 415-417.
Pollock, B. M., and Toole, V. K. (1966). *Plant Physiol.* **41**, 221-229.
Poulson, R., and Beevers, L. (1973). *Biochim. Biophys. Acta* **308**, 381-389.
Powell, A. A., and Matthews, S. (1977). *J. Exp. Bot.* **32**, 1045-1050.
Powell, A. A., and Matthews, S. (1978). *J. Exp. Bot.* **29**, 1215-1229.
Powell, A. A., and Matthews, S. (1979). *J. Exp. Bot.* **30**, 193-197.
Powell, A. A., and Matthews, S. (1981). *J. Exp. Bot.* **32**, 1045-1050.
Priestley, D. A., and Bruinsma, J. (1982). *Physiol. Plant.* **56**, 303-311.
Priestley, D. A., and de Kruijff, B. (1982). *Plant Physiol.* **70**, 1075-1078.
Priestley, D. A., and Leopold, A. C. (1980). *Physiol. Plant.* **49**, 198-204.
Retzlaff, G. (1970). *Protoplasma* **70**, 361-377.
Rodaway, S., Huang, B.-F., and Marcus, A. (1979). In 'The Plant Seed: Development, Preservation, and Germination' (I. Rubenstein, R. L. Phillips, C. E.. Green and B. S. Gengenbach, eds), pp. 203-239. Academic Press, New York.
Rogan, P. G., and Simon, E. W. (1975). *New Phytol.* **74**, 273-275.
Sargent, J. A., and Osborne, D. J. (1980). *Protoplasma* **104**, 91-103.
Sato, S., and Asahi, T. (1975). *Plant Physiol.* **56**, 816-820.
Seewaldt, V., Priestley, D. A., Leopold, A. C., Feigenson, G. W., and Goodsaid-Zalduondo, F. (1981). *Planta* **152**, 19-23.
Sen, S., Payne, P. I., and Osborne, D. J. (1975). *Biochem. J.* **148**, 381-387.
Shaykewich, C. F. (1973). *J. Exp. Bot.* **24**, 1056-1061.
Shaykewich, C. F., and Williams, J. (1973). *J. Exp. Bot.* **22**, 19-24.
Sherwin, T., and Simon, E. W. (1969). *J. Exp. Bot.* **20**, 776-785.
Shivanna, K. R., and Heslop-Harrison, J. (1981). *Ann. Bot.* **47**, 759-770.
Shull, C. A. (1913). *Bot. Gaz.* **61**, 169-199.
Siedow, J. N., and Girvin, M. E. (1980). *Plant Physiol.* **65**, 669-674.
Simon, E. W. (1974). *New Phytol.* **73**, 377-420.
Simon, E. W. (1978). In 'Dry Biological Systems' (J. H. Crowe and J. S. Clegg, eds), pp. 205-224. Academic Press, New York.
Simon, E. W., and Mills, L. (1983). *Recent Adv. Phytochem.* **17**, 9-27.

Simon, E. W., and Raja Harun, R. M. (1972). *J. Exp. Bot.* **23**, 1076-1085.
Simon, E. W., and Wiebe, H. H. (1975). *New Phytol.* **74**, 407-411.
Simon, E. W., Minchin, A., McMenamin, M. M., and Smith, J. M. (1976). *New Phytol.* **77**, 301-311.
Smith, C. A. D., and Bray, C. M. (1982). *Planta* **156**, 413-420.
Spedding, D. J., and Wilson, A. T. (1968). *Phytochemistry* **7**, 897-901.
Speer, H. L. (1973). *Plant Physiol.* **52**, 142-146.
Spiegel, S., and Marcus, A. (1975). *Nature* **256**, 228-230.
Spurný, M. (1964). *Flora* **154**, 167-180.
Spurný, M. (973). *In* 'Seed Ecology' (W. Heydecker, ed.), pp. 367-389. Butterworths, London.
Steponkus, P. L. (1970). *Cryobiology* **8**, 570-573.
Stewart, R. R. C., and Bewley, J. D. (1981). *Plant Physiol.* **68**, 516-518.
Stoddart, J. L., and Thomas, H. (1973). *Planta* **112**, 309-321.
Tao, K.-L., and Khan, A. A. (1976). *Plant Physiol.* **58**, 769-772.
Thompson, E. W., and Lane, B. C. (1980). *J. Biol. Chem.* **255**, 5965-5970.
Thomson, W. W., and Platt-Aloia, K. (1982). *Plant, Cell Environ.* **5**, 367-373.
Toivio-Kinnucan, M. A., and Stushnoff, C. (1981). *Cryobiology* **18**, 72-78.
Touzard, J. (1975). *In* 'La Germination des Semences' (R. Chaussat and Y. Le Deunff, eds), pp. 157-170. Gauthier-Villars, Paris.
Tully, R. E., Musgrave, M. E., and Leopold, A. C. (1981). *Crop Sci.* **21**, 312-317.
Vigil, E. L., Christiansen, M. N., Steare, R. L., and Wergin, W. P. (1982). *Plant Physiol.* **69**, Supp. 3.
Villanueva, V. R., Adlakha, R. C., and Cantera-Soler, A. M. (1978). *Phytochemistry* **17**, 1245-1249.
Waggoner, P. E., and Parlange, J.-Y. (1976). *Plant Physiol.* **57**, 153-156.
Walton, D. C. (1966). *Plant Physiol.* **41**, 298-302.
Waters, L. C., and Dure, L. S. (1966). *J. Mol. Biol.* **19**, 1-27.
Weeks, D. P., and Marcus, A. (1971). *Biochim. Biophys. Acta* **232**, 671-684.
Weil, J. H., and Parthier, B. (1982). *In* 'Nucleic Acids and Proteins in Plants II. Structure, Biochemistry and Physiology of Nucleic Acids' (B. Parthier and D. Boulter, eds), Encyclopedia of Plant Physiology, New Series, Vol. 14B, pp. 65-112. Springer-Verlag, Berlin.
Wilson, A. M. (1970). *Plant Physiol.* **45**, 524-526.
Wunderlich, F., Kreutz, W., Mahler, P., Ronai, A., and Heppeler, G. (1978). *Biochemistry* **17**, 2005-2010.
Yentur, Y., and Leopold, A. C. (1976). *Plant Physiol.* **57**, 274-276.

CHAPTER **4**

Mobilization of Polysaccharide Reserves from Endosperm

ANNE E. ASHFORD and FRANK GUBLER

I.	Introduction	117
II.	Starch Hydrolysis in Cereal Endosperm	118
	A. Breakdown of Starch Granules	118
	B. Effects of Gibberellin on α-Amylase Synthesis in Aleurone	121
	C. Isoenzymes of α-Amylase	127
	D. Secretion of Hydrolases from Aleurone	129
	E. Other Sites of Production of Starch-Hydrolysing Enzymes	140
	F. The Production of Gibberellins in Cereals	144
	G. The Whole Grain: A Synopsis	145
III.	Cell Wall Mobilization	146
	A. Cereal Endosperm Walls	146
	B. Mannan-rich Wall Deposits	147
IV.	Future Prospects	156
	References	157

I. INTRODUCTION

Mobilization of carbohydrates from the endosperm has probably received more attention than any other topic in seed physiology. This is a direct consequence of the role of endosperm carbohydrates in human affairs. Starch is the most common, major carbohydrate accumulated and its mobilization has been studied in the most detail in cereal grains. This results firstly from the central role of cereals as food crops (see Section 1.IV of

Volume 1) and secondly from the importance of barley (*Hordeum vulgare*) in the brewing of beer.

During germination and seedling establishment starch is rapidly mobilized but this is not the only hydrolytic event and the endosperm becomes completely liquefied. The cell walls, any remaining components of the cytoplasm and protein deposits are all dissolved, together with the starch. In cereal grains, the contribution of cell walls to the carbohydrate metabolism of the embryo is usually much less than that of starch, but nevertheless is still significant (Morrall and Briggs, 1978). In certain other species starch is not present and cell wall deposits constitute the major carbohydrate reserve of the endosperm. These often include mannans (Section 5.III,C of Volume 1) and their hydrolysis is considered here (Section III,B). The hydrolysis of proteins is the subject of the next chapter, but it must be emphasized that in addition to providing dipeptides and amino acids for embryo growth, protein breakdown supplies amino acids for the production of hydrolases necessary for carbohydrate breakdown.

II. STARCH HYDROLYSIS IN CEREAL ENDOSPERM

A. Breakdown of Starch Granules

Starch occurs in the endosperm of wheat (*Triticum aestivum*) and barley in two types of granule, which differ in size and shape (Fig. 1). The large granules, which are lenticular, make up about 90% of the total starch content and are considered to make the most important contribution (May and Buttrose, 1959). In other cereals, starch granules do not fall into two distinct size categories (Kent, 1975). Isolated starch granules contain amylose and amylopectin (see MacLeod, 1979). Amylose is a linear $(1\rightarrow 4)$-α-glucan, while amylopectin is a highly branched molecule, consisting of $(1\rightarrow 4)$-α-glucan with $(1\rightarrow 6)$-α-branch points to other glucan chains. Some authors consider that the separation of these two fractions is an artifact of extraction (see Williams, 1968) but most do not (Section 5.III,A,*1* of Volume 1).

Most studies on starch breakdown have been undertaken in wheat and barley and in these two species the pattern of hydrolysis is remarkably similar. However, there are differences in the hydrolysis of the large and small granules. In barley the small granules, which are spherical, appear to be degraded first and degradation occurs primarily at the granule surface (Palmer, 1972). In wheat they are less susceptible to digestion and granules showing interior breakdown have sometimes been found (MacGregor and

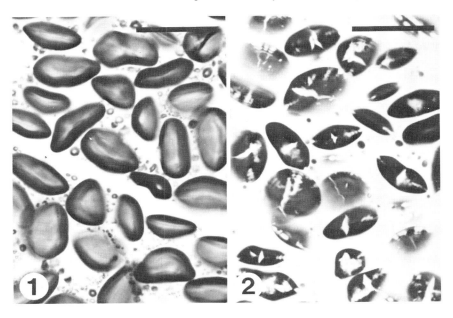

Fig. 1. Large and small starch grains *in situ* in the endosperm of unimbibed barley grains (*Hordeum vulgare*). Bar = 30 μm.

Fig. 2. Endosperm of six-day imbibed barley grains showing hydrolysis patterns in the large starch granules. Most of the small granules have already disappeared. Bar = 30 μm. The material for this Figure and for Figure 1 was embedded in glycol-methacrylate, sectioned and stained for carbohydrate with the periodic acid-Schiff (PAS) reaction.

Matsuo, 1982). Dissolution of the large granules also starts most commonly at their surface: many pits are formed and the equatorial groove becomes more prominent (Fig. 2). However, further corrosion converts the pits into a network of radial channels and hydrolysis progresses rapidly to the interior of the granule. This region appears to be degraded preferentially to produce a hollow shell (Fincher and Stone, 1974). This pattern of digestion has been interpreted to reflect preferential digestion either of less densely packed regions (Buttrose, 1960) or along microfractures in the granule (Palmer, 1972). In pearl millet (*Pennisetum americanum*) and oats (*Avena sativa*), starch granules are all of similar size and breakdown is predominantly at the surface (Lineback and Ponpipom, 1977).

Enzymes with a potential role in starch breakdown are listed together with their reactions in Table I. In the cereal endosperm the complete

Table 1. Enzymes implicated in starch hydrolysis

Enzyme	Type	Substrate preference
α-amylase [EC 3.2.1.1]	endohydrolase: cleaves $(1\rightarrow 4)$-α-links	large dextrins (inhibited near $(1\rightarrow 6)$-α-branch points)
β-amylase [EC 3.2.1.2]	exohydrolase: maltose released sequentially from non-reducing end	large dextrins (limited by $(1\rightarrow 6)$-α-branch points)
α-glucosidase [EC 3.2.1.20]	exohydrolase: glucose released sequentially from non-reducing end	maltose and small oligosaccharides (limited by $(1\rightarrow 6)$-α-branch points)
Limit dextrinase [EC 3.2.1.41]	hydrolyses $(1\rightarrow 6)$-α-branch points	limit dextrins
Starch phosphorylase [EC 2.4.1.1]	exohydrolase: in the presence of inorganic phosphate, releases glucose-1-phosphate sequentially from non-reducing end	degrades amylose, amylopectin and smaller dextrins (limited by $(1\rightarrow 6)$-α-branch points)

Data compiled from Greenwood and Milne, 1968; Manners, 1974; Preiss and Levi, 1980.

breakdown of starch is accomplished by a mixture of $(1\rightarrow 4)$-α-glucanases and $(1\rightarrow 6)$-α-glucanases. Barley and pearl millet α-amylases reported to be free from other glucan hydrolases readily degrade intact starch granules (Maeda et al., 1978; Beleia and Varriano-Marston, 1981). Alpha-amylase, an endohydrolase, is thought to catalyse the initial steps in depolymerisation of both the amylose and amylopectin and linear or branched dextrins are likely to be the major products of this action (see Greenwood and Milne, 1968). With prolonged incubation, purified cereal α-amylases can degrade amylose and amylopectin to low molecular weight oligosaccharides and glucose. Some of the oligosaccharides contain $(1\rightarrow 6)$-α-branch points, which are not attacked. *In vivo* it is unlikely that α-amylase would carry the reaction this far, because of the very slow rates of hydrolysis of malto-oligosaccharides and the presence of other enzymes. Therefore, following the initial step of starch hydrolysis by α-amylase, it seems likely that hydrolysis of the released dextrins is completed by a combination of β-amylase, limit dextrinase and α-glucosidase (Dunn, 1974).

Beta-amylase, which has been detected in a number of cereal grains during germination, catalyses the release of maltosyl residues from the non-reducing end of amylose and amylopectin (Greenwood and Milne, 1968). However, with amylopectin, the stepwise action of β-amylase is prevented when $(1\rightarrow 6)$-α-branch points are approached. Accumulation of maltose in barley grains during germination has been cited as evidence for the presence of both α-amylase and β-amylase in the endosperm (MacLeod, 1979). Furthermore, degradation of isolated starch granules by barley α-amylase is promoted by addition of β-amylase, although β-amylase alone has little effect (Maeda et al., 1978). Limit dextrinase is thought to be the sole debranching enzyme involved in starch hydrolysis in cereal endosperm. This catalyses the hydrolysis of $(1 \rightarrow 6)$-α-glucosidic linkages in branched dextrins and amylopectin (see Manners, 1974; MacLeod, 1979; Preiss and Levi, 1980). Alpha-glucosidase catalyses the stepwise release of glucosyl residues from the non-reducing end of dextrins, with a substrate preference for maltose and to a lesser extent other maltose oligosaccharides. Both these hydrolases have been detected in germinating cereal grains and show increases in activity in response to increasing gibberellic acid (GA_3) concentrations in barley half-grains in parallel with α-amylase (Manners and Yellowlees, 1973; Hardie, 1975). Although phosphorylase is considered to play an important role in starch degradation in legume cotyledons during germination (Section 7.II,A) there is so far little evidence for its involvement in degradation of cereal endosperm (Tsai and Nelson, 1968, 1969; Abbott and Matheson, 1972).

B. Effects of Gibberellin on α-Amylase Synthesis in Aleurone

1. Levels of α-Amylase Messenger RNA

Although it had been recognized for a long time that endosperm breakdown in barley does not occur unless the embryo is present, it was not at first clear whether the embryo produced the enzymes directly or secreted some factor which controlled their production elsewhere (Haberlandt, 1884, 1890; Brown and Morris, 1890). The only part of the endosperm considered to be living is the aleurone layer but evidence that enzymes for endosperm hydrolysis originate there was not obtained until the development of dissection techniques to separate aleurone tissue from the rest of the endosperm (see Briggs, 1973). A standard method was developed for isolation of aleurone layers from three-day imbibed half grains of barley (Chrispeels and Varner, 1967). Such layers always have a small and variable amount of starchy endosperm adhering to them. Contamination by microorganisms is kept to a minimum by using sterile technique and bacterial suppressants. The aleurone layers can be combined and incubated under controlled conditions. One of the early experiments (Yomo and Iinuma, 1964) showed that GA_3 caused a ten-fold increase in α-amylase activity: this was about 70% of the response elicited in endosperm halves.

Elegant experiments by Varner's group showed that the barley aleurone layer both synthesizes and secretes the major starch-hydrolysing enzyme α-amylase in response to GA_3 (see Yomo and Varner, 1971) and that in the

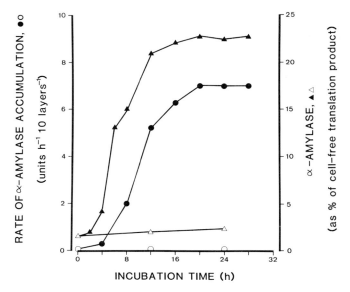

Fig. 3. Changes in the rate of α-amylase accumulation and levels of translatable α-amylase mRNA in barley aleurone layers in response to GA_3: ●▲ $+GA_3$; ○△ $-GA_3$ (after Higgins et al., 1982).

absence of the endosperm, α-amylase accumulates in the incubating medium. Sensitivity to cycloheximide and other inhibitors, and experiments with radioactive tracers, indicated that α-amylase production involves new protein synthesis (Varner and Chandra, 1964). Conclusive evidence of *de novo* synthesis was obtained by density labelling experiments, where newly synthesized proteins were distinguished from already existing ones by their increased density following incorporation of heavy (deuterium-containing) amino acids. All the protein with α-amylase activity produced in the presence of label was heavier than that made in its absence, indicating that all the enzyme molecules had been newly synthesized (Filner and Varner, 1967).

Four phases can be recognized during the time course of α-amylase synthesis (Fig. 3): a lag phase; a period when the rate of synthesis increases; a linear phase when the rate of synthesis is maximal; and a final stage when activity begins to decline. The lag phase has been defined as the time between application of GA_3 and the first appearance of α-amylase. Its perceived length has become shorter as more sensitive assays have been developed, and it has been measured at two to three hours after GA_3

addition (Higgins *et al.*, 1976). Little is known about what happens in the lag phase, but it has generally been assumed that most of it represents the time taken to induce the alterations of cell metabolism needed to develop the synthesis and secretion systems. However, Gibson and Paleg (1982) have shown that the time taken for the hormone to penetrate the tissue, and to associate irreversibly with it or activate it, is a significant component of the lag phase and cannot be ignored. The effects of inhibitors of RNA synthesis indicate that transcription occurs at the end of the lag phase and that α-amylase escapes from transcriptional control about 15 h after GA_3 addition, when the rate of α-amylase synthesis has just become constant (Goodwin and Carr, 1972; Jones and Jacobsen, 1978). The amino acids for this synthesis are produced by hydrolysis of protein stored in the aleurone grains, which are abundant throughout the cells and become progressively digested (Jones, 1969; Yomo and Varner, 1971; Briggs, 1973).

The mechanism of action of GA_3 in promoting α-amylase synthesis was controversial until Higgins *et al.* (1976) showed that GA_3 promotes an increase in a mRNA specific for α-amylase. The isolation of this mRNA from barley aleurone was made possible because, like much eukaryote mRNA, its 3-hydroxy end contains a poly(adenylic) acid sequence which allows it to be separated from other RNA by binding to columns containing a complementary base. Higgins *et al.* (1976) were able to show that this isolated poly(A)-containing RNA fraction included a mRNA specific for α-amylase, by translating it with a cell-free protein-synthesis system comprising wheat-germ polyribosomes. Alpha-amylase was identified among the translation products by its molecular weight and immunological properties. Its incorporation of radioactive [^{35}S]methionine could be used to determine the amount of translatable mRNA present. Alpha-amylase mRNA began to increase two to four hours after GA_3 addition and this increase continued until about 16 h, when it constituted 15%–25% of the total translatable mRNA (Jacobsen *et al.*, 1982). Other workers have obtained essentially similar results (Ho, 1979; Muthukrishnan *et al.*, 1979; Mozer, 1980a). The molecular weight of the barley α-amylase mRNA is estimated at about 580 000 daltons, giving it about 1600 nucleotides, sufficient to code for a protein of about 45 000 daltons such as α-amylase (Mozer, 1980a; Jacobsen *et al.*, 1982).

After GA_3 treatment, neither incorporation of labelled nucleotides into total RNA, nor amount of total RNA, increases. However, there is increased incorporation of labelled nucleotides into poly(A)RNA, indicating that the rate of formation of this fraction is increased (Ho and Varner, 1974; Jacobsen and Zwar, 1974a, b). There has been speculation as to whether GA_3 increases the level of this mRNA by increasing its synthesis (and therefore acting at the level of transcription) or by activation of an already existing precursor mRNA, which may be inactive in *in vitro* translation systems. Bernal-Lugo *et al.* (1981) tested these ideas by using a DNA-hydridization method which can measure mRNA whether it is translatable or not. They prepared radioactively labelled DNA complementary to a partially purified α-amylase mRNA, and used this DNA as a 'probe' to measure the mRNA levels in GA_3-treated barley aleurone layers. The assay indicated that α-amylase mRNA increases about forty-fold in GA_3-treated aleurone, supporting the idea of a true increase in the amount of messenger rather than merely its activation.

Gibberellic acid does not cause major changes in the amount of total RNA or protein synthesis, but redirects protein synthesis so that some proteins are produced in large amounts while synthesis of others is depressed. These effects may not be due simply to alterations of mRNA levels. Profiles of polypeptides translated in cell-free systems from GA_3-treated aleurone mRNA are much more complex than the spectrum of polypeptides obtained *in vivo*. Furthermore, the rate of increase in α-amylase synthesis *in vivo* is faster than the rise in translatable mRNA (Jacobsen *et al.*, 1982). Both these observations imply the existence of controls other than simply the level of mRNA. (It should perhaps be noted here that this level of mRNA *in vivo* at any time will result from the sum of its production and breakdown or deactivation, and there is as yet no direct information on the latter.)

The α-amylase produced from α-amylase mRNA in cell-free translation systems, although immunologically similar, is always larger than the α-amylase produced by aleurone layers (Okita *et al.*, 1979; Mozer, 1980a; Higgins *et al.*, 1982). This agrees with the concept that proteins destined for secretion contain a unique sequence of amino acids, which triggers attachment of the ribosomes to the membrane and directs the synthesized protein across it (signal hypothesis, see Section 3.V,A of Volume 1).

Further support for this hypothesis came from experiments in barley and wheat. *In vitro* translation systems were converted from producing the larger molecular weight form of α-amylase to producing the lower molecular weight, authentic α-amylase by adding membranes from pancreatic microsomes stripped of their ribosomes (Boston *et al.*, 1982; Higgins *et al.*, 1982). Such preparations can remove the signal sequence and vectorially transport precursors of secreted proteins from animal cells across membranes (Jackson and Blobel, 1977; Bielinska *et al.*, 1979). Furthermore, *Xenopus laevis* oocytes (used as a whole-cell translation system) produce authentic α-amylase, not the high molecular-weight form, on injection with the same wheat mRNA containing α-amylase messenger (Boston *et al.*, 1982). Therefore, it seems very likely that the signal hypothesis applies to aleurone cells: α-amylase is synthesized on membrane-bound polysomes and is vectorially transported into the lumen of the endoplasmic reticulum (ER).

2. Interaction of Other Growth Substances
Gibberellic acid is the gibberellin (GA) most commonly used in experiments with isolated aleurone layers, but other gibberellins (with the exception of GA_8) are also able to elicit the response. Kinetin and auxin are without direct effect (Clutterbuck and Briggs, 1973) and there is no evidence that cyclic-3,5-AMP is involved in any way (Keates, 1973).

Abscisic acid (ABA) inhibits GA_3-induced synthesis of α-amylase, when added at the same time as GA_3. It does not cause gross changes in metabolism, but antagonizes GA-enhanced processes specifically (Jacobsen, 1973). The mechanism by which this occurs has become the subject of controversy. Ho and Varner (1976) found that ABA continued to inhibit α-amylase synthesis after they considered that α-amylase mRNA synthesis was complete, and concluded that ABA acts via a post-transcriptional mechanism, probably the translation of α-amylase mRNA. A similar view was held by Mozer (1980b). This was based on observations that the levels of translatable mRNA extracted from aleurone treated with GA_3 and ABA were similar to those with GA_3 alone, while α-amylase synthesis was suppressed *in vivo*. In contrast, Jacobsen and co-workers reported that ABA strongly suppressed the amount of translatable mRNA but also affected the efficiency of mRNA translation, and these authors concluded that ABA exerts its effect both on the level of mRNA and its translatability (Jacobsen *et al.*, 1979).

More direct measurements of mRNA levels using complementary DNA support the view that ABA suppresses α-amylase mRNA levels (Chandler *et al.*, 1982), but this does not necessarily exclude an effect on translation. To what extent translation may be affected is at present unresolved. One feature which has confounded interpretation of the interactions between hormones and inhibitors in GA-induced responses is that net effects are being measured. Not all cells are positioned equally in relationship to the bathing medium and there is ample cytological evidence that the cells do not behave absolutely synchronously. Variability of aleurone response from experiment to experiment may explain some of the discrepancies. The inhibitory effect of ABA is not only on α-amylase synthesis. ABA appears also to prevent all of the other GA_3-promoted changes in mRNA levels and protein synthesis (Jacobsen *et al.*, 1979; Mozer, 1980b; Higgins *et al.*, 1982). In addition there is evidence that ABA, when added alone, stimulates the production of two or more specific proteins by increasing their mRNA levels, and that this effect is antagonized by GA_3. The situation as summed up by Jacobsen (1983) is that GA_3 and ABA appear to act in barley aleurone in similar ways, both modifying mRNA levels and protein synthesis and each antagonizing the other's action.

3. Effects of Gibberellin on the Endoplasmic Reticulum

Proliferation of the rough endoplasmic reticulum (ER) is commonly found in cells where large amounts of protein are synthesized and secreted (Gunning and Steer, 1975). Consequently, there has been much interest in the effect of GA on membrane biosynthesis and formation of the ER in aleurone. In barley, GA_3-induced proliferation of the rough ER has been inferred from many studies using electron microscopy. But unfortunately there are no quantitative data from the observations of fine structure to determine whether this reflects an increase in the rough ER or simply its reorganization (see Jones and Jacobsen, 1978). Some biochemical work supports the view that membrane biosynthesis occurs, such as an increase in the activity of enzymes of phospholipid biosynthesis in aleurone tissue, or increase in incorporation of radioactive choline and $^{32}P_i$ into aleurone microsomal fractions in response to GA_3 (Evins and Varner, 1971; Koehler and Varner, 1973), but other observations are contradictory. Firn and Kende (1974), for example, were unable to find net increase in total lipid, neutral lipids or total lipid phosphorus, and neither they, nor others, could demonstrate incorporation of acetate or glycerol into crude lipid fractions of GA_3-treated cells (Koehler and Varner, 1973). Varner and Ho (1976) interpret these apparently conflicting data to indicate that there is GA_3-enhanced lipid turnover, without net lipid synthesis. This would be expected if there were degradation of stored lipid in the aleurone spherosomes to produce new membrane components and neither argues for nor refutes the proposal that GA_3-induced ER synthesis occurs (Jones and Jacobsen, 1978).

Recently Jones (1980a, b) has evaluated the fractionation techniques used with barley aleurone layers and found most of them wanting. Using improved methods, which show a high recovery of recognizable ER, he has obtained evidence for a GA_3 stimulation in the amount of ER within the first 24 h of incubation with GA_3. But the enhancement was only by 80% and substantial increases in the ER also occurred during incubation of half grains and isolated aleurone layers without GA_3. Furthermore, later in incubation there was a sharp drop in the amount of ER in GA_3 treatments. This fall paralleled other GA_3-induced fine structural changes, such as the formation of a large central vacuole and reduction of the cytoplasmic volume of aleurone cells (Jones and Price, 1970; Jones, 1980b). The data

4. Endosperm Carbohydrate Mobilization   127

as a whole, therefore, show that the effect of GA₃ on the amount of ER may not be as great as was first thought and that the changes in fine structure observed are due to a major reorganization of the ER membranes rather than their increase.

It is difficult to make comparisons, because of the differences in treatment of the tissue, sampling times and techniques used (Jones and Jacobsen, 1978), but there is some evidence that the effects of GA may be similar in wheat aleurone. Colborne *et al.* (1976), on the basis of fine structural changes, reported that the ER increases in wheat aleurone without GA₃ and that there is also a GA₃-induced fall in the amount of ER, but they failed to find a GA₃ stimulation. Jelsema *et al.* (1982) found that GA₃ enhanced some of the enzymes of lipid biosynthesis but found that the major effect was on the timing of events.

The initial reports that GA₃ stimulates polysome formation (Ho, 1979) have also received criticism for reasons similar to those given above and, as yet, there is no compelling evidence that GA₃ stimulates either the number of ribosomes or polysomes. In fact, there is some evidence against this, namely, the failure of GA₃ to stimulate total protein synthesis, total RNA synthesis or an increase in ribosomal RNA (see Jacobsen, 1977). Gibberellic acid may increase the association of polysomes with the ER membranes, which would make the appearance of stacked rough ER cisternae more obvious. Some of the effects of GA₃ thought to be related to membrane biosynthesis appear to occur very early (i.e., within half an hour; Varner and Ho, 1977).

C. Isoenzymes of α-Amylase

Multiple forms of α-amylase have been demonstrated in isolated aleurone layers and whole germinated barley grains by a number of authors. The number of isoenzymes obtained varies according to the variety and species examined and the techniques used (see MacGregor, 1977; Jacobsen and Higgins, 1982 for references). However, it has become clear that in many barley varieties the isoenzymes can be separated into two major groups that have different physical and chemical properties, are immunologically distinct and behave differently in response to GA₃.

The α-amylase isoenzymes separated by Jacobsen and colleagues from GA₃-treated aleurone layers of Himalaya barley are shown in Table II. Traces of β-amylase are also released into the incubation medium, but cannot be detected inside aleurone layers and are thought to arise from adhering starchy endosperm. One α-amylase isoenzyme, amylase 5, is only found in the soluble fraction of aleurone tissue where it represents a large

Table II. Amylase isoenzymes in isolated aleurone layers of Himalaya barley treated with and without GA_3, for 22 h

	Total (0 h)	Medium		Extract organelles		Extract soluble	
		+GA	−GA	+GA	−GA	+GA	−GA
Group A							
1	+	+ +	+	+	+	+	+
2	+	+ + +	+	+	+	+ +	+
Group B							
3	−	+	−	+	−	+	−
4	−	+ + +	−	+	−	+ +	−
5	−	−	−	−	−	+ +	+
β-Amylase	trace	+ +	−	−	−	−	−

About 83% of enzyme extracted was in the soluble fraction. β-amylase is thought to arise from adhering starchy endosperm (−, enzyme not detected).
Data from Jacobsen and Higgins (1982); Jones and Jacobsen (1982).

proportion of enzyme activity. However, the status of this isoenzyme is not clear. It does not seem to occur under all extraction conditions, and appears to arise at the expense of group B isoenzymes (compare Jones and Jacobsen, 1982; Jacobsen and Higgins, 1982). It may therefore be produced by the extraction procedure.

Four α-amylase isoenzymes are released into the incubating medium. These can be separated into two distinct groups on the basis of their properties (Jacobsen and Higgins, 1982). Isoenzymes 3 and 4 belonging to group B behave as 'classical' aleurone α-amylases. They have a calcium requirement for activity and show the greatest response to GA_3, increasing 12-fold in the bathing medium. These isoenzymes are not produced at GA_3 concentrations lower than 0.1 mM and they cannot be detected until 8–13 h after GA_3 addition. However, they make an increasing contribution, so that by 24 h they may represent 60% of total α-amylase activity. Isoenzyme 4 is the major component. By contrast isoenzymes 1 and 2 belonging to group A have no calcium requirement and are sensitive to sulfhydryl group reagents. They are immunologically similar to each other, but differ immunologically from enzymes of group B. Their response to GA_3 is quite different. They are produced and appear in the incubating medium in small amounts without GA_3, mainly as isoenzyme 2. They appear in response to concentrations of GA_3 as low as 0.1 μM, and their presence in tissue incubated without GA_3 could be attributed to the low levels of endogenous gibberellins thought to occur in aleurone. They appear sooner in response to GA_3 and enhancement of their production and release by GA_3 is much less than for group B enzymes. Their increase in the bathing medium is only 2.5-fold, but nevertheless they contribute significantly (20%-25%) to the total amylase activity in $+GA_3$ incubating media at 24 h and obviously are important in endosperm hydrolysis.

Groups A and B can be distinguished by isoelectric focusing: their isoelectric points cluster into two distinct groups. Evidence is accumulating that these two groups may be produced from two different genes or gene clusters, situated on different chromosomes. Isoenzymes within each group have some peptides in common, but the polypeptides of one group differ from those of the other group (Jacobsen and Higgins, 1982). Other evidence that the two groups may arise from different genes comes from work on 'addition' lines, where single barley chromosomes have been inserted into the Chinese spring variety of wheat (Brown and Jacobsen, 1982). The products of the α-amylase genes situated on chromosomes 1 and 6 of Betzes barley can also be separated into two distinct groups by isoelectric focusing. The group from chromosome 1 clusters in the pH range 4.5-5.2 and that from chromosome 6 from pH 5.9-6.6, very similar to Groups A and B from Himalaya barley. Several other cultivars of barley and wild barley (*Hordeum spontaneum*) also contain α-amylase isoenzymes which separate into two similar, distinct groups. In some of these the two groups are immunologically different and arise asynchronously. This does not necessarily apply universally: other species of barley have multiple isoenzymes, but there is no evidence that they fall into distinct groups (Momotani and Kato, 1972, 1974).

In summary, the isoenzyme work on Himalaya barley tends to indicate that more than one gene may be involved in α-amylase production, that the expression of these genes is controlled differently by GA_3 and therefore the isoenzymes do not simply represent post-translational modifications. More than one α-amylase mRNA may be involved.

D. Secretion of Hydrolases from Aleurone

1. Gibberellin Control of Synthesis and Release of Other Hydrolases

Several other enzymes, many of which are not involved in carbohydrate mobilization are known also to be synthesized and secreted by barley aleurone layers in response to GA_3. Most of these are hydrolases. They include protease [EC 3.4], ribonuclease [EC 3.1.4], $(1\rightarrow 3)$-β-glucanase [EC 3.2.1.39], α-glucosidase, limit dextrinase (Varner and Ho, 1976), acid phosphatase [EC 3.1.3.2], esterase [EC 3.1.1.1] (Jacobsen and Knox, 1972;

Ashford and Jacobsen, 1974b), various cell wall hydrolases (Dashek and Chrispeels, 1977; Eastwell and Spencer, 1982) and peroxidase [EC 1.11.1.7] (Gubler and Ashford, 1983). Of these, only protease is known to be synthesized totally *de novo* and its response curve closely follows that of α-amylase (Jacobsen and Varner, 1967). The activities of all the other enzymes are enhanced by GA_3, but often only to a small degree. Furthermore, most of the enzymes are present already in quite large amounts at the time when aleurone layers are isolated, and they increase in activity in aleurone cells during incubation without GA_3. Ribonuclease and $(1\rightarrow 3)$-β-glucanase are in this category and for these two enzymes it has been shown that the increase in activity involves *de novo* synthesis, but not specifically of the GA_3-enhanced fraction (Bennett and Chrispeels, 1972). In wheat aleurone a similar spectrum of enzymes is enhanced in activity by GA_3, but it is not established whether any of them are synthesized *de novo* (Jones and Jacobsen, 1978). Therefore, hydrolase production in the aleurone layer comes under varying degrees of control by GA_3 and the situation is much more complex than was first thought.

Gibberellic acid also brings about enzyme release from isolated aleurone layers and appears to have a separate control over this process. We have found only two cases where appreciable amounts of an enzyme are released from isolated aleurone layers in the absence of GA_3, and there is evidence in one that this may be due to displacement of the enzyme from the cell walls by cations (see below). Gibberellic acid-induced release curves for several hydrolases are given in Figure 4. As with synthesis, four phases can be recognized; an initial lag phase is followed by a period when the rate of release increases until a maximum rate is achieved. Release is maintained at this rate for several hours, and after this, activity in the incubation medium either levels off or declines. It is clear that the enzymes are released in two distinct groups. In the first group, activity begins to increase in the medium 6–12 h after GA_3 addition. This group includes protease, esterase, $(1\rightarrow 3)$-β-glucanase, acid phosphatase, peroxidase and α-amylase (Jacobsen and Knox, 1972). The second group, which includes ribonuclease and endo-$(1\rightarrow 4)$-β-xylanase [EC 3.2.1.8], is released much later: these enzymes do not appear in the medium until 24 h after GA_3 addition.

It is probable that GA_3 exerts its effects on the rate of release differently in these two groups. Group 1 enzymes appear to be limited initially by their diffusion through the cell walls (Varner and Mense, 1972; Ashford and Jacobsen, 1974a, b). The lag phase for enzyme production (α-amylase 2–5 h; peroxidase 4–8 h) is much shorter than that for their release, and the enzymes may accumulate temporarily in the aleurone layer (Jacobsen, 1983; Gubler and Ashford, 1983). Later the maximal rate of release in some batches of barley appears to be limited by the availability of group 1 enzymes for secretion, since enzyme activities within the layer

4. Endosperm Carbohydrate Mobilization

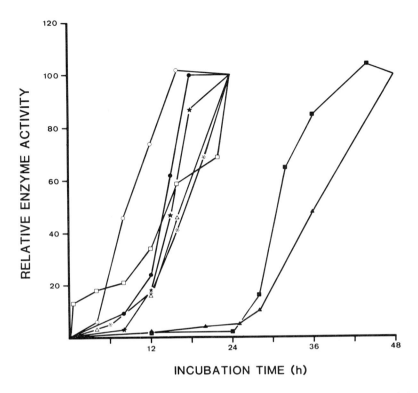

Fig. 4. Time course of enzyme release from barley aleurone layers in the presence of GA_3 showing that enzymes are released as two distinct groups. The first group is released 6–12 h after GA_3 addition: α-amylase ★ and protease ● (from Jacobsen and Varner, 1967); $(1\rightarrow3)$-β-glucanase ○ (from Jones, 1971); acid phosphatase * (from Ashford and Jacobsen, 1974b); β-xylopyranosidase △ (from Taiz and Honigman, 1976); peroxidase □ (from Gubler and Ashford, 1983). The second group are not released until 24 h after GA_3 addition: ribonuclease ▲ (from Chrispeels and Varner, 1967); endoxylanase ■ (from Dashek and Chrispeels, 1977). For enzymes of the first group a value of 100 was assigned for activity present in the medium after 24 h incubation and the second (ribonuclease and endoxylanase) were assigned a value of 100 for activity after 48 h incubation.

progressively fall to very low levels. In contrast, group 2 enzymes are released at a time when the walls obviously do not present a significant barrier to their release. Mozer (1980 b) has reported that a new group of proteins arises in GA_3-treated aleurone at 20 h: this is just before secretion of group 2 enzymes begins. These new proteins may include some of the enzymes of group 2 and this implies that some of them are released late because they are synthesized late, for example, the endoxylanase of Dashek and Chrispeels (1977). Synthesis does not provide the only control, because ribonuclease which is released with group 2 is synthesized earlier and temporarily accumulates in the tissue, where its activity falls concurrently with its release (Chrispeels and Varner, 1967). Some of the complexities in the GA_3-response of particular enzymes can probably be attributed to the occurrence of most of the hydrolases as multiple isoenzymes, which may be located at different cellular sites and may behave differently in response to GA_3 (Bailey et al., 1976; Jacobsen and Higgins, 1982; Gubler and Ashford, 1983).

It should also be realized that GA_3 may not exert its control only at these two points (Jones and Jacobsen, 1978; Tomos and Laidman, 1979). Gibberellic acid induces many changes in fine structure (see Yomo and Varner, 1971) and the time taken for induction of an intracellular transport system may not be insignificant (Obata and Suzuki, 1976; Obata, 1979). Finally, there are further complexities which must also be considered. Several hydrolases are present in the aleurone cell wall system prior to GA_3 treatment. Their release may be mediated by cation exchange, as seems to be the case with peroxidase (Gubler and Ashford, 1983) and possibly also acid phosphatase and esterase (Jacobsen and Knox, 1972; Ueki, 1979), or by wall hydrolysis (Fig. 5), if they are trapped in the wall or covalently bound. This appears to apply to esterase (compare Figs 6 and 7). Both these processes are influenced by GA_3 which is known to control wall hydrolysis (see Section II,D,*3*) and might influence the cation exchange of the wall by stimulation of ion release from aleurone cells (Eastwood and Laidman, 1971; Jones, 1973; Tomos and Laidman, 1979). The contribution of this component from cell walls to overall release curves has not been quantified.

2. Pathway of Secretion

Secretion of a newly synthesized protein from any plant cell can be considered as a three stage process involving: (i) intracellular transport from the site of synthesis to the plasma membrane; (ii) passage across the plasma membrane; and (iii) release through the cell walls. On the basis of current evidence, it is most probable that in aleurone cells the hydrolases are synthesized on the polysomes of the rough ER and passed into the ER lumen, where they are transported to the exterior of the cell either directly by fusion of the ER membranes with the plasmalemma, or via ER-derived vesicles, without the participation of Golgi bodies. They may accumulate temporarily in the periplasmic space between the plasmalemma and cell wall before they are released through the cell wall system. Release is facilitated and may indeed be controlled by digestion of the cell wall by hydrolases induced by GA_3. This proposed route is not proven: in fact, most aspects of it are controversial. The available evidence is evaluated below.

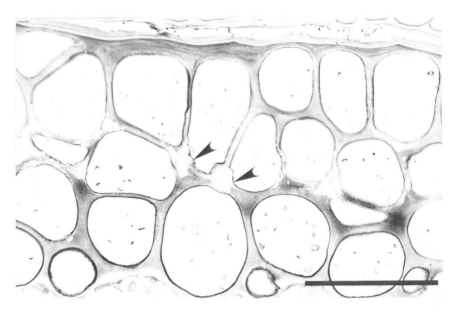

Fig. 5. Isolated barley aleurone layer treated for 24 h with GA_3. Wall hydrolysis indicated by loss of PAS staining is most pronounced on the endosperm side of the cells in the middle of the aleurone layer. The innermost wall region around each cell stains more intensely than the rest of the wall and resists hydrolysis (arrows) (from Ashford and Jacobsen, 1974b). Bar = 50 μm.

Pulse labelling experiments with radioactive amino acids have shown that newly synthesized proteins are associated with the microsomal fraction (see Jones and Jacobsen, 1982) and some α-amylase activity has also been shown to be associated with ER fractions isolated from aleurone layers, from both barley and wheat. This activity is latent but can be released by membrane disruption treatments, indicating that it is sequestered inside membranes of the ER (see Jones, 1980a, b). This agrees with older data on the association of α-amylase with a particulate fraction, showing properties of latency (see Chrispeels, 1976). Newly synthesized protein (as determined by incorporation of radioactive amino acids) associated with the ER and apparently in its lumen, has been shown to be α-amylase protein by immunological tests. Some of this protein is lost from the ER fraction if the pulse with radioactive amino acids is followed by a short chase period, indicating that the α-amylase is only in the ER temporarily (Jones and Jacobsen, 1982). The enzyme is fully active and has a molecular weight of about 44 000 daltons, which is the same as that of secreted amylase. The fact that this is smaller by about 1500 daltons than enzyme synthesized from barley α-amylase mRNA in a cell-free translation system is further circumstantial evidence that this is a secreted protein sequestered in a membrane bounded compartment (see Section II.B,*1*).

The results are equivocal because this ER-associated fraction represents only a small portion of the total α-amylase activity in the cells. Most of the enzyme ends up in the soluble fraction and the problem is its cellular location; there have been various suggestions that this may represent secreted but not yet released enzyme trapped in the periplasmic space and walls, or soluble cytoplasmic enzyme, or enzyme released from broken ER fragments or vesicles during extraction (Jones, 1980a). Furthermore, one of the five α-amylase isoenzymes reported by Jones and Jacobsen (1982) to be present in GA_3-treated barley aleurone was found exclusively in the soluble fraction, occured in aleurone tissue regardless of its treatment with or without GA_3 and did not appear in the medium. They interpreted this to indicate the production of a discrete α-amylase pool not associated with the ER and not released either. However, the status of this isoenzyme is not clear (Section II,C).

There is also evidence from enzyme histochemistry at the fine structural level that hydrolases are sequestered in the ER lumen during the GA_3-induced response of aleurone. Localization of acid phosphatase occurs in the ER lumen of GA_3-treated tissue but not incubated without GA_3 tissue (Pyliotis *et al.*, 1979). Such results unfortunately do not provide evidence that this is acid phosphatase *en route* to secretion. Gibberellic acid induces only a small increase in acid phosphatase in aleurone (Ashford and Jacobsen, 1974b; Bailey *et al.*, 1976). There is a large amount of phosphatase present initially and this is localized at several sites throughout the cells. There is no evidence that the acid phosphatase specifically localized in the ER is destined for secretion although this would be a reasonable assumption, since most of the acid phosphatase is released into the medium. Furthermore, phosphatase activity associated with the ER in fractionation studies does not show latency (Jones, 1980b), and this would be expected for an enzyme sequestered in the ER lumen.

The mode of transfer from the ER lumen to the outside of the cell is even more of a mystery. No obvious 'secretory organelle' has been found, although looked for many times. Golgi bodies are not prominent in aleurone cells as they are in some other protein or carbohydrate secreting systems although they are said to increase in number (see Chrispeels, 1976). For this reason, Jones and colleagues proposed that after α-amylase synthesis on the ER, it is secreted via the cytosol without participation of any organelle (see Chrispeels, 1976). In contrast, Gibson and Paleg (1975, 1976) thought that the secreted hydrolases of wheat were transported by lysosomes, while Vigil and Ruddat (1973) and Locy and Kende (1978) proposed that ER-derived vesicles mediate transfer. Asymmetric distribution of ER in aleurone cells especially near the cell periphery and the large numbers of vesicles or ER profiles near the plasmalemma during the phase of secretion (Vigil and Ruddat, 1973) favour the direct ER transfer hypothesis. Descriptions of vesiculation of ER or of vesicle fusion with the plasmalemma based solely on electron micrographs are equivocal, because of the known inability of modern fixatives to stabilize and immobilize membranes during the fixation process (Mersey and McCully, 1978). Secretion via the ER lumen does, however, seem an attractive hypothesis, if only because it would protect the contents of the aleurone cells from the potentially damaging effects of some of the hydrolases secreted.

The hydrolases, including α-amylase, accumulate in the periplasmic space prior to their release. Release is limited by diffusion (Varner and Mense, 1972) and appears to be impeded by the thick walls around the aleurone cells in tissue not treated with GA_3. Acid phosphatase, one of the enzymes released in response to GA_3, accumulates not only in the periplasmic space but also in the innermost regions of the wall in tissue incubated without GA_3, but it is not released into the medium (Ashford and Jacobsen, 1974a, b). This has led Jones and Jacobsen (1978) to suggest that movement of enzymes across the plasmalemma is not a GA_3 dependent process, although its rate may be enhanced in GA_3-treated tissue. Release through the cell walls does seem to be GA_3 controlled. Aleurone wall hydrolysis occurs exclusively in response to GA (e.g. Fig. 5; Jones, 1972) and secreted enzymes such as esterase (Fig. 8), peroxidase (Fig. 9) and acid phosphatase (Fig. 10) all appear in the cavities and channels that are formed in the wall system in GA_3-treated tissue. Cytochemical localization of enzymes indicates that these areas are preferential routes for enzyme release and that unhydrolysed wall is relatively impermeable to the enzymes (Ashford and Jacobsen, 1974a, b; A. E. Ashford and F. Gubler, unpublished data). Although a similar localization pattern applies to several GA-induced enzymes (Figs 8–10), this pattern has not yet been shown for α-amylase. Preferential release of enzymes via digested wall channels does not exclude the possibility that a slow release of enzymes from aleurone cells can occur in the absence of extensive wall hydrolysis. Traces of α-amylase protein have been detected in incubating media without added GA_3 (Higgins et al., 1982).

Gahan et al. (1979) showed very similar acid phosphatase localization patterns in wheat aleurone cell walls to those found in barley. In both dry and germinated grains enzyme activity was interpreted to be localized in digested wall channels, indicating that GA_3 was not necessary for wall hydrolysis. However, these authors did not show sections illustrating cell

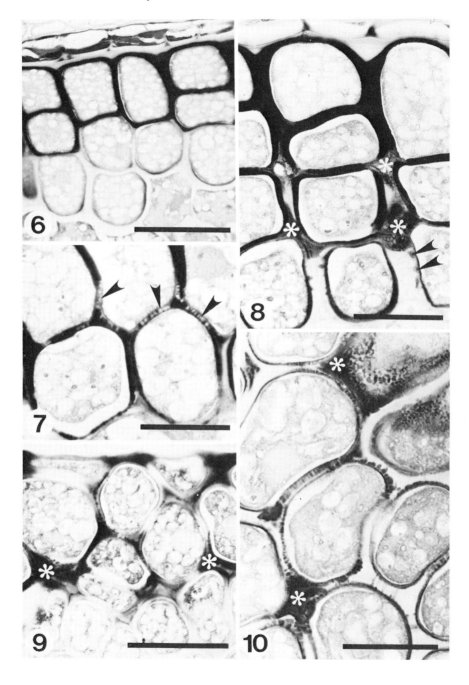

wall hydrolysis. The acid phosphatase localization that they obtained in dry grains is much more likely to represent enzyme activity associated with the resistant wall tube around plasmodesmata as it does in barley, rather than enzyme activity in hydrolysed wall areas (Ashford and Jacobsen, 1974a, b). The suggestion that some enzyme may be retained in the wall system of untreated aleurone layers by ionic attraction (Tomos and Laidman, 1979) also requires further study.

3. Gibberellin Control of Wall Hydrolysis and the Secretion of Wall-Hydrolysing Enzymes

Dissolution of the aleurone cell walls is of interest not only because these contribute carbohydrate in their own right, but also because wall hydrolysis appears to control the release of starch-hydrolysing enzymes from the aleurone layer. The GA_3-induced wall hydrolysis occurs first in the middle lamella region around groups of plasmodesmata and in patches spreading from the innermost region of the wall, particularly in the cell corners (Jones, 1972). It is similar in both wheat and barley, and occurs both in

---

Figs 6–10. Sections of barley aleurone embedded in glycol-methacrylate at low temperature to preserve enzyme activity and stained to show the distribution of various enzymes. Figures 6–8: esterase, substrate α-naphthyl acetate with tetrazotized o-dianisidine pH 8.0; Figure 9: peroxidase, substrate diaminobenzidine pH 5.0; Figure 10: acid phosphatase, substrate naphthol AS-BI phosphate with hexazotized pararosanilin pH 5.0. All show isolated aleurone layers incubated for 16 h + GA_3, except Figure 6, where incubation took place for 16 h without GA_3.

Fig. 6. Esterase reaction occurs in a band throughout the cell walls around the outermost one or two rows of aleurone cells in $-GA_3$ tissue. Bar = 50 μm.

Fig. 7. Esterase reaction is lost from regions of the cell wall around the outermost rows of cells: these are regions hydrolysed in response to GA_3 (arrows). Bar = 30 μm.

Fig. 8. Simultaneously, esterase reaction appears in and around other wall areas hydrolysed in response to GA_3 (asterisks). Bar = 30 μm.

Fig. 9. Peroxidase similarly appears in wall areas hydrolysed in response to GA_3 (asterisks). Bar = 40 μm.

Fig. 10. Acid phosphatase is also located in GA_3-induced hydrolysed wall regions. Reaction occurs in large patches (asterisks), strands and globular areas around the middle row of aleurone cells (from Ashford and Jacobsen, 1974a). Bar = 25 μm.

For all enzymes, reaction was shown to occur in digested wall regions by staining adjacent sections with PAS reagents. Controls (heat-inactivated or without substrate) are not similarly stained.

GA_3-treated aleurone and during normal germination (Fulcher et al., 1972; van der Eb and Nieuwdorp, 1967). In the three-layered barley aleurone, wall hydrolysis often begins first around the middle layer of cells (Fig. 5), especially on the endosperm side, dispelling the myth that all the cells of a GA-treated aleurone layer behave synchronously. As hydrolysis continues whole areas of wall material are removed, so that with time a continuum of connected space develops around all the aleurone cells, ultimately leaving only remnants of the wall in areas adjacent to the pericarp (Ashford and Jacobsen, 1974a, b).

Throughout the dissolution process the innermost region of the wall appears to resist hydrolysis and persists around individual aleurone cells. It can still be seen, although not necessarily intact, in aleurone layers incubated for 72 h with GA_3 (Gubler, 1983). There is also a resistant layer which may be of similar composition around plasmodesmata (Taiz and Jones, 1973). The function of the resistant inner wall layer is not clear, although there are several roles possible. It may prevent the aleurone cells from bursting as their contents become mobilized and osmotic pressure increases. It could regulate the flow of enzymes through the cell walls or simply prevent the aleurone layer from falling to pieces by maintaining cell connection via plasmodesmata, as the rest of the walls become digested. The composition of the inner wall layer is not known, but its staining properties indicate that this may be different from the rest of the wall (Taiz and Jones, 1970; Fulcher et al., 1972).

The major polysaccharides present in wheat and barley aleurone walls are heteroxylans and $(1\rightarrow3)(1\rightarrow4)$-$\beta$-glucans (Bacic and Stone, 1981b). Arabinoxylan is the main heteroxylan and this is thought to consist of a $(1\rightarrow4)$-β-xylan backbone, highly substituted with arabinofuranosyl residues. Wheat and barley aleurone walls contain unusually large amounts of the phenolic acid, ferulic acid (Fulcher et al., 1972; Bacic, 1979). This is likely to be esterified to the arabinosyl side chains of the arabinoxylan (Smith and Hartley, 1983; Gubler, 1983; F. Gubler and A. E. Ashford, unpublished data). It has been proposed that ferulic acid confers resistance to hydrolysis on the aleurone walls but the ferulic acid seems to be removed non-selectively along with the rest of the wall material (Fulcher et al., 1972). Analysis of ferulic acid-containing compounds in the incubation medium of GA_3-treated aleurone layers also indicates that ferulic acid does not prevent wall hydrolysis, and does not have to be modified before this can occur. Ferulic acid-containing compounds can be quantitatively recovered from the bathing medium. No free ferulic acid occurs, and all the ferulic acid detected in the incubating medium of GA_3-treated aleurone layers is still covalently bound to various hydrolysed wall fragments, containing both arabinose and xylose residues (Gubler, 1983; F. Gubler and A. E. Ashford, unpublished data).

4. Endosperm Carbohydrate Mobilization 139

Wall hydrolysis has been shown biochemically as well as structurally, by release of wall materials into the bathing medium. Arabinose and xylose-containing compounds and the free sugars first appear at about 10 h after GA_3 addition. There are also large amounts of glucose, much of which must be derived from the wall glucans (Dashek and Chrispeels, 1977). Wall hydrolysis appears to depend on the secretion of wall hydrolysing enzymes from the aleurone layer. Three arabinoxylan-degrading enzymes, endo-$(1\rightarrow 4)$-β-xylanase [EC 3.2.1.8]; β-xylopyranosidase [EC 3.2.1.72], and α-arabinofuranosidase [EC 3.2.1.55], are released into the medium, all in response to GA_3. Of these, only endoxylanase would be expected to degrade intact polymers. *In vitro*, the endoxylanase produces polymers of intermediate molecular weight, and low molecular weight products occur only after prolonged hydrolysis. So, on enzymological grounds, the most likely hydrolysis pattern for the arabinoxylans in the walls is initial degradation by an endoxylanase to produce oligosaccharides, followed by glycosidase action to produce monosaccharides (Taiz and Honigman, 1976). However, the endoxylanase produced by isolated aleurone layers is not released until 28-32 h after GA_3 addition. This is well after a major amount of wall hydrolysis has already occurred, whether this is measured biochemically or visually (Taiz and Honigman, 1976; Dashek and Chrispeels, 1977). The endoxylanase does not accumulate appreciably prior to secretion and cannot, therefore, be considered to be in the process of digesting its way through the cell walls. Nor, with a molecular weight of 29 000 daltons, would one expect its progress through the walls to be particularly impeded. Arabinofuranosidase and xylopyranosidase start to appear in the medium much earlier. They are released with group 1 enzymes at 12 h (Taiz and Honigman, 1976), but these enzymes would not be expected to release the large molecular weight compounds (over 10 000 daltons) found in the incubation medium at this time (Dashek and Chrispeels, 1977). Endoxylanase is a likely candidate for this function, but it does not seem to be present. Therefore the early hydrolysis of arabinoxylan in aleurone cell walls remains an enigma.

It has been reported that ethylene may facilitate enzyme release via the enhanced production of xylanase (Eastwell and Spencer, 1982). There are unresolved difficulties with this hypothesis. Firstly there are no detectable differences in xylanase activity, with or without ethylene, over the first 24 h of incubation, when wall hydrolysis would be expected to have the greatest effect on facilitating enzyme release. Secondly, although some enzyme accumulates in the tissue, as these authors point out, this does not establish that the enzyme is in contact with potential sites of wall hydrolysis. If 'total xylanase' activity represents mostly endoxylanase and it has been secreted into the walls, it is difficult to see how it would be retained when other enzymes of larger molecular weight are being released. Finally, measurement of total xylanase in terms of wall hydrolysis is difficult to interpret. Assays using increase of reducing sugars require total absence of contaminating substances, which is difficult to achieve in practice.

There have also been problems in interpreting the action of the $(1\rightarrow 3)$-β-glucanase. This enzyme increases in activity during incubation in the absence of GA_3 and there is little if any GA_3 stimulation of activity, but its release is under GA_3 control and it appears in the medium with α-amylase and other group 1 enzymes (Jones, 1971; Bennett and Chrispeels, 1972). It was first thought that $(1\rightarrow 3)$-β-glucanases were primarily responsible for hydrolysis of aleurone walls, but these contain mixed $(1\rightarrow 3)(1\rightarrow 4)$-$\beta$-glucans and it is now thought that $(1\rightarrow 3)(1\rightarrow 4)$-$\beta$-glucanases are the most important enzymes in their hydrolysis (Woodward and Fincher, 1982). The $(1\rightarrow 3)$-β-glucanase is probably involved in hydrolysing the $(1\rightarrow 3)$-β-glucan deposits that occur at the interface of the aleurone and sub-aleurone layers but may also contribute in a limited way to hydrolysis of $(1\rightarrow 3)(1\rightarrow 4)$-$\beta$-glucans (Moore and Stone, 1972; Fulcher *et al.*, 1977; Bacic and Stone, 1981b). Two endo-$(1\rightarrow 3)(1\rightarrow 4)$-$\beta$-glucanases [EC 3.2.1.73] have been found in germinating barley (Woodward and Fincher, 1982) and there is evidence of their release from aleurone tissue (G. B. Fincher, pers. comm. 1984).Cellulase activity [EC 3.2.1.4] was not detected in isolated aleurone layers of Himalaya barley either with or without GA_3 by Taiz and Honigman (1976). However, cellulase activity has been found in whole germinating barley grains. Most of this can be attributed to contaminating micro-organisms, but Hoy *et al.* (1981) found enzyme activity, which they believed to be of plant origin, in the endosperm of germinated, surface sterilized barley. There is little cellulose in endosperm cell walls but cellulase can depolymerize mixed linkage $(1\rightarrow 3)(1\rightarrow 4)$-$\beta$-glucans (see Woodward and Fincher, 1982) and may, therefore, also contribute to wall hydrolysis.

E. Other Sites of Production of Starch-Hydrolysing Enzymes

1. The Scutellum
The position of the scutellum between the reserves and embryo indicates that it plays a crucial role in the absorption of hydrolysis products. Glucose is absorbed from the endosperm, converted to sucrose and transported as such to the seedling (see Swing and O'Brien, 1970 for references). Initially there is no vascular connection across the scutellum and this must limit solute transfer to the embryo considerably. However, mature sieve tubes differentiate from pro-vascular tissue very rapidly and in wheat it has been shown that the connection of a bridge of phloem is more or less complete by 24 h after soaking (Swift and O'Brien, 1971). Accumulation of sugar has

been reported to inhibit enzyme secretion from the scutellum as with aleurone tissue (Brown and Morris, 1890; Jones and Armstrong, 1971) and this may regulate the amount of starch hydrolysis in early stages of germination. While the role of the scutellum in absorption and transport of the products of hydrolysis into the embryo is undisputed (see Swift and O'Brien, 1970), its function as a secretory tissue, releasing starch hydrolysing enzymes into the endosperm has been much more controversial. It now seems probable that the scutellum produces and secretes hydrolases and the problem has become not whether it contributes to the hydrolysis, but how much.

The evidence for secretion of enzymes from the scutellum arises from diverse techniques, such as hydrolysis patterns of the cell walls and starch in the endosperm, examination of the production and secretion of hydrolases from isolated embryos and scutella, and localization of hydrolases *in situ* in whole germinating grains. None of these approaches is without its critics, but taken together they provide a body of evidence, largely circumstantial, that the scutellum contributes to endosperm hydrolysis, at least in the early stages of germination. Endosperm cell wall hydrolysis precedes starch breakdown. It begins at the embryo end, in the immediate vicinity of the scutellum and spreads to the distal end of the grain (Briggs, 1972). There is some disagreement over whether breakdown occurs faster in the dorsal or ventral sides of the grain (Palmer, 1982), but the view that has the most support is that it progresses as a front parallel with the face of the scutellum (Briggs, 1972; Gibbons, 1981). Later, wall breakdown advances more rapidly adjacent to the aleurone layer. Starch breakdown similarly begins at the scutellum–endosperm junction (MacGregor and Matsuo, 1982). These patterns of hydrolysis indicate an involvement of the scutellum early in germination, but do not in themselves constitute conclusive evidence. The data have been interpreted differently by Palmer (1982) to reflect differential digestibility of endosperm or preferential pathways of enzyme diffusion. Moreover, as digestion progresses away from the scutellum surface, it becomes very difficult to differentiate between the effects of hydrolase release from the scutellum and adjacent aleurone tissue which, because of its position, would be expected to respond early. Gram (1982) has provided some evidence from electron micrographs that aleurone cells relatively close to the embryo do not show signs of secretory activity until late in germination, but more information is needed to clarify this problem fully.

There have been several reports of production and release of enzymes, including wall hydrolysing enzymes and amylase, from isolated barley embryos and pieces of scutellum (Brown and Morris, 1890; and see Briggs, 1972, 1973). It has been suggested that much of this activity may arise from aleurone tissue that continues to adhere to the embryo after isolation (MacLeod and Palmer, 1966). In recent experiments, Palmer (1982) showed that cores cut from the centre of the scutellum, and therefore free from aleurone tissue, produced very little α-amylase. This does not, however, resolve the problem, since only α-amylase in the tissue was measured and not that secreted. The scutellar epithelium in these experiments behaved abnormally and may have had a reduced response; isolated scutella may not follow the same metabolic patterns as intact embryos. Similar problems have been encountered in the assessment of GA production by isolated embryos (Briggs, 1973). Release of small amounts of α-amylase into the medium from isolated scutella with embryonic axis attached, but free from contaminating aleurone, has been reported in barley (Briggs, 1973) and rice (*Oryza sativa*).

Miyata et al. (1981) showed that rice scutellar α-amylase produced *in vivo* had a larger molecular weight than enzyme produced *in vitro* in cell free systems. They provided data to indicate that this amylase has a signal sequence of 1600 daltons, which is cleaved, and that the protein also becomes glycosylated with oligosaccharide portions estimated to be 2900 daltons (Miyata and Akazawa, 1982a). By adding polysomes carrying partially completed protein chains to cell free systems and monitoring the appearance of products, this group (Miyata and Akazawa, 1982b) was further able to show that cleavage of the signal sequence and glycosylation both occur while amino acids are still being added to the protein chain, and that peptide cleavage precedes glycosylation. Glycosylation is evidently not necessary to convert the protein into its enzymatically active form, while cleavage of the peptide sequence is. This was shown by using the substrate analogue β-cyclodextrin, which presumably only binds to the biologically active form of the enzyme (Miyata and Akazawa, 1982). Akazawa and Miyata (1982) interpret these data to indicate that scutellar secretion of α-amylase makes a very considerable contribution to hydrolysis of the starchy endosperm in rice, and they extrapolate to other cereals also. The major difficulty with this hypothesis is the apparently small amount of amylase released from the scutellum (Tanaka et al., 1970; Briggs, 1973; Miyata et al., 1981). In addition it is known that during germination starch is both deposited and removed in pro-amyloplasts of wheat scutellum (Swift and O'Brien, 1972), and starch deposition in the scutellum during germination also occurs in barley and rice (Gram, 1982; Okamoto et al., 1982). It is very likely that at least some of the α-amylase produced in the scutellum is related to starch turnover within the scutellum.

The most convincing demonstration of the scutellar production of hydrolases comes from enzyme histochemistry. Several enzymes have now been localized by different methods in a range of cereal grains. In all cases the enzyme reaction has been detected first associated with the scutellar epithelium or close to it. Activity appears to spread from here to the rest of the endosperm. Enzyme activites could not be detected in the aleurone layer until much later. The enzymes localized include α-amylase, detected either by substrate film methods or use of fluorescent antibodies in rice and barley, respectively (Okamoto and Akazawa, 1979a; Okamoto et al., 1980; Gibbons, 1980, 1981). However, from the enzyme histochemistry, it is not easy to be absolutely certain at what stage the aleurone layer begins to contribute to starch hydrolysis, or to assess its quantitative contribution. Impermeability of aleurone tissue may hinder histochemical staining within the layer initially. Also, if the embryo stimulus reaches the aleurone layer by diffusion, aleurone adjacent to digested endosperm could be stimulated preferentially. In this case, there may not be the characteristic localization pattern with digestion spreading along beneath the aleurone layer, that is usually taken to indicate aleurone involvement.

Structural changes accompany the changes in physiological activity of the scutellum during germination. The most obvious of these occur in the scutellar epithelium. These cells double in length and completely separate from one another so that their surface area is dramatically increased (see Swift and O'Brien, 1972 for references). Tubular evaginations of the plasmalemma, which would further enhance surface area have been reported by Harris et al. (1982). Cell wall hydrolysis is involved in cell separation, but as in the aleurone there is an inner region of the wall which stains differently and resists digestion (Swift and O'Brien, 1972; Smart and O'Brien, 1979a). Ferulic acid is a major component of the scutellum cell walls as it is in aleurone tissue. The intensity of wall fluorescence, attributed to ferulic acid, progressively increases during germination in some grasses, but not in wheat and barley, so the relationship between ferulic acid content and cell wall hydrolysis is not clear at present (Smart and O'Brien, 1979a, b). Within the cells several cytological changes occur which indicate alteration from a relatively quiescent to a very active metabolic state. These include increase in numbers of mitochondria and mitochondrial cristae, proliferation of rough ER, and increase in numbers of Golgi bodies with vesicles. Such changes have been described in wheat (Swift and O'Brien, 1972) and malted barley (Gram, 1982) and rice (Okamoto et al., 1982). Unfortunately, none of them can be related specifically to secretion, since there are so many other activities occurring in the cells at the same time.

2. Starchy Endosperm: The Origin of β-Amylases
Beta-amylase activity increases in the endosperm of whole grains during germination together with other starch hydrolases, but it does not originate from the aleurone layer. There is no evidence for *de novo* synthesis of

β-amylases in wheat and barley (Daussant and Corvazier, 1970; Hardie, 1975) and it is generally accepted that they arise from pre-existing forms, already present in the starchy endosperm. In wheat the latent form is associated with the glutenin fraction (Rowsell and Goad, 1962), and the enzymes are thought to be released from their bound form and activated by the GA-induced proteases secreted from the aleurone layer (Rowsell and Goad, 1964). In rice, Okamoto and Akazawa (1980) reported that small amounts of β-amylase are synthesized *de novo* in isolated embryos, but the amounts were insignificant and enzyme localization of β-amylase with fluorescent antibodies shows it to be present throughout the endosperm early in germination, as in other cereals (Okamoto and Akazawa, 1979b).

F. The Production of Gibberellins in Cereals

At about the same time that work on isolated aleurone layers was initiated, a search had begun for natural gibberellins in cereals. Gibberellins were first reported in germinated barley and malt by Radley (1959) using the Meteor pea (*Pisum sativum*) growth test as a bioassay. They were subsequently also found by other authors (see Paleg, 1965; Briggs, 1973). The early attempts to identify which gibberellins occur in germinating grains were limited by the techniques available, but they indicated that either GA_1 or GA_3 predominated (Cohen and Paleg, 1967; Radley, 1967; Groat and Briggs, 1969). Both GA_1 and GA_3 have now been identified in two varieties of malted barley by gas chromatography–mass spectrometry, and GA_1 found to be present in the higher amount (Yamada, 1982). The level of GA_1 rises rapidly during germination under malting conditions, reaches a maximum at two days, and thereafter decreases. This parallels the changes found for total GA-like compounds, measured in the earlier experiments by bioassay (Groat and Briggs, 1969). In contrast, GA_3 occurs in small amounts and changes little during germination (Weiler and Wieczorek, 1981; Yamada, 1982). Gibberellins GA_4 and GA_7 were not detected in germinating barley grains by Yamada (1982).

There is virtually no free GA present in dry barley grains (Groat and Briggs, 1969; Yamada, 1982) and the removal of the embryo prevents any increase in GA levels in the endosperm. This has led to the view that the embryo is the most likely source of GA during germination. GA-like compounds are known to be released from isolated embryos (Cohen and Paleg, 1967; Radley, 1967) and there is evidence that GA synthesis can occur in the embryo. Only the latter stages of synthesis, conversion of kaurenol and kaurenoic acid to GA, have been demonstrated. It may be that precursors of GA, which are known to be stored in mature grains (Murphy and Briggs, 1973), are converted into GA by the embryo, but the storage sites of such precursors of GA, which are known to be stored in mature grains (Murphy and Briggs, 1973), are converted into GA by the embryo, but the storage sites of such precursors are not known. Most of the evidence supports the view that the scutellum is the major site of GA synthesis (Radley, 1967, 1969), but MacLeod and Palmer (1966) found that for maximal GA production it was necessary for the embryonic axis to be left attached to the scutellum.

Although GA levels are low initially and the embryo has been shown to synthesize and secrete GA *in vitro*, this does not necessarily mean that the embryo provides all the GA during germination. An alternative view has been presented by Trewavas (1982) that free GA may arise in the endosperm by release from bound forms already present in the dry grain. In some seeds GA is known to occur in bound forms but there is no convincing evidence as yet that this is the case with cereals (Mounla and Michael, 1973; Jones and Stoddart, 1977).

Assuming that gibberellins are transported from the embryo to the aleurone layer *in vivo*, it is not known how or by what route they travel. It is tempting to speculate that they may travel symplasmically through the aleurone layer, via the plasmodesmata, and this may explain why the aleurone layer remains intact throughout germination. However, it is equally possible that the gibberellins may simply diffuse through the starchy endosperm.

G. The Whole Grain: A Synopsis

The level of free GA present in germinating grains is considered sufficient to trigger the secretory response of aleurone layers (Cohen and Paleg, 1967; Briggs, 1973; Yamada, 1982). It is known that many of the responses of isolated aleurone layers to added gibberellins are similar to those shown *in situ* in germinating seedlings (van der Eb and Nieuwdorp, 1967; Yomo and Varner, 1971; Fulcher, 1972). These and other observations that gibberellins can be synthesized in cereal embryos have led to the hypothesis that gibberellins emanating from the embryo control starch breakdown in the endosperm, via their action on the aleurone layer (see Fig. 11 of Chapter 5 for an illustration of the classical view). This is an attractive hypothesis and it is probably true, but there is no direct evidence that the processes are linked in this way. The evidence is entirely circumstantial.

Doubts have been expressed about the relationship between GA levels and α-amylase production in germinating whole grains by Trewavas (1982). He argues that if aleurone cells are the 'target cells' for the embryo 'signal' then there should be a lag between the appearance of GA-like compounds and the first detectable rise in α-amylase activity. This would allow for the time taken for GA to diffuse to the aleurone cells and induce a response. The failure to find any such lag phase in barley grains germinated at 25°C (Groat and Briggs, 1969) led Trewavas (1982) to suggest that the two events were concomitant and independent. However, it may be that some of the α-amylase measured early in germination arises from the scutellum and not the aleurone layer (see Section II,E,*1*) and in this case it would not be possible to detect a lag phase by measuring total levels of α-amylase activity in whole grains.

The aleurone response can vary in different seed batches, both between varieties and from year to year (Jackson, 1971), resulting in some in the production of large amounts of α-amylase even in the absence of GA, and in others in the classic all-or-nothing GA-induced response. Researchers interested in the GA response tend, of course, to select the latter type of seed for their work, and so this variability, which may be important in the biology of the seed, tends to be underemphasized. Variability in response, which may be due to different sensitivity to GA, or to varying endogenous levels of GA, may result from differences in environmental factors during seed growth and maturation. For example, the GA response of wheat aleurone layers has been found to depend on dehydration of the seed (Nicholls, 1979; Armstrong *et al.*, 1982), and temperature effects on the induction of aleurone sensitivity to GA_3 have been observed in wheat and barley (Nicholls, 1980, 1982; Norman *et al.*, 1982). Duration and conditions of storage also influence the GA response (Jackson, 1971).

Finally it should be remembered that not all events in cereal aleurone are controlled by GA (exceptions include lipid mobilization; Tomos and Laidman, 1979), and that events are not identical in different species. For example, added GA has little effect in sorghum (*Sorghum bicolor*) (see Palmer, 1982) and certain varieties of maize (*Zea mays*) (Harvey and Oaks, 1974; Goldstein and Jennings, 1975), perhaps because of high endogenous GA levels in these grains.

III. CELL WALL MOBILIZATION

A. Cereal Endosperm Walls

Dissolution of the starchy endosperm cell walls occurs together with starch hydrolysis and appears to be necessary for full access of the secreted amylases to the starch grains (Fincher and Stone, 1974). In starch-storing endosperms the cell walls are usually thin, although there are exceptions, such as *Bromus mollis* (Brown and Morris, 1890; Fincher and Stone, 1981). Even in thin-walled species such as barley, the non-starchy polysaccharides make a significant contribution. Morrall and Briggs (1978) estimated that the starchy endosperm and aleurone walls together provided 18.5% of the carbohydrate used by six-day old germinated barley seedlings. The composition of starchy endosperm walls in several grasses has been analysed by Stone and colleagues (see Fincher and Stone, 1981) and found to contain predominantly heteroxylans and $(1\rightarrow 3)(1\rightarrow 4)$-$\beta$-glucans, and some glucomannan, but there are only small amounts of cellulose (Fincher and Stone, 1981; Staudte *et al.*, 1983). They are therefore not basically different in overall composition from aleurone cell walls (Bacic and Stone, 1981a, b) and are most probably attacked by the same wall hydrolases. Enzymes arising from the aleurone layer are most certainly responsible, but dissolution patterns of endosperm walls indicate that wall hydrolysing enzymes may also originate from the scutellum (Section II,E,*1*).

B. Mannan-rich Wall Deposits

The replacement of starch as the major carbohydrate reserve by thick deposits on the endosperm cell walls is a widespread phenomenon, but the breakdown of such deposits on germination has been studied in detail in only a few species. Many of the deposits have been shown to be rich in mannan (Halmer and Bewley, 1982). Some of them are pure mannans, such as the deposits found in the date palm (*Phoenix dactylifera*), caraway (*Carum carvi*) and the perisperm of coffee bean (*Coffea arabica*) (see Meier and Reid, 1982). Others are known to consist of galactomannan, with their mannan chain substituted to varying degrees by α-galactosyl side chains. Galactomannans have been found in seeds of Annonaceae, Convolvulaceae, Ebenaceae, Leguminosae, Loganiaceae, Palmae and Rubiaceae (Dey, 1978; McCleary and Matheson, 1975). Their structure, synthesis (Section 5.III,C of Volume 1) and mobilization are best understood in Leguminosae.

1. Galactomannans and Their Mobilization in Legumes
Legume seeds that retain an endosperm at maturity invariably contain galactomannan. This consists of linear chains of $(1\rightarrow 4)$-β-linked mannosyl residues, with single α-galactosyl residues joined by $(1\rightarrow 6)$-linkages at intervals along the chain. The number of galactosyl residues varies from species to species (Reid and Meier, 1970). Galactomannan is deposited as a thick layer on the inner surface of the endosperm cell walls (Fig. 11). Oligosaccharides of the raffinose family often occur as well in these seeds and both are mobilized during germination (Dey, 1978).

Storage of galactomannan is thought to be an adaptation to semi-arid conditions (Reid, 1971). Because of its hydrophilic properties galactomannan is thought to play an important role in the water relations of seeds

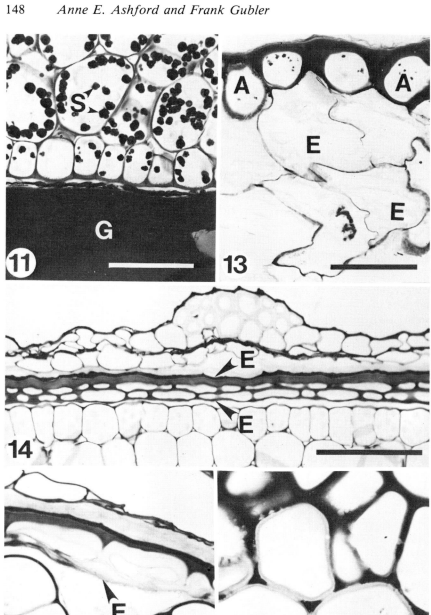

4. Endosperm Carbohydrate Mobilization

in dry environments, by improving their capacity for imbibition and water retention. Consequently there has been debate about its contribution as a reserve to overall embryo growth. Since galactomannan is broken down well after radicle emergence it is not considered to be important in germination *sensu stricto*. However, Reid and Bewley (1979) showed that there were differences in the growth of embryos of fenugreek (*Trigonella foenum-graecum*) seeds germinated intact compared with those where the endosperm was removed. By 55 h, the dry weight of a naked embryo was only about two-thirds that of an embryo from an intact seed. After 200 h the radicle and hypocotyl lengths of the seedlings germinated without endosperms were about 30% smaller. Therefore there is no doubt that, in fenugreek at least, galactomannan makes a significant contribution to seedling growth and establishment in the long run.

The oligosaccharides (usually either raffinose or stachyose and to a lesser extent, verbascose) are hydrolysed early in germination, releasing sucrose and galactose. However, in most of the seeds studied, galactomannan breakdown does not occur immediately, and the length of the lag phase differs in different species. Galactomannan breakdown was first

---

Fig. 11. Germinated fenugreek seed showing starch granules (S) in the cotyledon cells and unhydrolysed galactomannan deposits (G) in the adjacent endosperm. Bar = 50 μm.

Fig. 12. See page 152.

Fig. 13. Outer endosperm of the same section showing the aleurone-like layer (A) and adjacent endosperm cells (E), from which the galactomannan deposits have mostly been removed. A thin layer of the wall around all endosperm cells and the entire wall around the aleurone-like cells resist hydrolysis. Bar = 50 μm.

Fig. 14. Unimbibed lettuce seed showing the two layered endosperm (E-E), containing cells with thick walls. Bar = 50 μm.

Fig. 15. Hydrolysis pattern of the endosperm cell walls (E) in four-day imbibed lettuce seed. Bar = 30 μm.

Fig. 16. Wall hydrolysis in celery endosperm cells. There is an inner resistant wall layer as in barley aleurone. Bar = 40 μm.

The pattern of wall hydrolysis in both lettuce and celery supports the view that the enzymes involved arise from the endosperm cells themselves. For Figures 11, 13, 14, 15 and 16 the sections were all stained with PAS reagents.

studied extensively in fenugreek (Reid, 1971; Reid and Meier, 1972, 1973; Reid et al., 1977) but a few other legumes have also been examined in detail: guar (*Cyamopsis tetragonolobus*) (McCleary and Matheson, 1974; McClendon et al., 1976), carob (*Ceratonia siliqua*) (Seiler, 1977), honey locust (*Gleditsia triacanthos*), and lucerne (*Medicago sativa*) (McCleary and Matheson, 1974, 1975, 1976).

In fenugreek, galactomannan breakdown does not occur until well after germination has begun. Seeds germinated at 25°C in the dark show an initial swelling period, after which the radicle emerges (Reid, 1971). However, there is no change in endosperm polysaccharides for a further 18 h. After this, galactomannan breakdown begins and mobilization lasts for about 24 h. The sugars produced are rapidly transferred to the embryo: thus the major products of hydrolysis, mannose and galactose, accumulate in isolated endosperms but not in the endosperm of intact seeds (Reid and Meier, 1972; see also McCleary and Matheson, 1974, 1976 for data on other legumes). Transient accumulation of starch occurs in the cotyledons (Fig. 11) during the phase of galactomannan breakdown and a build up of sucrose has also been demonstrated here in some species (e.g., Reid, 1971; Seiler, 1977). Rapid uptake of D-galactose and D-mannose has been shown in isolated guar embryos and both these sugars were rapidly metabolized: other soluble sugars accumulated and about 25% of the labelled compounds were respired (McCleary and Matheson, 1976; McCleary, 1983). Significant levels of phosphomannoisomerase [EC 5.3.1.8], the enzyme which converts mannose-6-phosphate to fructose-6-phosphate so that it may enter the glycolytic pathway, were found in cotyledons of ten legume species, while seven species of non-legumes had small or undetectable amounts of this enzyme (McCleary and Matheson, 1976). (However, there was no particular correlation among the legumes between presence of this enzyme and storage of galactomannan in the seed.)

2. Enzymes of Galactomannan Breakdown

At least three enzymes are required for complete hydrolysis of galactomannan: an α-galactosidase [EC 3.2.1.22] which cleaves the $(1 \rightarrow 6)$-α-linked side chains, an endo-β-mannanase [EC 3.2.1.78] to cleave the mannan chain and a β-mannosidase [EC 3.2.1.25] to hydrolyse β-D-manno-oligosaccharides, such as mannobiose and mannotriose, to mannose (Reese and Shibata, 1965). All three enzymes have been found in germinating legume seeds and there is evidence that they are involved in galactomannan breakdown *in vivo*. A phosphorylase which can depolymerize manno-oligosaccharides and may therefore also be involved in breakdown has been detected in fenugreek seed extracts (Foglietti and Percheron, 1972) but its role *in vivo* is unknown. A generalized scheme for known enzyme interactions in galactomannan is given in Figure 12.

The activities of α-galactosidase and endo-β-mannanase both increase in the endosperm during germination in parallel with galactomannan breakdown. In contrast, β-mannosidase is already present throughout the endosperm in the resting stage in seeds of several legumes, and its activity actually falls during germination (McCleary, 1983). Previous inconsistencies in results, such as the failure to find β-mannosidase in several legumes by Somme (1970, 1971), while others found an apparent increase in activity during germination (Reid and Meier, 1973) were probably caused by problems with enzyme extraction. At the onset of germination quantitative recovery of β-mannosidase requires very specific conditions, since enzyme protein readily binds to other material in the extract. However, as the galactomannan deposits become solubilized the enzyme becomes progressively easier to extract (McCleary, 1983).

Purified α-galactosidase and endo-β-mannanase, when added together to galactomannan preparations, will break these down to produce galactose, mannotriose and mannobiose, and the β-mannosidases have been shown to be active towards the natural oligosaccharides, hydrolysing these to mannose. The β-mannosidase from guar seeds was found to be an exomannanase (McCleary, 1982). McCleary and Matheson (1975) reported that once scission of the mannan chain is initiated, breakdown to monomers rapidly follows. However, the details of hydrolysis still need to be worked out and they probably differ in different species. Since the activity of purified endo-β-mannanases towards various galactomannans becomes less the more highly substituted they are, it is thought that the α-galactosyl side chains protect the mannan backbone from scission. The α-galactosyl units must be at least partially removed before substantial chain breakage can occur and it has been proposed that hydrolysis may only be possible between two adjacent unsubstituted mannose residues (McCleary et al., 1976). On the other hand, α-galactosidase is unable to strip highly substituted galactomannans completely of their α-galactosyl side chains (Dey, 1978). Thus these two enzymes must act in concert, and their relative contributions to hydrolysis may differ. The modes of action of the enzymes from particular species may be matched closely to their own particular galactomannan.

3. Sources of Enzymes in Legumes
The outer layer of endosperm in fenugreek has been likened to the aleurone layer of cereal endosperm (Reid and Meier, 1972). The only cells of the endosperm that appear to be living in germinating fenugreek seeds are those belonging to this layer (Fig. 13), and galactomannan breakdown in the

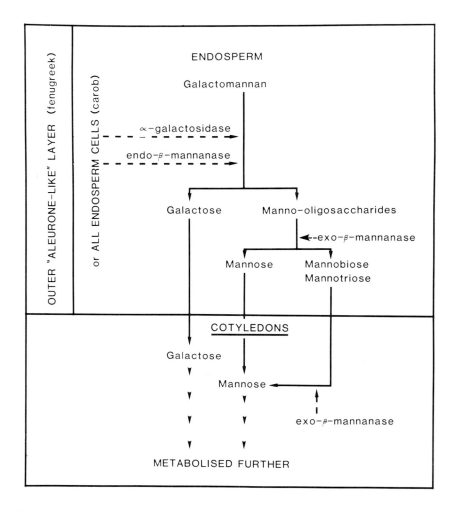

Fig. 12. A proposed scheme for the enzymic interaction in the degradation of endosperm galactomannan to monosaccharides (after McCleary, 1983).

remaining endosperm is reported to begin adjacent to them (Reid and Meier, 1972; Rijven, 1972). There is evidence that α-galactosidase and endo-β-mannanase are produced by cells of this layer.

In contrast to barley, there is no requirement for a factor from the embryo to stimulate enzyme production, nor does GA have any obvious effect (Rijven, 1972). These enzyme activities increase in endosperms incubated following isolation from dry seeds (Reid and Meier, 1973; Reid et al., 1977). Several of the fine structural changes of fenugreek aleurone-like cells parallel those in GA_3-stimulated barley aleurone cells. These include organization of polyribosomes and RER, digestion of cell walls and mobilization of protein body contents (Reid and Meier, 1972). Work with various inhibitors using isolated fenugreek endosperms indicates that galactomannan mobilization is dependent on protein synthesis and aerobic respiration, but there are still conflicting views about the requirement for DNA-dependent RNA synthesis (Reid and Meier, 1973). Abscisic acid is a potent inhibitor, another parallel with the barley aleurone system (Rijven, 1972), while ethylene has a stimulatory effect. Any role that these hormones might have in control processes in intact seeds is not clear.

Somewhat similar results have been obtained in carob seeds (Seiler, 1977). The endosperm cells are able to produce hydrolases independently of the embryo. Small quantities of α-galactosidase are already present in the endosperm of resting seeds. These presumably mobilize raffinose early in germination. An increase in α-galactosidase activity occurs coincident with galactomannan breakdown. The enzyme involved is very similar to the α-galactosidase described by McCleary and Matheson (1974), which is found exclusively in the endosperm in a number of legumes. The increase in its activity in carob is totally inhibited by cycloheximide and actinomycin D. Incorporation of [^{14}C]serine into enzyme protein has also been demonstrated, indicating that its synthesis is *de novo*.

In carob, in contrast to fenugreek, living cells do not appear to be restricted to the outer endosperm layer and mobilization is reported to occur more or less synchronously throughout the endosperm. Seiler (1977) resports that accumulation of free galactose, which is an inhibitor of α-galactosidase, appears to inhibit mobilization so that this progresses further only in regions adjacent to the embryo where the products of hydrolysis are readily absorbed. Seiler considers that in the absence of control by the embryo, inbibition is the trigger for hydrolase production. But he also reports that natural inhibitors are present, presumably in the seed coat, which depress hydrolase activity. These diffuse away during imbibition and their overall role in germination is not clear.

In several legume seeds a lack of correlation has been found in the timing of events. For example, galactomannan loss in isolated fenugreek endosperm occurs earlier than in intact seeds (Reid and Meier, 1972) and in several species increases in activities of the enzymes assumed to be involved in hydrolysis do not correspond well with the progress of galactomannan breakdown (see Bewley and Black, 1978). In some cases, the galactomannan appears to be lost before the enzymes are present in appreciable amounts, as, for example, occurs in isolated endosperms of fenugreek (compare Figs 1A and 1B of Reid and Meier, 1972, with Figs 2 and 3 of Reid et al., 1977). Some of these differences may result from the presence of inhibitors, or loss of the galactomannan without hydrolysis due to its high water solubility, but no explanation seems entirely satisfactory. There is little anatomical information for most species and it is therefore difficult to be sure to what extent the aleurone layer or the endosperm as a whole are involved in enzyme production.

Some lessons might also be learned from cereal grains. Demonstration that isolated leguminous endosperms are capable of total self digestion does not preclude a potential contribution of some galactomannan hydrolases from the embryo. Embryos are known to contain some though not all of the hydrolases involved in galactomannan breakdown. Alpha-galactosidases and β-mannosidases have both been recorded in the embryo of several leguminous seeds (McCleary and Matheson, 1974; 1975). There is a role for the α-galactosidase in oligosaccharide breakdown in the embryo but there is no known function for the embryo β-mannosidase. Whether or not isolated embryos secrete these enzymes is not known.

4. Lettuce Endosperm

In lettuce (*Lactuca sativa*) endosperm the major carbohydrate reserve is also deposited as thick cell walls which are hydrolysed during germination (Fig 14, 15; Jones, 1974). These contain primarily mannose (about 58%) with about 10% galactose, so it is likely that they also consist of galactomannan, although it is not yet known to what extent galactose substitution occurs (Halmer et al., 1975). The three enzymes expected to hydrolyse galactomannan occur in germinated lettuce, namely endo-β-mannanase, β-mannosidase and α-galactosidase. These originate in different regions of the seed and they all behave differently during germination (Bewley et al., 1983). The β-mannanase arises in the endosperm: its level is low initially and its increase in activity correlates fairly well with endosperm breakdown. In contrast, β-mannosidase is found exclusively in the cotyledons. It is present initially in large amounts in a bound form and its amount does not change during germination. Its behaviour parallels the β-mannosidase of legume seeds. Alpha-galactosidase is present in dry seeds but its activity increases in both endosperm and cotyledons early in germination, before radicle emergence and endosperm breakdown. Products of hydrolysis accumulate in isolated endosperm (Leung and Bewley, 1983) but their accumulation has not been detected in whole seeds, indicating rapid transfer to the embryo (Halmer et al., 1978) and subsequent metabolism. There is an increase in embryo dry weight equivalent to the loss in endosperm dry weight at the time of breakdown, further indicating transfer (Halmer et al., 1978). However, it should be realized that the cotyledons are the major storage area in lettuce: the endosperm in mature seeds consists of only two cell layers (Jones, 1974) and it makes only a minor contribution to embryo growth.

Gibberellin does not exert direct control over breakdown of lettuce endosperm but ABA has an inhibitory effect (Bewley and Halmer, 1980/1981). It is thought, however, that GA may act indirectly, by stimulating the cotyledons in some way to remove endogenous inhibitors such as ABA from the endosperm. In the Grand Rapids variety of lettuce, GA can partially overcome the red-light requirement for seed germination at supra-optimal temperatures, and it has been found in another lettuce variety that there is an inverse relationship between ABA content and percentage germination (McWha, 1976).

5. Gibberellin-controlled Cell Wall Hydrolysis in Celery Endosperm

The only well documented case of direct GA control of endosperm wall breakdown in dicotyledons is in celery (*Apium graveolens*, Umbelliferae), although the phenomenon may be more widespread (see Jacobsen et al., 1976). Celery has a seed rich in oil and protein, but the endosperm walls are very thick and obviously supply a significant amount of carbohydrate to the embryo (Figs 16, 17). In fact, all of the cells of the endosperm, which are filled with protein bodies and spherosomes, are living, and are remarkably similar to cereal aleurone cells. Evidence for the GA control of endosperm breakdown was obtained by Jacobsen et al. (1976), who found that endosperms isolated from the embryo do not undergo modification unless gibberellins are added. In this case the best response is elicited by GA_{4+7}, not GA_3 (Palevitch and Thomas, 1976). Endosperm breakdown includes wall hydrolysis and cell separation. As in barley aleurone, the pattern of wall hydrolysis indicates that cell-wall hydrolysing enzymes originate from the protoplast of the same cell and that there is an inner region of the wall that resists digestion (Fig. 16). The cell wall composition is not known, but in caraway, a close relative, it is mostly mannan (Hopf and Kandler, 1977). The overall pattern of endosperm mobilization in celery indicates that the stimulus for breakdown comes from the embryo (Jacobsen and Pressman, 1979) since cells of the central endosperm, which are closest to the embryo, respond earlier than those at the periphery. Earlier breakdown of central cells in GA-treated isolated endosperms (Jacobsen et al., 1976) similarly reflects their position in relationship to the embryo (Dwarte and Ashford, 1982).

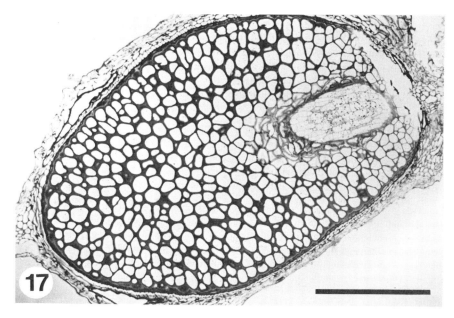

Fig. 17. Section through ungerminated celery fruit stained with PAS reagents to show appearance of thick-walled cells throughout the endosperm. Bar = 300 μm (material was provided by Mr Dennis Dwarte).

IV. FUTURE PROSPECTS

As is apparent, most of our information about the endosperm is derived from work on cereals. We know very little about the fine structure of endosperm or the physiology of its breakdown in most other species. In all but a few seeds nothing is known about the controls operating during germination to initiate the hydrolysis of reserves, or the origin of the hydrolytic enzymes involved. The role of the embryo in these processes is also unclear. Sufficient differences have emerged, however, in the few species so far studied to prevent extrapolation of results from cereals to other species. Examination of the structure and germination processes in a much wider range of species is needed to separate fundamental controls common to all seeds from species differences.

In spite of the intensive studies on barley and wheat endosperm these systems are still by no means fully understood. The properties of GA receptors, primary sites of interaction of GA with aleurone cells, have yet to be determined. This is fundamental to understanding hormone action in plants. Primary GA-induced processes should be separated from secondary events. Furthermore, the data obtained in isolated endosperm systems need to be better related to events in whole seeds during normal germination. This is illustrated by the controversy, still very much alive, about the relative contributions of the scutellum and aleurone layer in endosperm hydrolysis during germination. Clarification of the relationship between GA levels and aleurone response, pathways of GA movement and the mechanism of GA transport in whole seeds is also needed.

ACKNOWLEDGEMENTS

The authors wish to thank Dr W. G. Allaway for his helpful advice and criticism of the manuscript, and Dr J. V. Jacobsen for making available unpublished data and for valuable discussions. We also thank Mrs H. Grunsett for typing the manuscript. Support from the Australian Research Grants Scheme in the form of a Post Doctoral Fellowship to Frank Gubler is gratefully acknowledged.

REFERENCES

Abbott, I. R., and Matheson, N. K. (1972). *Phytochemistry* **11**, 1261-1272.
Akazawa, T., and Miyata, S. (1982). *In* 'Essays in Biochemistry' Vol. **18** (P. N. Campbell and R. D. Marshall, eds), pp. 40-78. Academic Press, London.
Armstrong, C., Black, M., Chapman, J. M., Norman, H. A., and Angold, R. (1982). *Planta* **154**, 573-577.
Ashford, A. E., and Jacobsen, J. V. (1974a). *In* 'Mechanisms of Regulation of Plant Growth' (R. L. Bieleski, A. R. Ferguson and M. M. Cresswell, eds), pp. 519-599. Bulletin 12, The Royal Society of New Zealand, Wellington.
Ashford, A. E., and Jacobsen, J. V. (1974b). *Planta* **120**, 81-105.
Bacic, A. (1979). 'Biochemical and Ultrastructural Studies on Endosperm Walls'. Ph.D. Thesis, La Trobe University, Melbourne.
Bacic, A., and Stone, B. A. (1981a). *Aust. J. Plant Physiol.* **8**, 453-474.
Bacic, A., and Stone, B. A. (1981b). *Aust. J. Plant Physiol.* **8**, 475-495.
Bailey, K. M., Phillips, I. D. J., and Pitt, D. (1976). *J. Exp. Bot.* **27**, 324-336.
Beleia, A., and Varriano-Marston, E. (1981). *Cereal Chem.* **58**, 437-440.
Bennett, P. A., and Chrispeels, M. J. (1972). *Plant Physiol.* **49**, 445-447.
Bernal-Lugo, I., Beachy, R. N., and Varner, J. E. (1981). *Biochem. Biophys. Res. Commun.* **102**, 617-623.
Bewley, J. D., and Black, M. (1978). 'Physiology and Biochemistry of Seeds in Relation to Germination. Vol. 1. Development, Germination, and Growth'. Springer-Verlag, Berlin.
Bewley, J. D., and Halmer, P. (1980/1981). *Isr. J. Bot.* **29**, 118-132.
Bewley, J. D., Leung, D. W. M., and Oullette, F. B. (1983). *Rec. Adv. in Phytochem.* **17**, 137-152.

Bielinska, M., Rogers, G., Rucinsky, T., and Boime, I. (1979). *Proc. Nat. Acad. Sci. USA* **76**, 6152-6156.
Boston, R. S., Miller, T. J., Mertz, J. E., and Burgess, R. R. (1982). *Plant Physiol.* **69**, 150-154.
Briggs, D. E. (1972). *Planta* **108**, 351-358.
Briggs, D. E. (1973). *In* 'Biosynthesis and its Control in Plants' (B. V. Milborrow, ed.), pp. 219-277. Academic Press, London.
Brown, A. H. D., and Jacobsen, J. V. (1982). *Genetical Research Cambridge* **40**, 315-324.
Brown, H. T., and Morris, G. H. (1890). *Journal of the Chemical Society, Transactions* **57**, 458-528.
Buttrose, M. S. (1960). *J. Ultrastructural Res.* **4**, 231-257.
Chandler, P., Higgins, T. J. V., Zwar, J. A., and Jacobsen, J. V. (1984). *Plant Mol. Biol.* In press.
Chrispeels, M. J. (1976). *Annu. Rev. Plant Physiol.* **27**, 19-38.
Chrispeels, M. J. and Varner, J. E. (1967). *Plant Physiol.* **42**, 398-406.
Clutterbuck, V. J., and Briggs, D. E. (1973). *Phytochemistry* **12**, 537-546.
Cohen, D., and Paleg, L. G. (1967). *Plant Physiology* **42**, 1288-1296.
Colborne, A. J., Morris, G., and Laidman, D. L. (1976). *J. Exp. Bot.* **27**, 759-767.
Dashek, W. V., and Chrispeels, M. J. (1977). *Planta* **134**, 251-256.
Daussant, J., and Corvazier, P. (1970). *FEBS Lett.* **7**, 191-194.
Dey, P. M. (1978). *Advances in Carbohydrate Chemistry and Biochemistry* **35**, 341-376.
Dunn, G. (1974). *Phytochemistry* **13**, 1341-1346.
Dwarte, D., and Ashford, A. E. (1982). *Bot. Gaz.* **143**, 164-175.
Eastwell, K. C., and Spencer, M. S. (1982). *Plant Physiol.* **69**, 563-567.
Eastwood, D., and Laidman, D. L. (1971). *Phytochemistry* **10**, 1275-1284.
Evins, W. H., and Varner, J. E. (1971). *Proc. Nat. Acad. Sci. USA* **68**, 1631-1633.
Filner, P., and Varner, J. E. (1967). *Proc. Nat. Acad. Sci. USA* **58**, 1520-1526.
Fincher, G. B., and Stone, B. A. (1974). *Aust. J. Plant Physiol.* **1**, 297-311.
Fincher, G. B., and Stone, B. A. (1981). *In* 'Encyclopedia of Plant Physiology New Series, Volume 13B. Plant Carbohydrates II Extracellular Carbohydrates'(W. Tanner and F. A. Loewus, eds), pp. 68-132. Springer-Verlag, Berlin.
Firn, R. D., and Kende, H. (1974). *Plant Physiol.* **54**, 911-915.
Foglietti, M. J., and Percheron, F. (1972). *Comptes Rendus des seances de l'Academie de Science Paris Series D* **274**, 130-132.
Fulcher, R. G. (1972). 'Observations on the Aleurone Layer with Emphasis on Wheat'. Ph.D. Thesis, Monash University, Clayton, Australia.
Fulcher, R. G., O'Brien, T. P., and Lee, J. W. (1972). *Aust. J. Biol. Sci.* **25**, 23-34.
Fulcher, R. G., Setterfield, G., McCully, M. E., and Wood, P. J. (1977). *Aust. J. Plant Physiol.* **4**, 917-928.
Gahan, P. B., Sierakowska, H., and Dawson, A. L. (1979). *Planta* **145**, 159-166.
Gibbons, G. C. (1980). *Carlsberg Res. Commun.* **45**, 177-184.
Gibbons, G. C. (1981). *Carlsberg Res. Commun.* **46**, 215-225.
Gibson, R. A., and Paleg, L. G. (1975). *Aust. J. Plant Physiol.* **2**, 41-49.
Gibson, R. A., and Paleg, L. G. (1976). *J. Cell Sci.* **22**, 413-425.
Gibson, R. A., and Paleg, L. G. (1982). *Aust. Plant Physiol.* **9**, 623-628.
Goldstein, L. D., and Jennings, P. H. (1975). *Plant Physiol.* **55**, 893-898.
Goodwin, P. B., and Carr, D. J. (1972). *Planta* **106**, 1-12.
Gram, N. H. (1982). *Carlsberg Res. Commun.* **47**, 143-162.
Greenwood, C. T., and Milne, E. A. (1968). *Advances in Carbohydrate Chemistry* **23**, 281-366.

4. Endosperm Carbohydrate Mobilization 159

Groat, J. I., and Briggs, D. E. (1969). *Phytochemistry* **8**, 1615–1627.
Gubler, F. J. (1983). 'Peroxidase and Wall Phenolics in Barley Aleurone'. Ph.D. Thesis, University of New South Wales, Kensington, Australia.
Gubler, F., and Ashford, A. E. (1983). *Aust. J. Plant Physiol.* **10**, 87–97.
Gunning, B. E. S., and Steer, M. W. (1975). 'Ultrastructure and the Biology of Plant Cells'. Edward Arnold, London.
Haberlandt, G. F. J. (1884). 'Physiological Plant Anatomy'. Translated from the German edition by M. Drummond. Reprint edition, 1965. Today and Tomorrow Book Agency, New Delhi.
Haberlandt, G. (1890). *Berichte der deutsches botanischen Gesellschaft* **8**, 40–48.
Halmer, P. and Bewley, J. D. (1982). *In* 'Encyclopedia of Plant Physiology New Series. Volume 13A. Plant Carbohydrates I. Intracellular Carbohydrates' (F. A. Loewus and W. Tanner, eds), pp. 748–793. Springer-Verlag, Berlin.
Halmer, P., Bewley, J. D., and Thorpe, T. A. (1975). *Nature* **258**, 716–718.
Halmer, P., Bewley, J. D., and Thorpe, T. A. (1978). *Planta* **139**, 1–8.
Hardie, D. G. (1975). *Phytochemistry* **14**, 1719–1722.
Harris, N., Oparka, K. J., and Walker-Smith, D. J. (1982). *Planta* **156**, 461–465.
Harvey, B. M. R., and Oaks, A. (1974). *Planta* **121**, 67–74.
Higgins, T. J. V., Zwar, J. A., and Jacobsen, J. V. (1976). *Nature* **260**, 166–169.
Higgins, T. J. V., Jacobsen, J. V., and Zwar, J. A. (1982). *Plant Mol. Biol.* **1**, 191–215.
Ho, D. T. H. (1979). *In* 'Molecular Biology of Plants' (I. Rubenstein, R. L. Phillips, C. E. Green and B. G. Gengenbach, eds), pp. 217–240. Academic Press, London.
Ho, D. T. H., and Varner, J. E. (1974). *Proc. Nat. Acad. Sci. USA* **71**, 4783–4786.
Ho, D. T. H., and Varner, J. E. (1976). *Plant Physiol.* **57**, 175–178.
Hopf, H., and Kandler, O. (1977). *Phytochemistry* **16**, 1715–1717.
Hoy, J. L., Macauley, B. J., and Fincher, G. B. (1981). *Inst. Brewing* **87**, 77–80.
Jackson, D. I. (1971). *J. Exp. Bot.* **22**, 613–619.
Jackson, R. C., and Blobel, G. (1977). *Proc. Nat. Acad. Sci. USA* **74**, 5598–5602.
Jacobsen, J. V. (1973). *Plant Physiol.* **51**, 198–202.
Jacobsen, J. V. (1977). *Annu. Rev. Plant Physiol.* **28**, 537–564.
Jacobsen, J. V. (1983). *In* 'The Biochemistry and Physiology of Gibberellins' Volume 2 (A. Crozier, ed.), pp. 159–187. Praegar Publishers, New York.
Jacobsen, J. V., and Higgins, T. J. V. (1982). *Plant Physiol.* **70**, 1647–1653.
Jacobsen, J. V., and Knox, R. B. (1972). *In* 'Plant Growth Substances 1970' (D. J. Carr, ed.), pp. 344–351. Springer-Verlag, Berlin.
Jacobsen, J. V., and Pressman, E. (1979). *Planta* **144**, 241–248.
Jacobsen, J. V., and Varner, J. E. (1967). *Plant Physiol.* **42**, 1596–1600.
Jacobsen, J. V., and Zwar, J. A. (1974a). *Aust. J. Plant Physiol.* **1**, 343–356.
Jacobsen, J. V. and Zwar, J. A. (1974b). *Proc. Nat. Acad. Sci. USA* **71**, 3290–3293.
Jacobsen, J. V., Pressman, E., and Pyliotis, N. A. (1976). *Planta* **129**, 113–122.
Jacobsen, J. V., Higgins, T. J. V., and Zwar, J. A. (1979). *In* 'The Plant Seed: Development, Preservation, and Germination' (I. Rubenstein, R. L. Phillips, C. E. Green and B. G. Gengenbach, eds), pp. 241–262. Academic Press, New York.
Jacobsen, J. V., Chandler, P. M., Higgins, T. J. V., and Zwar, J. A. (1982). *In* 'Plant Growth Substances, 1982' (P. F. Wareing, ed.), pp. 111–119. Academic Press, London.
Jelsema, C. L., Morré, D. J., and Ruddat, M. (1982). *Bot. Gaz.* **143**, 26–31.
Jones, R. L. (1969). *Planta* **87**, 119–133.
Jones, R. L. (1971). *Plant Physiol.* **47**, 412–416.
Jones, R. L. (1972). *Planta* **103**, 95–109.

Jones, R. L. (1973). *Plant Physiol.* **52**, 303-308.
Jones, R. L. (1974). *Planta* **121**, 133-146.
Jones, R. L. (1980a). *Planta* **150**, 58-69.
Jones, R. L. (1980b). *Planta* **150**, 70-81.
Jones, R. L., and Armstrong, J. E. (1971). *Plant Physiol.* **48**, 137-142.
Jones, R. L., and Jacobsen, J. V. (1978). *Bot. Mag. Tokyo Special Issue* **1**, 83-99.
Jones, R. L., and Jacobsen, J. V. (1982). *Planta* **156**, 421-432.
Jones, R. L., and Price, J. M. (1970). *Planta* **94**, 191-202.
Jones, R. L., and Stoddart, J. L. (1977). *In* 'The Physiology and Biochemistry of Seed Dormancy and Germination' (A. A. Khan, ed), pp. 77-110. Elsevier/North Holland Biomedical Press, Amsterdam.
Keates, R. A. B. (1973). *Nature* **244**, 355-357.
Kent, N. L. (1975). 'Technology of Cereals with Special Reference to Wheat'. Pergamon Press, Oxford.
Koehler, D. E., and Varner, J. E. (1973). *Plant Physiol.* **52**, 208-214.
Leung, D. W. M., and Bewley, J. D. (1983). *Planta* **157**, 274-277.
Lineback, D. R., and Ponpipom, S. (1977). *Die Stärke* **29**, 52-60.
Locy, R., and Kende, H. (1978). *Planta* **143**, 89-99.
MacGregor, A. W. (1977). *J. Inst. Brewing* **83**, 100-103.
MacGregor, A. W., and Matsuo, R. R. (1982). *Cereal Chem.* **59**, 210-216.
McCleary, B. V. (1982). *Carbohydrate Research* **101**, 75-92.
McCleary, B. V. (1983). *Phytochemistry* **22**, 649-658.
McCleary, B. V., and Matheson, N. K. (1974). *Phytochemistry* **13**, 1747-1757.
McCleary, B. V., and Matheson, N. K. (1975). *Phytochemistry* **14**, 1187-1194.
McCleary, B. V., and Matheson, N. K. (1976). *Phytochemistry* **15**, 43-47.
McCleary, B. V., Matheson, N. K., and Small, D. M. (1976). *Phytochemistry* **15**, 1111-1117.
McClendon, J. H., Nolan, W. G., and Wenzler, H. F. (1976). *Am. J. Bot.* **63**, 790-797.
MacLeod, A. M. (1979). *In* 'Brewing Science, Vol. 1' (J. R. A. Pollock, ed.), pp. 145-232. Academic Press, London.
MacLeod, A. M., and Palmer, G. H. (1966). *J. Inst. Brewing* **72**, 520-589.
McWha, J. A. (1976). *Aust. J. Plant Physiol.* **3**, 849-851.
Maeda, I., Kiribuchi, S., and Nakamura, M. (1978). *Agr. Biol. Chem.* **42**, 259-267.
Manners, D. J. (1974). *In* 'Essays in Biochemistry' Vol. 10 (P. N. Campbell and F. Dickens, eds), pp. 37-71. Academic Press, London.
Manners, D. J., and Yellowlees, D. (1973). *J. Inst. Brewing* **79**, 377-385.
May, L. H., and Buttrose, M. S. (1959). *Aust. J. Biol. Sci.* **12**, 146-159.
Meier, H., and Reid, J. S. G. (1982). *In* 'Encyclopedia of Plant Physiology' New Series. Volume 13A. Plant Carbohydrates I. Intracellular Carbohydrates (F. A. Loewus and W. Tanner, eds), pp. 418-471. Springer-Verlag, Berlin.
Mersey, B., and McCully, M. E. (1978). *J. Microscopy* **114**, 49-76.
Miyata, S., and Akazawa, T. (1982a). *Plant Physiol.* **70**, 147-153.
Miyata, S., and Akazawa, T. (1982b). *Proc. Nat. Acad. Sci. USA* **79**, 6566-6568.
Miyata, S., and Akazawa, T. (1982c). *Proc. Nat. Acad. Sci. USA* **79**, 7792-7795.
Miyata, S., Okamoto, K., Watanabe, A., and Akazawa, T. (1981). *Plant Physiol.* **68**, 1314-1318.
Momotani, Y., and Kato, J. (1972). *In* 'Plant Growth Substances 1970' (D. J. Carr, ed.), pp. 352-355. Springer-Verlag, Berlin.
Momotani, Y., and Kato, J. (1974). *In* 'Plant Growth Substances 1973), pp. 633-637. Hirokawa Publishing Co., Tokyo.

Moore, A. E., and Stone, B. A. (1972). *Biochim. Biophys. Acta* **258**, 248-264.
Morrall, P., and Briggs, D. E. (1978). *Phytochemistry* **17**, 1495-1502.
Mounla, M. A. Kh., and Michael, G. (1973). *Physiol. Plant.* **29**, 274-276.
Mozer, T. J. (1980a). *Plant Physiol.* **65**, 834-837.
Mozer, T. J. (1980b). *Cell* **20**, 479-485.
Murphy, G. J. P., and Briggs, D. E. (1973). *Phytochemistry* **12**, 2597-2605.
Muthukrishnan, S., Chandra, G. R., and Maxwell, E. S. (1979). *Proc. Nat. Acad. Sci. USA* **76**, 6181-6185.
Nicholls, P. B. (1979). *Aust. J. Plant Physiol.* **6**, 229-240.
Nicholls, P. B. (1980). *Aust. J. Plant Physiol.* **7**, 645-653.
Nicholls, P. B. (1982). *Aust. J. Plant Physiol.* **9**, 373-383.
Norman, H. A., Black, M., and Chapman, J. M. (1982). *Planta* **154**, 578-586.
Obata, T. (1979). *Ann. Bot.* **44**, 333-337.
Obata, T., and Suzuki, H. (1976). *Plant, Cell Physiol.* **17**, 63-71.
Okamoto, K., and Akazawa, T. (1979a). *Plant Physiol.* **63**, 336-340.
Okamoto, K., and Akazawa, T. (1979b). *Plant Physiol.* **64**, 337-340.
Okamoto, K., and Akazawa, T. (1980). *Plant Physiol.* **65**, 81-84.
Okamoto, K., Kitano, H., and Akazawa, T. (1980). *Plant, Cell Physiol.* **21**, 201-204.
Okamoto, K., Murai, T., Eguchi, G., Okamoto, M., and Akazawa, T. (1982). *Plant Physiol.* **70**, 905-911.
Okita, T. W., Decaleya, R., and Rappaport, L. (1979). *Plant Physiol.* **63**, 195-200.
Paleg, L. G. (1965). *Annu. Rev. Plant Physiol.* **16**, 291-322.
Palevitch, D., and Thomas, T. H. (1976). *Physiol. Plant.* **37**, 247-252.
Palmer, G. H. (1972). *J. Inst. Brewing* **78**, 326-332.
Palmer, G. H. (1982). *J. Inst. Brewing* **88**, 145-153.
Preiss, J., and Levi, C. (1980). In 'The Biochemistry of Plants. Volume 3. Carbohydrates: Structure and Function' (J. Preiss, ed.), pp. 371-423. Academic Press, New York.
Pyliotis, N. A., Ashford, A. E., Whitecross, M. I., and Jacobsen, J. V. (1979). *Planta* **147**, 134-140.
Radley, M. (1959). *J. Chem. Ind.*, 877-878.
Radley, M. (1967). *Planta* **75**, 164-171.
Radley, M. (1969). *Planta* **86**, 218-223.
Reese, E. T., and Shibata, Y. (1965). *Can. J. Microbiol.* **11**, 167-183.
Reid, J. S. G. (1971). *Planta* **100**, 131-142.
Reid, J. S. G., and Bewley, J. D. (1979). *Planta* **147**, 145-150.
Reid, J. S. G., and Meier, H. (1970). *Z. Pflanzenphysiol.* **62**, 89-92.
Reid, J. S. G., and Meier, H. (1972). *Planta* **106**, 44-60.
Reid, J. S. G., and Meier, H. (1973). *Planta* **112**, 301-308.
Reid, J. S. G., Davies, C., and Meier, H. (1977). *Planta* **133**, 219-222.
Rijven, A. H. G. C. (1972). *Acta Bot. Neerl.* **21**, 381-386.
Rowsell, E. V., and Goad, L. J. (1962). *Biochem. J.* **84**, 73p.
Rowsell, E. V., and Goad, L. J. (1964). *Biochem. J.* **90**, 12p.
Seiler, A. (1977). *Planta* **134**, 209-221.
Smart, M. G., and O'Brien, T. P. (1979a). *Aust. J. Bot.* **27**, 403-411.
Smart, M. G., and O'Brien, T. P. (1979b). *Aust. J. Plant Physiol.* **6**, 485-491.
Smith, M. M., and Hartley, R. D. (1983). *Carbohydrate Research* **118**, 65-80.
Somme, R. (1970). *Acta Chem. Scand.* **24**, 72-76.
Somme, R. (1971). *Acta Chem. Scand.* **25**, 759-761.
Staudte, R. G., Woodward, J. R., Fincher, G. B., and Stone, B. A. (1983). *Carbohydrate Polymers* **3**.

Swift, J. G., and O'Brien, T. P. (1970). *Aust. J. Bot.* **18**, 45-53.
Swift, J. G., and O'Brien, T. P. (1971). *Aust. J. Bot.* **19**, 63-71.
Swift, J. G., and O'Brien, T. P. (1972). *Aust. J. Biol. Sci.* **25**, 469-486.
Taiz, L., and Honigman, W. A. (1976). *Plant Physiol.* **58**, 380-386.
Taiz, L., and Jones, R. L. (1970). *Planta* **92**, 73-84.
Taiz, L., and Jones, R. L. (1973). *Am. J. Bot.* **60**, 67-75.
Tanaka, Y., Ito, T., and Akazawa, T. (1970). *Plant Physiol.* **46**, 650-654.
Tomos, A. D., and Laidman, D. L. (1979). *In* 'Recent Advances in the Biochemistry of Cereals' (D. L. Laidman and R. G. Wyn Jones, eds), pp. 119-146. Academic Press, London.
Trewavas, A. J. (1982). *Physiol. Plant.* **55**, 60-72.
Tsai, C. Y., and Nelson, O. E. (1968). *Plant Physiol.* **43**, 103-112.
Tsai, C. Y., and Nelson, O. E. (1969). *Plant Physiol.* **44**, 159-167.
Ueki, K. (1979). *Plant, Cell Physiol.* **20**, 789-796.
van der Eb, A. A., and Nieuwdorp, P. J. (1967). *Acta Bot. Neerl.* **15**, 690-699.
Varner, J. E., and Chandra, G. R. (1964). *Proc. Nat. Acad. Sci. USA* **52**, 100-106.
Varner, J. E., and Ho, D. T. H. (1976). *In* 'The Molecular Biology of Hormone Action' (J. Papaconstantinou, ed.), pp. 173-194. Academic Press, New York.
Varner, J. E., and Ho, D. T. H. (1977). *In* 'Regulation of Enzyme Synthesis and Activity in Higher Plants' (H. Smith, ed.), pp. 83-92. Academic Press, London.
Varner, J. E., and Mense, R. M. (1972). *Plant Physiol.* **49**, 187-189.
Vigil, E. L., and Ruddat, M. (1973). *Plant Physiol.* **51**, 549-558.
Weiler, E. W., and Wieczorek, U. (1981). *Planta* **152**, 159-167.
Williams, J. M. (1968). *In* 'Starch and Its Derivatives' (J. A. Radley, ed.), pp. 91-138. Chapman and Hall, London.
Woodward, J. R., and Fincher, G. B. (1982). *Carbohydrate Research* **106**, 111-122.
Yamada, K. (1982). *J. Am. Soc. Brewing Chemists* **40**, 18-25.
Yomo, H., and Iinuma, H. (1964). *Proc. Am. Soc. Brew. Chemists*, 97-102.
Yomo, H., and Varner, J. E. (1971). *Current Topics in Developmental Biology* **6**, 111-144.

CHAPTER 5

Mobilization of Nitrogen and Phosphorus from Endosperm

MICHAEL J. DALLING and PREM L. BHALLA

I.	Introduction	163
II.	Mobilization of Nitrogen	164
	A. Nature and Localization of Nitrogenous Reserves	164
	B. Changes in Nitrogenous Compounds during and after Germination	165
	C. Role of Peptide Hydrolase Enzymes during and after Germination	165
	D. Role of Scutellum or Cotyledons in the Transfer of Nitrogen from Endosperm to Embryo	178
III.	Mobilization of Phosphorus	182
	A. Phytin	184
	B. Nucleic Acids	190
	C. Phospholipids	192
IV.	Conceptual Model of Nitrogen and Phosphorus Mobilization	193
	References	196

I. INTRODUCTION

When resting seeds are exposed to appropriate conditions of temperature, moisture and aeration it is usual for the seeds to germinate. However simple this process may appear, the physiological and biochemical processes underyling it are still intriguing. In essence all the activities of the seed are directed towards the growth of the embryo and a little later, the developing seedling. During this early period all the nutrients, including carbon, necessary for growth and development of the axis are derived from various storage compounds within the seed. Later, the seedling is totally dependent on nutrient acquisition from its aerial and edaphic environment.

This chapter is concerned with the mobilization of stored forms of nitrogen and phosphorus from the endosperm. Its emphasis is on cereals, where the endosperm is differentiated into an outer aleurone cell layer (or layers) distinct from the larger starch-storing cells of the interior, which are non-living in many species. On germination, the endosperm functions as a secondary lysosome, allowing various hydrolytic enzymes to degrade its contents, but simultaneously protecting the embryo from the action of the same hydrolytic enzymes.

II. MOBILIZATION OF NITROGEN

A. Nature and Localization of Nitrogenous Reserves

Proteins comprise the most abundant nitrogenous reserves in seeds with endosperm, accounting for about 8%–20% of seed weight. The other compounds that may contribute to the nitrogen nutrition of the embryonic axis are RNA and soluble amino acids, but their contribution is relatively minor compared to that of the reserve proteins.

In cereal grains it is well known that the aleurone layer is richer in protein than the bulk of the endosperm, although the latter contains the major proportion of the total. Proteins of the aleurone cells are localized in protein bodies originally termed 'aleurone grains' or 'aleurone vacuoles' (for discussion see Ashton, 1976; Bewley and Black, 1978). These protein bodies are characterized by the presence of globoid inclusions that correspond to phytin, the main reserve form of phosphorus, and sometimes other inclusions, as illustrated in Chapter 4 of Volume 1. The proteins that accumulate in the vacuolar protein bodies of aleurone cells are distinct in their amino acid composition and properties. As shown for wheat (*Triticum aestivum*) (Table I of Chapter 1, Volume 1), the aleurone proteins contain about three times the arginine content of the proteins of the starchy endosperm. The latter comprise a mixture of lysine-deficient gliadins (wheat prolamins, Section 1.II,C of Volume 1), together with glutenins (wheat glutelins). Each of these groups possesses relatively high contents of proline and glutamine (Wall, 1979). In wheat and barley (*Hordeum vulgare*) the prolamins and glutelins share the same intracellular location, the matrix surrounding the starch grains, but in many other cereals they occupy different protein bodies with distinct ontogeny (Chapter 3 of Volume 1).

B. Changes in Nitrogenous Compounds during and after Germination

A decline in the amount of endosperm nitrogen and a concomitant increase in axis nitrogen can be detected within 48 h of the beginning of imbibition (Fig. 1). Once the nitrogen reserves of the endosperm are depleted, any further increase in axis nitrogen must occur by absorption from the growth medium.

Perhaps the earliest study of the breakdown of individual endosperm protein fractions is the work of Folkes and Yemm (1958), who measured changes in the hordein (glutelin), hordenin (prolamin), albumin and globulin fractions of barley grains during the first ten days of seedling development. Hordein, the most abundant fraction, declined most rapidly, so that by day six only 10% remained. At the same time 23% of the hordenin remained. This association between relative abundance and susceptibility to degradation has been confirmed in several subsequent studies.

Harvey and Oaks (1974b) found that the maize (*Zea mays*) glutelin, representing 43% of total endosperm nitrogen, was degraded more rapidly than zein, 30% of the total (Fig. 2). Metivier and Monteiro (1981) compared the mobilization of endosperm protein in normal and zein deficient *opaque-2* maize. Glutelin depletion was faster than zein depletion in both. Prior to day three the rate of glutelin breakdown was 1670 µg/day in normal, and 1970 µg/day in *opaque-2* compared with only 421µg/day and 209 µg/day respectively for zein. In wheat, Preston et al. (1978) found that gliadin (47% of total endosperm protein) declined more rapidly than glutenin (40% of total endosperm protein). Therefore, it would appear that regardless of identity, the most abundant protein fraction is utilized most rapidly.

C. Role of Peptide Hydrolase Enzymes during and after Germination

According to Mikola (1983) there are three distinct stages of protoeolysis in germinating seeds. In the first stage, hydrolysis provides amino acids for the synthesis of those hydrolytic enzymes which will achieve degradation of the insoluble reserves of the endosperm. This hydrolysis may be confined to the cells of the aleurone layer, more specifically, the protein bodies. The second stage accounts for the hydrolysis of the main reserve proteins and results in the provision of amino acids to the growing seedling. The third and final stage includes the senescence of the reserve-depleted storage tissue. During this stage the mobilization system is broken down to provide

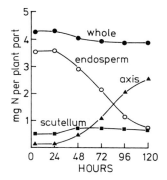

Fig. 1. Changes in the total nitrogen content of the endosperm, scutellum and embryonic axis of maize seedlings (data of Ingle et al., 1964).

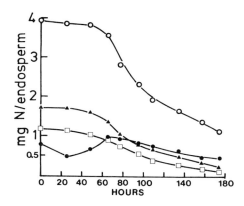

Fig. 2. Changes in maize endosperm proteins following imbibition: ○, total nitrogen; ▲, glutelin; □, zein; ●, albumin + globulin + dialysable nitrogen (data of Harvey and Oaks, 1974b).

the last ration of amino acids to the seedling before the onset of autotrophic growth. It is highly likely that in most seeds the populations of peptide hydrolases and their interactions with regulatory factors such as endogenous inhibitors are different in each stage.

1. Nomenclature and Classification of Peptide Hydrolases

The peptide hydrolases have been classified into two major sub-groups, based on the position of the peptide bond hydrolysed. The proteinases, or endopeptidases [EC 3.4.21–EC 3.4.24] cleave bonds distant from the ends of the polypeptide, whereas exopeptidases [EC 3.4.11–EC 3.4.17] cleave bonds only near (one or two residues from) the end of the polypeptide.

a. Exopeptidases
The exopeptidases can be categorized into six major classes on the basis of their substrate specificity (Table I). Exopeptidases acting at the N-terminal end of peptides require a free α-amino group (or α-imino group in the case of iminopeptidases). The N-terminal exopeptidases that release free amino acids comprise the class called α-aminoacyl peptide hydrolases (aminopeptidases), while those that release intact dipeptides belong to the dipeptidylpeptide hydrolases. The C-terminal exopeptidases can likewise be designated as peptidylamino-acid hydrolases (carboxypeptidases) or peptidyldipeptide hydrolases, depending on whether a free amino acid or an intact dipeptide is released. The dipeptide hydrolases, or dipeptidases, cleave dipeptides into their constituent amino acids.

b. Endopeptidases
Because of the complex substrate specificity of the endopeptidases, they have been classified according to the nature of their catalytic site (Table I).

Care must be taken while interpreting the effects of the 'group-specific' inhibitors used in the classification of endopeptidases. Most of the inhibitors lack complete selectivity, and are not sufficiently effective to allow clear cut demarcation in every case (Barrett, 1977). Similarly, caution should be taken while distinguishing between endopeptidase and exopeptidase activities, especially in crude extracts. The most commonly used measure of proteinase activity against protein substrates has been the increase in trichloroacetic acid (TCA)-soluble α-amino groups. The release of α-amino groups resulting from proteinase action is then usually detected by reaction with ninhydrin reagent. However, as pointed out by Ashton (1976), this method is generally inadequate for measuring proteinase activity. In cases where large peptide fragments are released as a result of proteinase action, activity may be underestimated because the reaction products are still large enough to be TCA-insoluble. Also, in many cases the proteinase substrates are unsuitable, since carboxypeptidases have been demonstrated to act on a number of intact proteins (Mikola and Pietilä, 1972; Kruger and Preston, 1977). In addition, carboxypeptidases are able to attack the numerous C-terminals produced as a result of proteinase action. Clearly, proteinase activity will be overestimated in these situations. Some indication of the nature of the peptide hydrolases present can be gauged from the ratio of total-N:α-amino-N of the breakdown products. If this ratio is two, then degradation is predominantly due to exopeptidase activity. The higher the ratio, the more likely it is that degradation is due

Table I. Peptide hydrolase classification

EC Number	Class
	Exopeptidases
3.4.11	α-aminoacyl peptide hydrolases (aminopeptidases)
	The N-terminal amino acid is cleaved.
3.4.13	dipeptide hydrolases (dipeptidases)
	Dipeptides are cleaved into their constituent amino acids.
3.4.14	dipeptidyl peptide hydrolases
	Dipeptides are cleaved from the N-terminal end.
3.4.15	peptidyldipeptide hydrolases
	Dipeptides are cleaved from the C-terminal end.
3.4.16	serine carboxypeptidases (peptidylamino-acid hydrolases)
	The C-terminal amino acid is cleaved.
3.4.17	metallo-carboxypeptidases (peptidylamino-acid hydrolases)
	The C-terminal amino acid is cleaved.
	Proteinases
3.4.21	serine proteinases
3.4.22	thiol proteinases
3.4.23	acid (carboxyl) proteinases
3.4.24	metallo-proteinases
3.4.99	proteinases of unknown catalytic mechanism

to endopeptidase activity. More recently sodium dodecyl sulphate (SDS)-polyacrylamide gel electrophoresis has been employed to visualize the products of peptide hydrolase action (Ragster and Chrispeels, 1981; Thomas and Huffaker, 1981).

2. Peptide Hydrolases Present in Quiescent Seed

Low peptide hydrolase activity is generally detectable in quiescent seed of most species. Salmia (1981a) has described a proteinase with acid pH optimum from resting seed of Scots pine (*Pinus sylvestris*) which is inhibited by pepstatin-A, a specific inhibitor of the acid proteinase group [EC 3.4.23]. The pepstatin-sensitive enzyme activity does not change during the course of germination, whereas proteinases I and II increase substantially (Fig. 3). This enzyme is similar to that isolated from resting seeds of rice (*Oryza sativa*) by Doi et al. (1980a).

Such enzymes are obvious candidates for the role of initiator proteinase. However, before any conclusions can be drawn in this regard, more information is required about their *in vivo* localization and regulation. Because of their low pH optimum it is highly likely that they may be associated with a protein body or vacuole where the pH is in the range 5.7–5.9 (Nishimura, 1982). The nature of the mechanism which

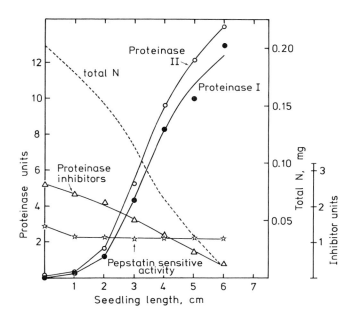

Fig. 3. Changes in the activities of proteinases and proteinase inhibitors in the storage tissue of Scots pine during early seedling growth (data of Salmia, 1980, 1981a).

prevents enzyme action prior to germination is open to speculation. Compartmentation is one possibility, however this necessitates a mechanism to bring the substrate and proteinase together. Inhibitors are another possibility, but in Scots pine the pepstatin-sensitive acid proteinase is not affected by the endogenous inhibitors so far characterized (Salmia et al., 1978; Salmia and Mikola, 1980).

Carboxypeptidase activity has been reported in resting seeds of Scots pine (Salmia and Mikola, 1976a), rice (Doi et al., 1980a) and barley (Mikola and Kolehmainen, 1972). In resting wheat grain, high levels of carboxypeptidase activity were reported by Preston and Kruger (1976a). In a subsequent study (Preston and Kruger, 1976b) it was shown that this activity could be resolved into two enzymes, each with very broad specificity, including the ability to hydrolyse the storage protein, gluten. These authors suggest that the carboxypeptidase activity may be extremely important in the early stage of germination, prior to the induction and secretion of further hydrolytic enzymes by the aleurone cells.

Aminopeptidase activities have generally been found to be high in resting seeds, for example, barley (Mikola and Kolehmainen, 1972), Scots pine (Salmia and Mikola, 1976b), maize (Feller et al., 1978) and wheat (Kruger and Preston, 1978; Waters and Dalling, 1979). An exception is castor bean (*Ricinus communis*), where a four-fold increase in activity occurs during the most rapid period of protein degradation in the endosperm (Tully and Beevers, 1978).

3. Peptide Hydrolases Appearing in the Endosperm during Germination

a. Wheat

There have been several reports which indicate that hydrolysis of the endosperm storage proteins during germination is associated with an increase in the general level of peptide hydrolase activity. Redman (1971) demonstrated rapid softening of gluten due to the action of peptide hydrolase in sprouted wheat. Hwang and Bushuk (1973) observed a 17-fold increase in peptide hydrolase activity eight days after germination of a Canadian hard red spring wheat. Associated with the *in vitro* activity was a decrease in endosperm protein of high molecular weight and a concomitant increase in free amino acids. Hanford (1967) measured both exopeptidase and endopeptidase activity in wheat flour and concluded that because of a lack of correlation between gluten softening and the accumulation of TCA-soluble nitrogen, gluten hydrolysis was most likely due to endopeptidase activity. Preston *et al.* (1978) measured endopeptidase activity with the synthetic substrate azocasein and concluded that enhancement of endopeptidase activity was required for extensive storage protein hydrolysis. Although the role of exopeptidases was recognized in terms of secondary processing of peptides, Preston *et al.* (1978) observed only a small increase in exopeptidase activity during germination. This increase was probably due to carboxypeptidase because aminopeptidase activity declines in parallel with protein loss during germination (Waters and Dalling, 1979).

b. Scots pine

The so-called endosperm of Scots pine is actually haploid megagametophyte storage tissue. It has the same function as true endosperm. As already indicated (Section I.B,*2*), it contains a pepstatin-A sensitive acid proteinase. Once germination commences two other proteinases, I and II, increase in activity (Fig. 3). Salmia (1981a) has separated proteinases I and II and shown them to be sensitive to inhibitors of sulfhydryl groups and also to the endogenous inhibitors of the endosperm. Proteinase I has a molecular weight (MW) of about 42 000 daltons and exhibits high activity at pH 3.7 against haemoglobin, but much lower activity against casein at pH 5.4 and

7.0. Proteinase II has a MW of about 30 000 daltons and exhibits a broad pH optimum against haemoglobin and casein. Proteinase II is much more stable than proteinase I.

The changes in activity of exopeptidase enzymes have also been well characterized. Salmia and Mikola (1975) detected two different alkaline dipeptidases in resting seeds, both of which increase in activity during germination. On the basis of observations that similar activities are associated with most plant tissues, the authors were unable to conclude that the dipeptidases are uniquely associated with the germination process. Resting seeds also contain enzymes with aminopeptidase activity. During germination, however, there is little change in their activity until the senescence phase, when their activity declines (Salmia and Mikola, 1976b). Carboxypeptidase activity is very low in resting seed but increases markedly (40-fold) during germination (Salmia and Mikola, 1976a). However, the highest carboxypeptidase activity is observed after the major portion of the reserve proteins has been degraded, that is, during the senescence stage.

c. Rice

Doi *et al.* (1980b) have partially characterized two acid endopeptidases from germinating rice seeds. One is present in the resting seed, has a MW of 65 000–70 000 daltons and is inhibited by pepstatin-A. The other enzyme appears during germination, has a MW of 20 000–25 000 daltons and is inhibited by the sulfhydryl group inhibitor *p*-chloromercuribenzoic acid and leupeptin, a tripeptide inhibitor of thiol-endopeptidases (Umezawa and Aoyagi, 1977).

d. Castor bean

Although castor bean is a dictotyledon, the major storage organ of the seed is endosperm. Tully and Beevers (1978) have made a comprehensive study of the relation between protein mobilization and the activity of a range of exopeptidases and endopeptidases. Endopeptidase activity was assumed from activity against haemoglobin at pH 3.5. This enzyme activity increased during the period of rapid protein mobilization and significant activity persisted until the later stage of endosperm senescence. The endopeptidase was inhibited by the sulfhydryl-group inhibitors, iodoacetamide and N-ethylmaleimide. This enzyme, together with the other acidic hydrolytic enzymes, carboxypeptidase, phosphodiesterase [EC 3.1.4.1], RNase [EC 3.1.4.22], phytase [EC 3.1.3.26] and β-glucosidase [EC 3.2.1.21], was shown by Nishimura and Beevers (1978) to be localized in vacuoles that arise from protein bodies during the course of germination.

More recently, Alpi and Beevers (1981) have shown that careful introduction of leupeptin into whole castor bean seeds inhibits germination.

In vitro degradation of the water-soluble and crystalloid-storage proteins of the protein bodies of the sulfhydryl-sensitive endopeptidase is completely inhibited by 10 μM leupeptin. Furthermore, application of leupeptin to endosperm halves from germinating seeds prevents the normal development of enzymes concerned with gluconeogenesis and severely limits production of sugar from oil reserves (Table II). This is an excellent example of how other enzymes essential for germination are reliant on protein mobilization as a source of amino acids for their own synthesis. The effect of leupeptin on germination is arguably the most compelling evidence in support of the hypothesis that the sulfhydryl-sensitive proteinase plays an essential role in protein mobilization.

e. Maize

Degradation of zein and glutelin, the major storage proteins in maize endosperm, begins during the second day of germination (Fig. 2). Peptide hydrolase activity, assayed with either zein or gliadin as substrate, increases as protein is lost (Harvey and Oaks, 1974a). Although exopeptidase activity was not measured directly, it appeared to be low because the ratio of total-N:α-amino-N released was always in the range 8–40. Activity was enhanced by the addition of β-mercaptoethanol to the assay medium. Calculations by Harvey and Oaks (1974a) indicate that the *in vitro* activities can readily account for the rate of protein hydrolysis *in vivo*.

Perhaps the most comprehensive study of the inter-relations between exopeptidase and endopeptidase enzymes during maize germination is that of Feller *et al.* (1978). Endopeptidase activity was measured by following the degradation of casein at pH 3.8, 5.4 and 7.5. Endopeptidase activity increased as endosperm nitrogen was lost (Fig. 4). Mercaptoethanol increased the endopeptidase activity several-fold, especially at pH 5.4. Aminopeptidase activity was high in the endosperm (including aleurone layer) of resting seed and during the course of germination this activity declined with the loss of endosperm nitrogen. Carboxypeptidase activity exhibited the same pattern of activity as the endopeptidase. However, some distinction could be made between the two, as the endopeptidase activity was sensitive to mercaptoethanol, which had no effect on carboxypeptidase activity.

The maize endosperm endopeptidase has been purified and characterized by Abe *et al.* (1977). The enzyme was found to have a MW of 21 000 daltons and an isoelectric point of 2.3 or less. The optimum pH for degradation of denatured haemoglobin was 3.0. The endopeptidase was completely inhibited by 1 mM *p*-chloromercuribenzoic acid and 70% inhibited by 25 μM leupeptin; pepstatin had no effect.

Table II. Influence of leupeptin[a] on growth, sugar accumulation and enzyme activities in excised endosperm halves incubated for 4 days at 30°C

	Without Leupeptin	With Leupeptin
Increase in length, cm	0.74	0.25
Increase in weight, mg	381	226
Catalase, mmol/min	2.70	0.30
Isocitrate lyase, μmol/min	0.58	0.13
3-Hydroxy-acyl CoA dehydrogenase, μmol/min	1.92	0.67
Cyt c oxidase, μmol/min	1.92	0.02
FDPase, μmol/min	0.52	0.17
Sucrose, mg	33.0	9.2
Glucose, mg	0.66	0.10

[a] Leupeptin was supplied at 1 μg per half endosperm. Values are expressed per whole endosperm.
Data of Alpi and Beevers (1981).

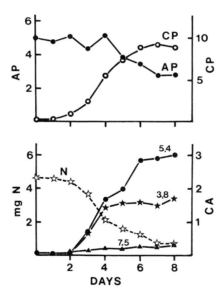

Fig. 4. Changes in proteolytic enzyme activities and nitrogen content (N) of the endosperm of maize following imbibition: AP, aminopeptidase activity as increase in absorbance (A) 410 nm per hour following release of p-nitrophenol; CP, carboxypeptidase activity as μmoles amino N/h; CA, caseolytic activity as μmoles amino-N/h, measured at the three pH values indicated (data of Feller et al., 1978).

f. Sorghum

Glennie *et al.* (1983) have described the ultrastructural changes that occur during germination of sorghum. Modification of the endosperm commences at the endosperm–scutellum interface and moves subsequently into the starchy endosperm. The matrix protein begins to disappear first and then simultaneous degradation of the starch granules and protein bodies takes place.

Acid peptide hydrolase activity increases markedly in endosperm of germinating sorghum grain (Aisien, 1982). An acid endopeptidase has been purified from germinated sorghum seed by Garg and Virupaksha (1970a, b). This enzyme was not sensitive to sulfhydryl-group inhibitors. The pH optimum, using bovine serum albumin as substrate, was 3.6 and the MW was estimated to be 80 000 daltons. The enzyme has a narrow specificity, cleaving only the peptide bond between the α-carboxyl group of aspartate or glutamate and the amino group of the adjacent amino acid. The prolamin and glutelin protein fractions of sorghum endosperm contain large amounts of glutamic acid and aspartic acid (Virupaksha and Sastry, 1968).

g. Barley

During germination endopeptidase activity increases markedly (Mikola and Enari, 1970). This activity is attributed to the secretion by the aleurone cells of at least two sulfhydryl-sensitive endopeptidases with acid pH optima (Sundbolm and Mikola, 1972). Starchy endosperm seems devoid of both alkaline dipeptidase and aminopeptidase activity (Fig. 5). These enzymes are associated with the aleurone cells, the scutellum and the developing axis. Carboxypeptidase activity in the starchy endosperm, however, increases substantially during germination.

Fractionation of extracts of starchy endosperm by chromatography on diethylaminoethyl (DEAE)-cellulose reveals the presence of five enzymes with carboxypeptidase activity using carbobenzoxy (Z-)-dipeptides as substrates. One of these enzymes exhibits a high specificity for peptides of at least three residues and a C-terminal proline residue (Mikola and Mikola, 1980). This enzyme must play a significant role in the degradation of the barley protein reserves. In barley grains, proline is one of the most abundant amino acids, comprising about 12% of the amino acids in the total protein and about 20% in the major reserve protein, hordein (Shewry *et al.*, 1978). Degradation of these proteins could be achieved through the co-operative action of the acid endopeptidases, the carboxypeptidases that show broad specificity, and the proline-specific carboxypeptidase (Mikola and Mikola, 1980).

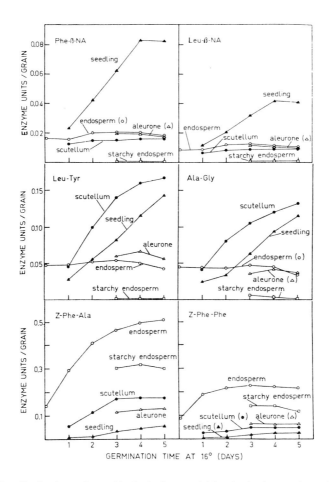

Fig. 5. The distribution of peptide hydrolase activities in barley seeds and seedlings. Aminopeptidases were measured using aminoacyl-β-naphthylamides (Phe-β-NA and Leu-β-NA), dipeptidases using Ala-Gly and Leu-Tyr, and carboxypeptidases using Z-Phe-Ala and Z-Phe-Phe as substrates (from Mikola and Kolehmainen, 1972).

4. Regulation of Peptide Hydrolases

a. Protein synthesis

The ability of gibberellic acid to increase the level of peptide hydrolase activity of barley grains has been well established. A response to gibberellic acid was observed in the wort of malted barley (Ault, 1961; Kringstad et al., 1960; Macey and Stowell, 1961), in de-embryonated barley seeds (Briggs, 1963; MacLeod et al., 1964) and in isolated aleurone tissue (MacLeod et al., 1964). The physiological basis of this phenomenon was substantially resolved by the study of Jacobsen and Varner (1967). Through the use of density labelling, they were able to demonstrate unequivocally that the gibberellic acid-induced increase in peptide hydrolase activity was due to *de novo* protein synthesis. In addition, they concluded that gibberellic acid has an additional avenue for control, whereby the hormone stimulates secretion of the peptide hydrolase from the aleurone layer (Section 4.II).

Control of the level of carboxypeptidase, however, is not clear. Mikola and Kolehmainen (1972) concluded that the aleurone layer was unlikely to synthesize and secrete the carboxypeptidases found in the starchy endosperm. They found that incubation of whole endosperms in the presence of gibberellic acid did not increase carboxypeptidase activity. More recently, Schroeder and Burger (1978) have reported that carboxypeptidase was produced by aleurone cells of germinating barley. Production of the enzyme was inhibited by 6-methylpurine, cycloheximide and cordycepin. Activity as measured against the synthetic peptides Z-Phe-Ala and Z-Phe-Phe was not due to endopeptidase activity, as is evident from the inhibition by phenylmethylsulfonyl fluoride. Gibberellic acid accelerated the rate of enzyme release from the aleurone layer, but was not essential for release, and did not influence the ultimate amount of carboxypeptidase produced.

Preston and Kruger (1979) have suggested that germinating cereals may contain two distinct systems for controlling peptide hydrolase activity. They report also that the increase in exopeptidase (presumably carboxypeptidase) observed in the endosperm of germinating wheat is largely independent of gibberellic acid. Inhibitors of RNA synthesis also had no effect on the development of exopeptidase activity. Therefore, it was concluded that the increase in activity of endosperm exopeptidase activity accompanying germination may be due to removal of endogenous inhibitors, rather than *do novo* synthesis.

b. Compartmentation

As a form of regulation, compartmentation has two aspects. Firstly, compartmentation can give rise to discrete and different locations for particular enzymes and their respective substrates at both the cellular and tissue level. Secondly, separation of enzymes and substrates is invariably associated with marked differences in the environment of these sites, particularly in cellular or organelle pH. Clearly, the two phenomena are unlikely to be independent.

Germinating barley seeds provide the best example of regulation by compartmentation (Enari and Mikola, 1977). Protein hydrolysis in the starchy endosperm is promoted by its low pH, which is usually close to pH 5 during the course of germination (Mikola and Mikola, 1980). This is close to the pH optimum for the sulfhydryl-sensitive endopeptidase and the carboxypeptidases secreted from the aleurone layer (Section II.C,3). No neutral exopeptidases are associated with the endosperm (Enari and Mikola, 1977). Mikola and Virtanen (1980) consider that the acidity of the endosperm is maintained by the active secretion of malic acid by the aleurone layer.

The pH in the endosperm also facilitates the uptake of amino acids and small peptides by the scutellum. Dipeptide uptake by germinating barley embryos has a pH optimum of about 4 (Higgins and Payne, 1977b; Sopanen, 1979). Leucine uptake has a pH optimum between 3.5 and 5.0, depending on the experimental conditions (Sopanen et al., 1980).

The endosperm, therefore, can be considered to function as a secondary lysosomal compartment. The primary lysosome is the vacuole and the role of this intracellular compartment has been documented for castor bean endosperm. Hydrolysis of the storage proteins at a pH optimum of 5.0 was demonstrated by Nishimura and Beevers (1979) and more recently Nishimura (1982) confirmed that the intravacuolar pH is in the range 5.7–5.9.

In contrast to the endosperm, the scutellum has a pH of 7–8. This pH facilitates the activity of the scutellar exopeptidases, aminopeptidases and dipeptidases, all of which have neutral to alkaline pH optima.

c. Endogenous inhibitors

Proteins that inhibit endopeptidases have been described in detail in Section 3.III of Volume 1. The physiological functions of these inhibitors are not clear. Two important functions might be: (i) protection of other cellular proteins against the marked increase in proteinase activity that occurs during germination; or, (ii) inhibition of proteinases already present in mature seed. Any regulatory role for these inhibitors in resting seed would seem unlikely, in view of the finding that the major endopeptidase stored in Scots pine seed is a pepstatin-A sensitive acid proteinase which is unaffected by the endogenous inhibitors (Salmia, 1981b).

Protection of proteins from indiscriminate proteinase attack during germination seems possible. However, in all tissues so far studied, the amount of inhibitor in the resting seed (the maximum amount) is always less than the endopeptidase activity of the germinating seed; for example, 5%–15% for barley (Mikola and Enari, 1970) and 30%–50% in the case of Scots pine (Salmia, 1980). Since these endogenous inhibitors are located in the cytoplasm (Section 3.III,B), it would seem that they could function primarily to protect cytoplasmic proteins from attack by endopeptidases released accidentally or prematurely from protein bodies. This protective role would be short-lived, however, as the amount of inhibitor usually declines rapidly once germination commences (Fig. 3; Mikola and Enari, 1970).

D. Role of Scutellum or Cotyledons in the Transfer of Nitrogen from Endosperm to Embryo

1. Scutellum

It is clear from the many studies of enzymatic hydrolysis of endosperm storage proteins that the concerted action of the endopeptidases and carboxypeptidases is most likely to result in a mixture of amino acids and small peptides. Complete hydrolysis is unlikely because of the apparent inability of the carboxypeptidases to act on dipeptides and tripeptides (Visuri *et al.*, 1969). The scutellum, an absorptive organ responsible for the transfer of nutrients between the endosperm and seedling, shows high levels of exopeptidase activity, especially aminopeptidases and alkaline dipeptidases (Mikola and Kolehmainen, 1972). This suggests that the scutellum may have a dual function: absorption of amino acids and peptides, and hydrolysis of the absorbed peptides.

The earliest study of amino acid absorption by scutellum was that of Stewart (1971). He observed that the uptake of glutamine by maize scutellum had a broad pH optimum (4.3–5.2) and was inhibited by several monovalent cations. Uptake was also inhibited by several amino acids and was sensitive to the addition of respiratory inhibitors.

Sopanen *et al.* (1980) have described the characteristics of a leucine transport system of barley scutellum. Scutella isolated from three-day old seedlings took up L-leucine from an aerated incubation medium at rates varying between 4 and 14 μmol g^{-1} h^{-1}. Uptake took place against a concentration gradient (estimated ratio concentration inside:outside was 270:1) and was strongly inhibited by dinitrophenol (Table III). The optimum pH for uptake was in the range 3.5–5 and was dependent on the buffer composition and pre-washing time. The relationship between uptake rate and L-leucine concentration was complex, with Michaelis constant, K_m, values of 3.4 mM and 15.5 mM below and above 3 mM leucine, respectively.

Development of the leucine uptake system during germination exhibited a complex pattern, with a rapid response within the first two days, constant activity for another two days, followed by a further late increase

Table III. Uptake of leucine by scutella, root tips and coleoptile slices of Pirkka barley

Organ	Uptake of leucine	
	Without DNP	+0.25 mM DNP
	μmol g^{-1} h^{-1}	
Scutella, 0 days	1.33 ± 0.09	0.38 ± 0.01
Scutella, 6 days	10.64 ± 0.30	1.49 ± 0.09
Roots	1.84 ± 0.10	0.25 ± 0.01
Coleoptiles	2.13 ± 0.22	0.44 ± 0.03

DNP, dinitrophenol.
From Sopanen et al. (1980).

(Sopanen et al., 1980). Scutella separated from seeds which had been imbibed for four hours took up leucine from a 1 mM solution at a rate of 3.7 h^{-1} per scutellum, and about 70% of the uptake was inhibited by dinitrophenol. During the course of germination, the rate of uptake increased over 10-fold to 45 nmol h^{-1} per scutellum. Addition of inhibitors of RNA and protein synthesis significantly arrested the increase of uptake activity, suggesting that the carrier protein or system was synthesized *de novo* in the scutellum. The presence of the embryo axis was not necessary for development of the uptake system and neither gibberellic acid nor its antagonist, abscisic acid, had any effect (Table IV). The endosperm thus has a strong regulatory role and this may be manifested to some extent through glutamine or other amino acids (Nyman et al., 1983).

Transfer of nitrogen from endosperm to embryo can also be facilitated by peptide uptake by the scutellum. Higgins and Payne (1981) indicate that in terms of absolute amounts of nitrogen transferred, peptides play at least as important a role as amino acids. The first studies of peptide uptake by the scutellum of barley used the peptides glycylsarcosine (Gly-Sar), Gly-Sar-Sar and Gly-Sar-Sar-Sar, which are very resistant to hydrolysis owing to N-methylation of their peptide bonds (Higgins and Payne, 1977a, b; Sopanen et al., 1977). Collectively, these studies show that peptide uptake occurs against a concentration gradient (inside > outside) by a saturable mechanism that is dependent on metabolic energy. Uptake is not affected by glycine, but is strongly inhibited by competing peptides (Sopanen et al., 1977). Thus peptide transport occurs by a separate mechanism to amino acid transport.

Peptides as large as pentaalanine are absorbed intact and undergo subsequent hydrolysis in the scutellum (Higgins and Payne, 1978a). Studies to date indicate that all peptides share a common carrier mechanism

Table IV. Effect of various treatments on development of leucine uptake activity by scutella from Pirkka barley. Measurements were made 15 h after the beginning of imbibition.

Germination conditions	Uptake of leucine	
	nmol h^{-1} scut^{-1}	% of control
Whole grain	19.5 ± 0.9	100
Endosperm removed	43.4 ± 1.5	249
Embryo axis removed	17.0 ± 1.2	84
Endosperm and axis removed	39.3 ± 1.5	224
Endosperm removed + 5 mM glutamine	22.3 ± 1.4	117
Endosperm removed + 50 mM glutamine	16.7 ± 0.7	83
Endosperm removed + 500 µg/mL GA$_3$	36.3 ± 0.1	204
Endosperm removed + 0.1 mg/mL CCC	42.8 ± 1.2	244
Endosperm removed + 10 mg/mL CCC	14.2 ± 1.2	67
Endosperm removed + 10 mg/mL CCC + 500 µg/mL GA$_3$	14.0 ± 0.2	66
Endosperm removed + 10 µM ABA	39.6 ± 2.4	224
Endosperm removed + 100 µM ABA	36.5 ± 2.1	205

ABA, abscisic acid; CCC, 2-chloroethyltrimethylammonium chloride; GA$_3$, gibberellic acid. From Sopanen et al. (1980).

(Higgins and Payne, 1978b; Sopanen et al., 1978). This is best illustrated through studies of Gly-Gly uptake (Sopanen et al., 1978). Scutella take up [^{14}C]Gly-Gly rapidly with a pH optimum of about 4.5. Uptake conforms to Michaelis–Menten kinetics with a K_m of 2.3 mM and V_{max} of 41 µmol g^{-1} h^{-1} and is strongly inhibited by dinitrophenol, cyanide and lack of oxygen. The non-hydrolysed dipeptide Gly-Sar is a competitive inhibitor of Gly-Gly uptake and 10 other dipeptides tested also inhibit Gly-Gly uptake (Table V). The carrier mechanism shows a stereochemical preference for L-amino acid residues (Higgins and Payne, 1978c). The extent to which transport is inhibited depends on the position of the D residue in the primary sequence, with D residues at the C-terminal end of the peptide having the greatest effect. Normally L-dipeptides are hydrolysed immediately upon uptake by the scutellum. However, the dipeptide L-Leu-D-Leu was accumulated against a concentration gradient, indicating that the active transport mechanism has a function independent of peptide hydrolysis within the scutellum. Inhibitors of RNA and protein synthesis block any development of the peptide transport mechanism during the course of germination. The embryo axis, gibberellic acid, abscisic acid and endosperm are without effect once imbibition has started (Sopanen, 1979). Some of the characteristics of the transporter responsible for peptide uptake have recently been determined by Walker-Smith and Payne (1983). On the basis of sensitivity

Table V. Effects of dipeptides on the uptake of Gly-Gly by barley scutella

Peptide	Uptake	Apparent K_i
	% of control value	mM
Ala-Ala	27	0.4
Val-Val	16	0.2
Leu-Leu	17	0.2
Pro-Pro	33	0.5
Glu-Glu	55	1.3
Glu-Gly	55	1.3
Ala-Gly	40	0.7
Pro-Gly	72	2.7
Lys-Gly	16	0.2
Gly-Pro	54	1.2

From Sopanen *et al.* (1978).

to various thiol-group inhibitors, these authors concluded that peptide transport into the scutellum is conducted by a redox-sensitive, dithiol-dependent protein.

2. Cotyledons

The cotyledons of castor bean and Scots pine function in an analagous manner to the scutellum of the cereals. They are the site of absorption of sucrose and nitrogen-containing compounds from the endosperm and facilitate transfer of these compounds throughout the developing axis. Robinson and Beevers (1981) have studied the uptake of amino acids by isolated cotyledons of castor bean. They selected glutamine because examination of the exudate from severed hypocotyls showed that 35%–40% of the total amino nitrogen was glutamine (Fig. 6). Glutamine was also the dominant amino acid in extracts of the cotyledons. Uptake was linear for at least 120 min and glutamine accumulated against a concentration gradient. Active transport was implied as uptake was inhibited by respiratory inhibitors and uncouplers of phosphorylation. There was no distinct pH optimum. The K_m for glutamine uptake was 11.8 mM. Though high, this value should be compared with the concentration of glutamine in the endosperm which was 15.2 mM. Other amino acids, especially those transported at high rates, also reduced glutamine uptake. The rate of glutamine uptake per cotyledon was greatest at the time of maximum export of nitrogen from the endosperm.

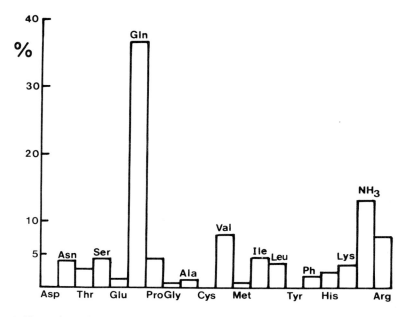

Fig. 6. The amino acid composition of the exudate from four-day old castor bean seedlings. Values are as % of total amino nitrogen (data of Robinson and Beevers, 1981).

III. MOBILIZATION OF PHOSPHORUS

The phosphorus-containing constituents of seed can be conveniently described in terms of four groups of compounds: phytic acid, nucleic acid, lipid and protein. Phytic acid (*myo*-inositol hexakis dihydrogen phosphoric acid) is always the most abundant phosphorus-containing compound. For example, in oats (*Avena sativa*) 53% of the seed phosphorus is associated with phytic acid, whereas the nucleic acid fraction accounts for 12% and the lipid and protein fractions 6.5%–8.5% and 6%, respectively (Hall and Hodges, 1966). In many species the proportion of seed phosphorus associated with phytic acid can be at least 80% (Table I of Chapter 4, Volume 1).

A detailed account of the changes in phosphorus constituents during germination of oats is summarized in Figure 7. In each group of reserve compounds a marked decline in total amount is observed, with a concomitant increase in the total phosphorus content of the developing axis. Though phytic acid is initially the dominant phosphorus reserve (53% of total),

5. *Endosperm N and P Mobilization*   183

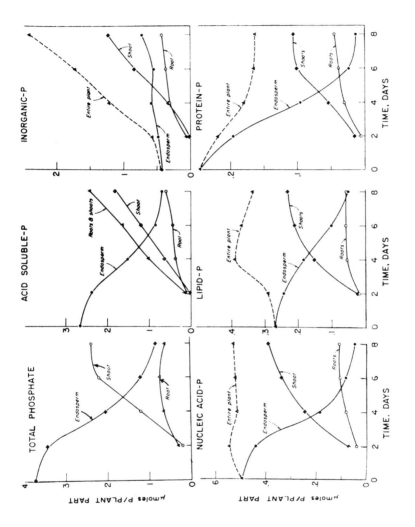

Fig. 7. Changes in the phosphorus constituents of axial and non-axial tissues of oat seedlings. Note that the 'endosperm' values are endosperm + scutellum (from Hall and Hodges, 1966).

thereafter this proportion declines, suggesting a preferential degradation of phytic acid compared to other phosphorus-containing compounds (Eastwood and Laidman, 1971).

A. Phytin

Phytin is the insoluble complex of phytic acid with a range of cations. The proportions of the associated cations vary with tissue and with species (Maga, 1982; Chapter 4 of Volume 1) and may depend on pH, temperature and ionic strength (Graf, 1983). In the phytin of rice, potassium and magnesium predominate, with smaller quantities of calcium: 18.9%, 5.3% and 0.5% respectively (Ogawa et al., 1977, 1979). In barley, the proportion of calcium is higher (Liu and Pomeranz, 1975). The proportion of phytin that remains as free acid is also known to vary. In rice aleurone grains there is an excess of cations over available phytic acid indicating very few, if any, free phosphoric acid residues (Tanaka et al., 1973). In cotton seed about 50% of the phosphoric acid groups are not associated with cations and hence the amount of phytic acid or partially associated phytin is correspondingly high (Lui and Altschul, 1967).

1. Occurrence

In general, phytates account for about 1% by weight of cereal grains, although as much as 5% was reported by de Boland et al. (1975). In barley, for example, a range of 0.97%–1.08% was observed over 18 varieties by Lolas et al. (1976). No significant difference in phytin content has been found between normal maize and high-lysine maize (O'Dell et al., 1972; de Boland et al., 1975; O'Dell and de Boland, 1976).

2. Localization

Phytin is found as globoid inclusions within protein bodies of vacuolar origin (Chapters 3 and 4 of Volume 1). In most cereals, most of the phytin is present in the protein bodies of the aleurone layer. For example, O'Dell et al. (1972) found that 85% of wheat kernel phytin is in the aleurone layer, with 13% in the embryo and only 2% in the starchy endosperm. A similar distribution is found in rice (Tanaka et al., 1974; Kennedy and Schelstraete, 1975). Maize is an exception in that almost 90% of the kernel phytate is associated with the embryo (O'Dell et al., 1972). This observation was supported by Kyle and Styles (1977), who were unable to detect phytate deposits in the protein bodies of developing aleurone cells of maize. Phytin is distributed throughout the endosperm of castor bean (Fig. 7 of Chapter 4 of Volume 1) but in lettuce (*Lactuca sativa*), another dicotyledon, 80% of the phytin is present in the cotyledons (Halmer et al., 1978).

3. Phytase

a. Function

Phytase, *myo*-inositol hexakis (dihydrogen phosphate) phosphohydrolase is a phosphomonoesterase which can hydrolyse phytin to yield inorganic orthophosphate and an initial series of lower phosphate esters of *myo*-inositol. *Myo*-inositol is the ultimate product derived from continued hydrolysis of these lower phosphate esters.

In the dephosphorylation of phytin by mung bean (*Vigna radiata*) phytase, the first phosphate liberated is from position 6, followed by those at 5 and 4, or 1 and 3 (Maiti *et al.*, 1974). The fate of position 2 phosphate was not determined. *Myo*-inositol phosphate esters with fewer than six phosphate groups do not accumulate within wheat grain during phytin hydrolysis (Matheson and Strother, 1969). This observation suggests that hydrolysis of the lower order esters is rapid and that the initial events of hydrolysis are probably rate limiting. As well as the orthophosphate released, the *myo*-inositol derived from phytin hydrolysis is likely to be of considerable importance to the developing axis. *Myo*-inositol is a precursor of all pentosyl and uronosyl sugar residues and so contributes directly to the synthesis of pectin and other polysaccharides associated with developing cell walls (Roberts *et al.*, 1968; Loewus and Loewus, 1982). It also contributes to certain lipids (Matheson and Strother, 1969).

Most of the phytases of plant origin reported so far are non-specific acid phosphomonoesterases [EC 3.1.3.2] rather than phytate-specific phosphatases (Yamagata *et al.*, 1980). Whether this biochemical distinction is relevant in relation to phytin degradation *in situ* is difficult to decide. Some distinction can be made, however, by examining phytase activity isolated from the sites of phytin storage. Acid phosphatase activity has been shown by histochemical techniques to be associated with aleurone grains of barley seed (Ashford and Jacobsen, 1974; Fig. 10 of Chapter 4). Phytin is also located in this organelle. In addition, Yamagata *et al.* (1979) were able to demonstrate that aleurone grains isolated from rice grains contained four isoenzymes of acid phosphatase. The major isoenzyme was subsequently purified and shown to be capable of degrading phytin together with the non-specific model substrate, *p*-nitrophenyl phosphate (Yamagata *et al.*, 1980). The existence of phytate-specific phosphatase activity associated with mung bean cotyledons has been demonstrated by Maiti and Biswas (1979). While this enzyme showed no activity towards substrates indicative of non-specific phosphatase action, it had very low activity towards lower order *myo*-inositol phosphates. This may suggest that in this tissue, such an enzyme is an initiator of phytin degradation and that complete hydrolysis is the function of other non-specific phosphatases. Although no distinction was made between either aleurone or endosperm as the primary location,

Ory and Henningsen (1969) reported that protein bodies isolated from ungerminated barley seeds contain an acid phosphatase with very high specificity for phytin (Table VI). This enzyme appears similar to that of the rice aleurone grain described by Yamagata et al. (1980) as it shows high activity towards phytin compared to a range of other phosphatase substrates.

b. *Biochemical characteristics*

Definitive studies of the nature and properties of seed phytases were difficult to find. The reason for this lies, perhaps, with a preoccupation of the investigators with the use of artificial substrates (e.g., *p*-nitrophenyl phosphate) during isolation and characterization. This weakness in methodology is compounded by the existence of multiple forms or possibly oligomeric series of acid phosphatases (Goel and Sharma, 1979; Yamagata et al., 1979). In order to reduce the confusion as to the identification of the enzyme(s) responsible for phytin degradation, we have only considered those studies in which the enzyme was isolated from phytin-containing storage bodies (e.g., aleurone grains). With this arbitrary distinction in mind, the study of Yamagata et al. (1980) seems to be the most appropriate. They isolated an acid phosphatase from rice aleurone grains. The enzyme is a glycoprotein of MW 70 000 daltons. It probably contains manganese and has functional sulfhydryl groups. Fluoride is a non-competitive inhibitor. The enzyme hydrolysed all inositol phosphates together with a range of other substrates. These properties are shared to varying degrees by several other phytases isolated from homogenates of whole grain or different grain tissues (Table VII).

4. *Regulation of Phytin Degradation*

In germinating maize seeds, increased phytase activity is readily detected within 24 h of imbibition (Fig. 8). Removal of the embryo abolishes this

Table VI. Acid phosphatase activity of barley grain protein bodies

Substrate	Activity measured (at pH 5.0)	
	Protein bodies	Crude extract
	$\times 10^{-4}$ μmoles Pi per min	
β-Glycerol-phosphate	0.87	11.5
Sodium phytate	58.0	8.14
Glucose-1-phosphate	0	0
Fructose-1,6-bisphosphate	0	5.2
ATP	0	31.4

From Ory and Henningsen (1969).

Table VII. Summary of biochemical properties of seed phytase

Source of enzyme	K_m	Optimum pH	Molecular weight	Inhibitors	Stabilizers	Optimum temperature	Reference
Rice							
(a) aleurone grains (dry seed)	—	4.5	—	—	—	—	Yoshida et al. (1975)
(b) aleurone grains (dry seed)	0.43 mM	—	approx. 70 000	F^- — non-competitive; L-tartrate, PO_4^{2-}, molydate, arsenate — competitive; Ag^+, Zn^{2+}, Hg^{2+}, Cu^{2+}, Fe^{2+}, EDTA	Triton X-100 lysolecithin	—	Yamagata et al. (1980)
(c) aleurone grains (dry seed)	—	4.5	—	—	—	45°C	Tanaka et al. (1973)
Triticale partially purified whole kernel extract (dry seed)	0.22 mM	5.4	—	p-CMB; Ag^+, Cu^{2+}, Fe^{2+}, Ni^{2+}, Co^{2+}	—	45°C	Singh and Sedetti (1979)
Maize endosperm and scutellum extract (4 days germination)	9 mM	5.6	—	F^-	—	50°C	Chang (1967)
Wheat wheat bran F_1 enzyme fraction	0.02 mM	5.6	47 000	PO_4^{2-} — competitive	lysolecithin	—	Lim and Tate (1973)
Mung Bean cotyledon (3 days germination)	0.65 mM	7.5	160 000	F^- — non-competitive; PO_4^{2-} — competitive, EDTA, citrate, mercaptoethanol, p-CMB	—	57°C	Mandal et al. (1972); Maiti and Biswas (1979)

increase in phytase activity (Eastwood and Laidman, 1971). This observation gave rise to a further series of experiments, all designed to identify the nature of the influence of the embryo. Applications of exogenous gibberellic acid, kinetin, auxin or various combinations of these hormones could not restore the influence of the embryo and increase *in vitro* phytase activity in embryo-free seeds (Eastwood and Laidman, 1971). Gibberellic acid, however, could partly substitute for the embryo in stimulating the release of the associative cations, potassium, calcium and magnesium, presumably due to its stimulation of phytin degradation *in situ*. An alternative explanation is that the level of free phytic acid increases as the associative cations are removed. *In vitro* phytase activity could be increased through the addition of glutamine to embryo-free grains, but added glutamine had little effect on *in vitro* phytase activity of isolated aleurone layers.

The role of glutamine *in vivo* is unclear. Gibberellic acid treatment of embryo-free seeds induces the release of glutamine into the endosperm. However, this glutamine is unable to induce increased phytase activity. The problem is not one of inadequate availability of glutamine, since Chittenden (1976) has shown that only 40 nmol of glutamine per aleurone layer is required to induce phytase activity in embryo-free grain. Dry grain already contains 62.5 nmol of glutamine per grain and after 24 hours of germination this approaches 100 nmol. Whether the glutamine released *in situ* has access to the same intracellular sites as the glutamine added *in vitro* is an important distinction which has not been resolved.

Eastwood and Laidman (1971) concluded from their studies that gibberellic acid was able to increase the rate of phytin degradation *in vivo* through its promotion of the rate of secretion of phosphate, a strong inhibitor of phytase (Table VII). This conclusion is supported by the studies of Jones (1973), Clutterbuck and Briggs (1974) and more recently, Katayama and Suzuki (1980). Indirect evidence which supports the role of phosphate in controlling phytin degradation is the report of Baset and Sutcliffe (1975). They observed a reduction in phytin mobilization as the exogenous supply of phosphorus to germinating oat seeds was increased.

In contrast to its apparent lack of direct effect on phytase activity, gibberellic acid has been shown to increase the amount of acid phosphatase in wheat half-seeds (Akiyama and Suzuki, 1980; 1981), barley aleurone cells (Ashford and Jacobsen, 1974) and embryo-free wheat seeds (Eastwood and Laidman, 1971). The increase in activity of acid phosphatase is apparently due to *de novo* synthesis (Akiyama *et al.*, 1981), but gibberellic acid also affects secretion of this enzyme(s) across the plasmalemma — a phenomenon which is well characterized for α-amylase (Moll and Jones, 1982). In the case of the embryo-free wheat seeds studied by Eastwood and Laidman (1971) the increase in phosphatase was not matched by increased phytase activity.

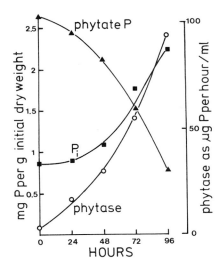

Fig. 8. Changes in the content of phytin, inorganic phosphate (Pi) and phytase activity of the endosperm + scutellum of maize seedlings (data of Chang, 1967).

The nature of the mechanisms controlling the induction of increased phytase activity observed with intact seeds is not clear. Kuvaeva and Kretovich (1978) separated two phytases from germinating pea cotyledons, one of which was also found in dry seed. The increase in cotyledon phytase associated with germination was due entirely to the enzyme not found in dry seed. Synthesis of this enzyme was inhibited by cycloheximide, and seeds imbibed in the presence of [^{14}C]-amino acids showed the label associated with the phytase formed during germination. That control of phytase synthesis is by inorganic phosphate is suggested by the work of Bianchetti and Satirana (1967). They observed that orthophosphate, added early during germination, was able to repress phytase synthesis. They conclude that orthophosphate exerts control at the transcription stage. Alternatively, the increased phytase activity may be due to activation of existing enzyme or zymogens (Eastwood *et al.*, 1969). Induction of aleurone cell phytase by glutamine and hydroxylamine was found to be unaffected by inhibitors of protein and RNA synthesis. Furthermore, incubation of cell-free extracts of aleurone cells with glutamine and hydroxylamine caused an increase in phytase activity. It is disappointing that this work was not continued and the mechanism of induction elucidated.

In summary, we conclude that phytase plays an important role in phytin degradation. However, it is apparent that *in vitro* phytase activity does not reflect the *in vivo* rate of phytin degradation (Eastwood and Laidman, 1971). Dry seed contains more than enough phytase activity to account for the rate of phytin breakdown observed during germination. The four-fold increase in phytase during germination seems superfluous. In addition, phytase activity continues to increase and reaches a maximum sometime after the phytin reserves have been depleted (Mukherji *et al.*, 1971). The lack of a close correlation between changes in the *in vitro* phytase activity and *in vivo* phytin degradation has also been well documented by Sutcliffe and Baset (1973) in the endosperm of germinating oat seeds. However, in the same tissue these authors observed a very close correlation between *in vivo* protein degradation and *in vitro* peptide hydrolase activity and *in vivo* starch degradation and *in vitro* α-amylase activity. Rather, phytin degradation appears to be modulated *in vivo* by the effect of inorganic phosphate — the product of phytase activity and an inhibitor of phytase. In this instance, gibberellic acid exerts control of the phosphate and associated cation released, primarily through the induction of secretory mechanisms across the plasmalemma (Section 4.II,D).

B. Nucleic Acids

1. Change in Nucleic Acid Content of Seed during Germination
Hall and Hodges (1966) reported that 12% of the phosphorus in oat seed is present in the form of nucleic acid. During germination the nucleic acid phosphorus of the endosperm declines with a concomitant increase in amount of phosphorus of the developing axis (Fig. 7). These data show that the total nucleic acid phosphorus of the seedling as a whole does not change during the first eight days of germination. One interpretation of this observation is that synthesis of the nucleic acids of the axis is entirely dependent upon the pool of nucleic acid in the rest of the seedling. This is not always the case, as the study of Ingle and Hageman (1965) clearly demonstrates: the nucleic acid content of the whole maize seedling increases dramatically during the first 121 h of germination (Table VIII). Examination of the absolute changes in nucleic acid of the non-axis tissue indicates that this increase must result from *de novo* synthesis.

2. Degradation of Nucleic Acids: Release of Phosphorus
Phosphate release from RNA requires the concerted action of RNase and a 3′-nucleotidase [EC 3.1.3.32]. Endosperm RNase activity increased to a maximum 1–2 days after germination of oat seed (Sutcliffe and Baset, 1973), and thereafter remained essentially constant while the major portion of RNA was being degraded. The difficulty of relating *in vitro* RNase activity with the content of RNA *in situ* is even more apparent in the developing axis. In this organ, total RNA content increases, but so does

Table VIII. The soluble nucleotide, RNA and DNA content of the axis, scutellum, endosperm and whole maize seedling during a 5-day period of germination (µg per plant part)

	Age of seedling (h)					
	4	23	48	71	95	121
	Axis					
Soluble nucleotide	18.2	24.3	97.1	206.0	413.0	585.0
RNA	63.3	66.6	142.5	323.0	485.0	710.0
DNA	6.3	6.4	17.2	39.6	63.5	106.5
	Scutellum					
Soluble nucleotide	53.0	74.0	114.0	138.0	141.0	138.0
RNA	105.0	95.0	117.5	136.3	146.0	114.0
DNA	12.7	11.4	15.0	15.3	15.1	15.8
	Endosperm					
Soluble nucleotide	92.2	107.0	107.0	116.5	121.5	100.0
RNA	47.5	25.4	28.5	31.7	25.4	25.4
DNA	21.4	21.0	14.0	11.9	11.9	9.1
	Whole seedlings[a]					
Soluble nucleotide	163.4	205.3	318.1	460.5	675.5	823.0
RNA	215.8	187.0	288.5	491.0	656.4	849.4
DNA	40.4	38.8	46.2	66.8	90.5	131.4

[a] Obtained by summation of the individual parts.
Data of Ingle and Hageman (1965).

RNase activity (Ingle and Hageman, 1965). Ingle and Hageman (1965) also observed a marked increase in 3'-nucleotidase activity in endosperm and scutellum of corn during germination. This enzyme had previously been shown to increase in germinating wheat seeds (Shuster and Gifford, 1962).

Wilson (1971) considers that the enzyme designated RNase I is the major enzyme responsible for the degradation of maize endosperm RNA. RNase I (ribonucleate nucleotido-2'-transferase (cyclizing); [EC 3.1.4.22]) is a soluble enzyme with a pH optimum of about 5 and produces 2':3'-cyclic nucleotides. This enzyme had also been isolated by Wilson (1967), shown to have a MW of 23 000 daltons and to be inhibited by dithiothreitol.

3. Regulation of Synthesis of RNase

Chrispeels and Varner (1967) were the first to show that the formation of RNase by barley aleurone layer is enhanced by gibberellic acid. The effect of gibberellic acid was inhibited by inhibitors of protein and RNA synthesis. During the first 24 h of induction the RNase was retained by the aleurone cells, thereafter, synthesis and release were coincident (Fig. 4 of Chapter 4). The role of the aleurone cells in undertaking *de novo* synthesis of the hydrolases responsible for degradation of the endosperm reserves seems to be well established (Chapter 4). This concept, however, has recently been questioned by Okamoto et al. (1980). These authors maintain that the initial site of formation of RNase is the epithelium of the scutellum. They do not question the ability of the aleurone cells to synthesize RNase, only the sequence of events — the scutellum epithelium producing enzyme before the aleurone cells. More recently, Glennie et al. (1983) have presented ultrastructural evidence which shows that modification of the endosperm protein bodies and starch granules of sorghum starchy endosperm begins at the scutellum–endosperm interface. Such evidence is consistent with the concept of synthesis and secretion of hydrolytic enzymes by the scutellum.

C. Phospholipids

1. Occurrence

About 80% of the phospholipid of whole wheat grains is associated with the starch granules (Morrison, 1979). Meredith et al. (1978) found that the starch lipid content increases as starch granule size increases and is apparently related to the surface area of the large granules (A-type, diameter > 20 μm), and the volume of the smaller granules (B-type, diameter $< 4\mu$m). These differences in size are illustrated in Figure 1 of Chapter 4. Starch-bound phospholipid is in the form of lysophosphatidylcholine (LPC), and it is thought the LPC occurs as an inclusion complex within the amylose helix (Baisted, 1981). Degradation of amylose is not affected by the inclusion of LPC (Fujikura and Baisted, 1983).

2. Degradation of Starch-bound Lysophosphatidylcholine

During germination of barley seed, the starch-bound LPC is released and amylase activity increases (Fig. 9). Between day four and day six the starch-bound LPC falls from about 0.4 to 0.1 μmol per seed, while the unassociated LPC in the seed remains unchanged at about 0.02–0.04 μmol per seed (Baisted, 1981). Endosperm tissue contains a soluble and a particulate form of the enzyme lysophospholipase (LPL, EC 3.1.1.5). The soluble enzyme activity peaks after day three and then declines rapidly (Fig. 10). However, starch-bound LPC is not released until after day four, which suggests that the soluble LPL activity has no role in the release of starch-bound LPC. The particulate-LPL activity increases five-fold between day four and five and this coincides remarkably well with the loss of starch-bound LPC.

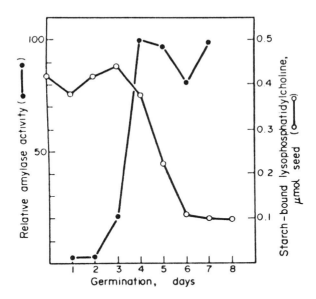

Fig. 9. Loss of starch-bound lysophosphatidylcholine compared with the increase in amylase activity of germinating barley grains (from Baisted and Stroud, 1982a).

3. Regulation of Lysophospholipase

Baisted and Stroud (1982b) have compared the development of acidic LPL in aleurone and starchy endosperm fractions of embryo-containing and embryo-free half seeds of barley (Table IX). Most of the LPL activity was in the embryo-containing half seed and its activity was almost entirely associated with the aleurone layer. Exposure of embryo-free seeds to 1 μM gibberellic acid caused enhancement of acidic and alkaline LPL activities in the aleurone cells and the bathing medium. Gibberellic acid had little effect on LPL activity in the endosperm.

IV. CONCEPTUAL MODEL OF NITROGEN AND PHOSPHORUS MOBILIZATION

Figure 11 summarizes the essential features of mobilization of nitrogen and phosphorus from the endosperm during seed germination. The model incorporates the important inter-relation between the embryo, aleurone layer, starchy endosperm and scutellum. Each has a specific function with

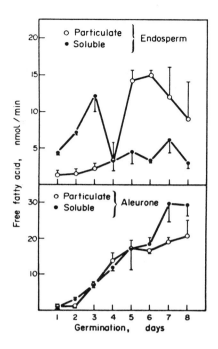

Fig. 10. Changes in the particulate and soluble lysophospholipase activities in the aleurone and starchy endosperm of germinating barley grains (from Baisted and Stroud, 1982b).

Table IX. Lysophospholipase (pH 5) activities of fractions from embryo-containing (+) and embryo-free (−) barley half-seeds

Time of imbibition (days)	Amount of palmitate released (nmol min^{-1} per total fraction)							
	Aleurone				Starchy endosperm			
	Soluble		Particulate		Soluble		Particulate	
	+	−	+	−	+	−	+	−
4	49.5	6.2	22.2	1.8	11.1	<0.2	4.5	<0.2
6	42.2	9.2	27.1	3.1	23.8	2.3	11.7	0.4

Data of Baisted and Stroud (1982b).

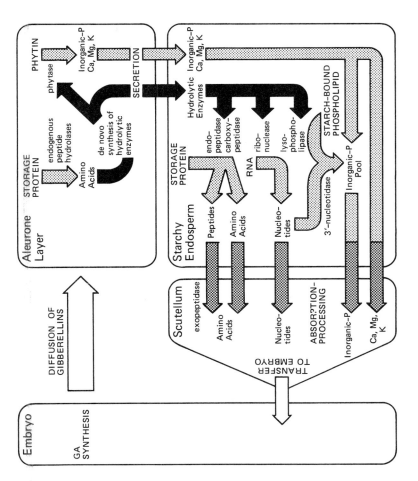

Fig. 11. Conceptual model of nitrogen and phosphorus mobilization.

respect to mobilization, transport, further processing and transfer to the growing seedling.

The sequence of events important to mobilization is as follows. Firstly, imbibition is closely followed or coincident with an increase in the *in situ* activity of the latent peptide hydrolases. The endopeptidase appears to be a pepstatin-A sensitive acid proteinase and is associated with the aleurone layer protein bodies. The characteristics of any of the exopeptidases are not so clear, although carboxypeptidase activity is also detectable at this time. Unfortunately, we know little of the mechanism whereby these enzymes, which are most likely located within the same organelle as their protein substrates, are kept inactive in the resting seed but become active as a consequence of imbibition. It seems that the amino acids for the next phase of germination — the gibberellic acid-induced synthesis and secretion of a range of hydrolytic enzymes, arise as a result of the activity of those enzymes detected in the resting seed.

Apart from phytase, which degrades the aleurone layer phytin, the other hydrolytic enzymes must be secreted into the endosperm. These enzymes are all acid peptide hydrolases, a property consistent with the low pH of the endosperm. Through their concerted action, amino acids and small peptides are produced. Degradation of RNA and LPC occurs also by action of enzymes synthesized and secreted by the aleurone layer. The scutellum (or the cotyledons of species like castor bean and Scots pine) absorbs the products of these degradative reactions. Absorption is rapid and the measured *in vitro* rates appear more than adequate to cope with the *in situ* flux of nitrogen from the endosperm. The scutellum also functions as a secondary processing tissue, a function well defined for nitrogen. This tissue is a rich source of aminopeptidase and alkaline dipeptidase activity, the enzymes most likely to be responsible for hydrolysis of those small peptides absorbed from the endosperm.

REFERENCES

Abe, M., Arai, S., and Fujimaki, M. (1977). *Agr. Biol. Chem.* **41**, 893-899.
Aisien, A. O. (1982). *J. Sci. Food. Agr.* **33**, 754-759.
Akiyama, T., and Suzuki, H. (1980). *Biochem. Physiol. Pflanzen.* **175**, 436-446.
Akiyama, T., and Suzuki, H. (1981). *Plant Sci. Lett.* **21**, 137-143.
Akiyama, T., Uchimiya, H., and Suzuki, H. (1981). *Plant, Cell Physiol.* **22**, 1029-1034.
Alpi, A., and Beevers, H. (1981). *Plant Physiol.* **68**, 851-853.
Ashford, A. E., and Jacobsen, J. V. (1974). *Planta* **120**, 81-105.
Ashton, F. M. (1976). *Annu. Rev. Plant Physiol.* **27**, 95-117.
Ault, R. G. (1961), *J. Inst. Brew.* **67**, 405-416.
Baisted, D. J. (1981). *Phytochemistry* **20**, 985-988.
Baisted, D. J., and Stroud, F. (1982a). *Phytochemistry* **21**, 29-31.

5. Endosperm N and P Mobilization 197

Baisted, D. J., and Stroud, F. (1982b). *Phytochemistry* **21**, 2619–2623.
Barrett, A. J. (1977). In 'Research Monographs in Cell and Tissue Physiology' (J. T. Dingle, ed.) Vol. 2, pp. 1–55. Elsevier/North-Holland Biomedical Press, Amsterdam.
Baset, Q. A., and Sutcliff, J. F. (1975). *Ann. Bot.* **39**, 31–41.
Bewley, J. D., and Black, M. (1978). 'Physiology and Biochemistry of Seeds in Relation to Germination' Vol. I. Springer-Verlag, Berlin.
Bianchetti, R., and Satirana, M. L. (1967). *Biochim. Biophys. Acta* **145**, 485–490.
Briggs, D. E. (1963). *J. Inst. Brew.* **69**, 13–19.
Chang, W. C. (1967). *Cereal Chem.* **44**, 129–142.
Chittenden, C. G. (1976). 'Amino Acid Pools in Germinating Wheat Grains'. Ph.D. Thesis, University of Wales, U.K.
Chrispeels, M. L., and Varner, J. E. (1967). *Plant Physiol.* **42**, 398–406.
Clutterbuck, V. J., and Briggs, D. E. (1974). *Phytochemistry* **13**, 45–54.
de Boland, A. R., Garner, G. B., and O'Dell, B. L. (1975). *J. Agr. Food Chem.* **23**, 1186–1189.
Doi, E., Komori, T., Matoba, T., and Morita, Y. (1980a). *Agr. Biol. Chem.* **44**, 77–83.
Doi, E., Shibata, D., Matoba, T., and Yonezawa, D. (1980b). *Agr. Biol. Chem.* **44**, 435–436.
Eastwood, D., and Laidman, D. L. (1971). *Phytochemistry* **10**, 1275–1284.
Eastwood, D., Tavener, R. J. A., and Laidman, D. L. (1969), *Biochem. J.* **113**, 32–33.
Enari, T.-M., and Mikola, J. (1977). In 'Ciba Foundation Symposium: Peptide Transport and Hydrolysis', pp. 335–352. Elsevier/North-Holland, Amsterdam.
Feller, U., Soong, T.-S.T., and Hageman, R. H. (1978). *Planta* **140**, 155–162.
Folkes, B. F., and Yemm, E. W. (1958). *New Phytol.* **57**, 106–131.
Fujikura, Y., and Baisted, D. (1983). *Phytochemistry* **22**, 865–868.
Garg, G. K., and Virupaksha, T. K. (1970a). *Eur. J. Biochem.* **17**, 4–12.
Garg, G. K., and Virupaksha, T. K. (1970b). *Eur. J. Biochem.* **17**, 13–18.
Glennie, C. W., Harris, J., and Leidenberg, N. V. D. W. (1983). *Cereal Chem.* **60**, 27–31.
Goel, M., and Sharma, C. B. (1979). *Phytochemistry* **18**, 1939–1942.
Graf, E. (1983). *J. Agr. Food Chem.* **31**, 851–855.
Hall, J. R., and Hodges, T. K. (1966). *Plant Physiol.* **41**, 1459–1464.
Halmer, P., Bewley, J. D., and Thorpe, T. A. (1978). *Planta* **139**, 1–8.
Hanford, J. (1967). *Cereal Chem.* **44**, 499–511.
Harvey, B. M. R., and Oaks, A. (1974a). *Plant Physiol.* **53**, 449–452.
Harvey, B. M. R., and Oaks, A. (1974b). *Plant Physiol.* **53**, 453–457.
Higgins, C. F., and Payne, J. W. (1977a). *Planta* **134**, 205–206.
Higgins, C. F., and Payne, J. W. (1977b). *Planta* **136**, 71–76.
Higgins, C. F., and Payne, J. W. (1978a). *Planta* **138**, 211–215.
Higgins, C. F., and Payne, J. W. (1978b). *Planta* **138**, 217–221.
Higgins, C. F., and Payne, J. W. (1978c). *Planta* **142**, 299–305.
Higgins, C. F., and Payne, J. W. (1981). *Plant Physiol.* **67**, 785–792.
Hwang, P., and Bushuk, W. (1973). *Cereal Chem.* **50**, 147–160.
Ingle, J., and Hageman, R. H. (1965). *Plant Physiol.* **40**, 48–53.
Ingle, J., Beevers, L., and Hageman, R. H. (1964). *Plant Physiol.* **39**, 735–740.
Jacobsen, J. V., and Varner, J. E. (1967). *Plant Physiol.* **42**, 1596–1600.
Jones, R. L. (1973). *Plant Physiol.* **52**, 303–308.
Katayama, N., and Suzuki, H. (1980). *Plant, Cell Physiol.* **21**, 115–123.
Kennedy, B. M., and Schelstraete, M. (1975). *Cereal Chem.* **52**, 173–182.
Kringstad, H., Busengdal, H., and Rasch, S. (1960). *J. Inst. Brew.* **66**, 477–480.
Kruger, J. E., and Preston, K. R. (1977). *Cereal Chem.* **54**, 167–174.

Kruger, J. E., and Preston, K. R. (1978). *Cereal Chem.* **55**, 360-372.
Kuvaeva, E. B., and Kretovich, V. L. (1978). *Sov. Pl. Physiol.* **25**, 290-295.
Kyle, D. J., and Styles, E. D. (1977). *Planta* **137**, 185-193.
Lim, P. E., and Tate, M. E. (1973). *Biochim. Biophys. Acta* **302**, 316-328.
Liu, D. J., and Pomeranz, Y. (1975). *Cereal Chem.* **52**, 620-629.
Loewus, M. W., and Loewus, F. A. (1982). *Plant Physiol.* **70**, 765-770.
Lolas, G. M., Palamidas, N., and Markakis, P. (1976). *Cereal Chem.* **53**, 867-871.
Lui, N. S. T., and Altschul, A. M. (1967). *Arch. Biochem. Biophys.* **121**, 678-684.
Macey, A., and Stowell, K. C. (1961). *J. Inst. Brew.* **67**, 396-404.
MacLeod, A. M., Duffus, J. H., and Johns, C. S. (1964). *J. Inst. Brew.* **70**, 521-528.
Maga, J. A. (1982). *J. Agr. Food Chem.* **30**, 1-9.
Maiti, I. B., and Biswas, B. B. (1979). *Phytochemistry* **18**, 316-318.
Maiti, I. B., Majumder, A. L., and Biswas, B. B. (1974). *Phytochemistry* **13**, 1047-1051.
Mandal, N. C., Burman, S., and Biswas, B. B. (1972). *Phytochemistry* **11**, 495-502.
Matheson, N. K., and Strother, S. (1969). *Phytochemistry* **8**, 1349-1356.
Meredith, P., Dengate, H. N., and Morrison, W. R. (1978). *Stärke* **30**, 119-125.
Metivier, J., and Monteiro, A. M. (1981). *J. Exp. Bot.* **32**, 1321-1332.
Mikola, J. (1983). In 'Seed Proteins' (J. Daussant, J. Mossé and J. G. Vaughan, eds), pp. 35-52. Phytochemical Society of Europe Symposia No. 20, Academic Press, London.
Mikola, J., and Enari, T.-M. (1970). *J. Inst. Brew.* **76**, 182-188.
Mikola, J., and Kolehmainen, L. (1972). *Planta* **104**, 167-177.
Mikola, L., and Mikola, J. (1980). *Planta* **149**, 149-154.
Mikola, J., and Pietilä, K. (1972). *Phytochemistry* **11**, 2977-2980.
Mikola, J., and Virtanen, M. (1980). *Plant Physiol.* **66**, S783.
Moll, B. A., and Jones, R. L. (1982). *Plant Physiol.* **70**, 1149-1155.
Morrison, W. R. (1979). In 'Recent Advances in the Biochemistry of Cereals' (D. L. Laidman and R. G. Wyn Jones, eds), pp. 313-335. Academic Press, London.
Mukherji, S., Dey, B., Paul, A. K., and Sircar, S. M. (1971). *Physiol. Plantarum* **25**, 94-97.
Nishimura, M. (1982). *Plant Physiol.* **70**, 742-744.
Nishimura, M., and Beevers, H. (1978). *Plant Physiol.* **62**, 44-48.
Nishimura, M., and Beevers, H. (1979). *Nature* **277**, 412-413.
Nyman, S., Sopanen, T., and Mikola, J. (1983). *Plant Physiol.* **73**, 135-141.
O'Dell, B. L., and de Boland, A. (1976). *J. Agr. Food Chem.* **24**, 804-808.
O'Dell, B. L., de Boland, A. R., and Koirtyohann, S. R. (1972). *J. Agr. Food Chem.* **20**, 718-721.
Ogawa, M., Tanaka, K., and Kasai, Z. (1977). *Cereal Chem.* **54**, 1029-1034.
Ogawa, M., Tanaka, K., and Kasai, Z. (1979). *Plant, Cell Physiol.* **20**, 19-27.
Okamoto, K., Kitano, H., and Akazawa, T. (1980). *Plant, Cell Physiol.* **21**, 201-204.
Ory, R. L., and Henningsen, K. W. (1969). *Plant Physiol.* **44**, 1488-1498.
Preston, K., and Kruger, J. (1976a). *Can. J. Plant Sci.* **56**, 217-223.
Preston, K. R., and Kruger, J. E. (1976b). *Plant Physiol.* **58**, 516-520.
Preston, K. R., and Kruger, J. E. (1979). *Plant Physiol.* **64**, 450-454.
Preston, K. R., Dexter, J. E., and Kruger, J. E. (1978). *Cereal Chem.* **55**, 877-888.
Ragster, L. E., and Chrispeels, M. J. (1981). *Plant Physiol.* **67**, 104-109.
Redman, D. G. (1971). *J. Sci. Food. Agr.* **22**, 75-78.
Roberts, R. M., Deshusses, J., and Loewus, F. (1968). *Plant Physiol.* **43**, 979-989.
Robinson, S. P., and Beevers, H. (1981). *Plant Physiol.* **68**, 560-566.
Salmia, M. A. (1980). *Physiol. Plant.* **48**, 266-270.
Salmia, M. A. (1981a). *Physiol. Plant.* **51**, 253-258.

Salmia, M. A. (1981b). *Physiol. Plant.* **53**, 39-47.
Salmia, M. A., and Mikola, J. J. (1975). *Physiol. Plant.* **33**, 261-265.
Salmia, M. A., and Mikola, J. J. (1976a). *Physiol. Plant.* **36**, 388-392.
Salmia, M. A., and Mikola, J. J. (1976b). *Physiol. Plant.* **38**, 73-77.
Salmia, M. A., and Mikola, J. J. (1980). *Physiol. Plant.* **48**, 126-130.
Salmia, M. A., Nyman, S. A., and Mikola, J. J. (1978). *Physiol. Plant.* **42**, 252-256.
Schroeder, R. L., and Burger, W. C. (1978). *Plant Physiol.* **62**, 458-462.
Shewry, P. R., Hill, J. M., Pratt, H. M., Leggatt, M. M., and Miflin, B. J. (1978). *J. Exp. Bot.* **29**, 677-692.
Shuster, L., and Gifford, R. H. (1962). *Arch. Biochem. Biophys.* **96**, 534-540.
Singh, B., and Sedetti, H. G. (1979). *Cereal Chem.* **56**, 267-272.
Sopanen, T. (1979). *Plant Physiol.* **64**, 570-574.
Sopanen, T., Burston, D., and Matthews, D. M. (1977). *FEBS Lett.* **79**, 4-7.
Sopanen, T., Burston, D., Taylor, E., and Matthews, D. M. (1978). *Plant Physiol.* **61**, 630-633.
Sopanen, T., Uuskallio, M., Nyman, S., and Mikola, J. (1980). *Plant Physiol.* **65**, 249-253.
Stewart, C. R. (1971). *Plant Physiol.* **47**, 157-161.
Sundbolm, N.-O., and Mikola, J. (1972). *Physiol. Plant.* **27**, 281-284.
Sutcliffe, J. F., and Baset, Q. A. (1973). *Plant Sci. Lett.* **1**, 15-20.
Tanaka, K., Yoshida, T., Asada, K., and Kasai, Z. (1973). *Arch. Biochem. Biophys.* **155**, 136-143.
Tanaka, K., Yoshida, T., and Kasai, Z. (1974). *Plant, Cell Physiol.* **15**, 147-151.
Thomas, H., and Huffaker, R. C. (1981). *Plant Sci. Lett.* **20**, 251-262.
Tully, R. E., and Beevers, H. (1978). *Plant Physiol.* **62**, 746-750.
Umezawa, H., and Aoyagi, T. (1977). *In* 'Proteinases in Mammalian Cell and Tissues' (A. J. Barrett, ed.), pp. 637-662. Elsevier/North-Holland, Amsterdam.
Virupaksha, T. K., and Sastry, L. V. S. (1968). *J. Agr. Food Chem.* **16**, 199-203.
Visuri, K., Mikola, J., and Enari, T.-M. (1969). *Eur. J. Biochem.* **7**, 193-199.
Walker-Smith, D. J., and Payne, T. W. (1983). *FEBS Lett.* **160**, 25-30.
Wall, J. S. (1979). *In* 'Recent Advances in the Biochemistry of Cereals' (D. L. Laidman and R. G. Wyn Jones, eds), pp. 275-311. Academic Press, London.
Waters, S. P., and Dalling, M. J. (1979). *Aust. J. Plant Physiol.* **6**, 595-606.
Wilson, C. M. (1967). *J. Biol. Chem.* **242**, 2260-2263.
Wilson, C. M. (1971). *Plant Physiol.* **48**, 64-68.
Yamagata, H., Tanaka, K., and Kasai, Z. (1979). *Agr. Biol. Chem.* **43**, 2059-2066.
Yamagata, H., Tanaka, K., and Kasai, Z. (1980). *Plant, Cell Physiol.* **21**, 1449-1460.
Yoshida, T., Tanaka, K., and Kasai, Z. (1975). *Agr. Biol. Chem.* **39**, 289-290.

CHAPTER **6**

Mobilization of Oil and Wax Reserves

RICHARD N. TRELEASE and DIANE C. DOMAN

I.	Introduction	202
II.	Oil and Wax Composition of Mature Seeds	202
	A. Structure and Fatty Acid Composition of Triacylglycerols	204
	B. Structure and Composition of Wax Esters	204
III.	Developmental Patterns of Glyoxysomes and Their Enzymes	207
	A. The Period of Lipid Degradation	207
	B. Metabolic Pathways: An Overview	210
	C. Glyoxysomes	211
IV.	Lipolysis of Triacylglycerols	217
	A. General Considerations	217
	B. Time Courses	218
	C. Subcellular Compartmentation of Lipases	218
	D. Fate of Storage Lipid Bodies	222
V.	Lipolysis of Wax Esters	223
VI.	β-Oxidation of Fatty Acids	224
	A. Enzyme Sequence and Compartmentation	224
	B. Time Courses	228
VII.	Metabolism of Acetyl Coenzyme A	228
	A. Glyoxylate Cycle	228
	B. Gluconeogenesis from Oxaloacetate	231
	C. Experiments with Radioactive Precursors	234
	D. Fate of Products	235
VIII.	Control of Lipid Mobilization	237
	A. General Considerations	237
	B. Effect of the Axis	238
	C. Control by the Availability of Oxygen	239
IX.	Summary and Perspectives	240
	References	242

I. INTRODUCTION

Plant species that store lipids in their seeds are more numerous than those that store mainly carbohydrates. About 40 different species provide seeds that can be harvested for commercial oil production. However, only nine species account for about 90% of total oil production. These are soybean (*Glycine max*), sunflower (*Helianthus annuus*), peanut (*Arachis hypogaea*), cotton (*Gossypium hirsutum*), rape (*Brassica napus*), coconut (*Cocos nucifera*), palm (*Elaeis guineensis*), linseed (*Linum usitatissimum*) and castor bean (*Ricinus communis*) (Duffus and Slaughter, 1980). Safflower (*Carthamus tinctorius*), maize (embryo) (*Zea mays*), olive seed (*Olea europaea*), sesame (*Sesamum indicum*) and tung (*Aleurites fordii*) also provide useful oils, but these are produced on a much smaller scale.

The impetus to find seed oils with characteristics suitable for a variety of human needs has resulted in considerable information on the oil content of seeds and on the fatty acid composition of the triacylglycerols (formerly known as triglycerides) that comprise the main category of seed reserve lipid (Earle *et al.*, 1959; Earle and Jones, 1962; Hilditch and Williams, 1964; Butt and Beevers, 1966; Hitchcock and Nichols, 1971; Barclay and Earle, 1974). The biosynthetic pathways leading to the accumulation of lipids during seed development are discussed in Chapter 6 of Volume 1. Various aspects of their metabolism following seed germination have been considered by Butt and Beevers (1966), Ching (1972), Mayer and Shain (1974), Galliard and Mercer (1975), Bewley and Black (1978), Mazliak *et al.* (1980), Stumpf (1980) and Mayer and Poljakoff-Mayber (1982). Relationships between lipid mobilization and cellular metabolism are discussed by Beevers (1975, 1979, 1980), Appleqvist (1975), Galliard (1975, 1980), Gurr (1980), Cioni *et al.* (1981), Huang (1984) and Huang *et al.* (1983). We intend to provide a comprehensive account of oil and wax mobilization that integrates the anatomical, physiological and biochemical information presently available.

II. OIL AND WAX COMPOSITION OF MATURE SEEDS

In Table I plants are ranked in order of their seed oil content (or wax content in the case of jojoba, see Section II.B). It is apparent from Table I that although different parts of the seed store lipid, depending on species, cotyledons are the predominant storage organ. There is considerable variation in lipid content, extending to almost 80% in the Queensland nut, *Macadamia ternifolia*. The range of values given for each species may partly reflect different methods of extraction and analysis, but seasonal, geographic and genetic factors and even the position of the fruit bearing the

Table I. Lipid and protein content of selected oil seeds

Species		Major storage organ	Lipid[a]	Protein[a]
Queensland nut	(*Macadamia ternifolia*)	cotyledon	75–79	9
Brazil nut	(*Bertholletia excelsa*)	radicle/hypocotyl	65–70	17–18
Coconut	(*Cocos nucifera*)	endosperm	63–72	18–21
Hazel nut	(*Corylus avellana*)	cotyledon	60–68	18–20
Pecan	(*Carya illinoensis*)	cotyledon	65	12
Jojoba	(*Simmondsia chinensis*)	cotyledon	41–57	20–38
Castor bean	(*Ricinus communis*)	endosperm	40–64	30–35
Sesame	(*Sesamum indicum*)	cotyledon	44–54	20–25
Watermelon	(*Citrullus vulgaris*)	cotyledon	46–52	38
Crambe	(*Crambe abyssinica*)	cotyledon	50	30
Oil palm	(*Elaeis guineensis*)	endosperm	48–50[b]	9–19[b]
Peanut	(*Arachis hypogaea*)	cotyledon	38–50	20–31
Pine	(*Pinus pinea*)	megagametophyte	45–48	34–35
Pumpkin	(*Cucurbita pepo*)	cotyledon	47–48	35
Rape	(*Brassica napus*)	cotyledon	35–48	20–35
Nigerseed	(*Guizotia abyssinica*)	cotyledon	40–45	37
Linseed, flax	(*Linum usitatissimum*)	cotyledon	24–43	23–26
Hemp	(*Cannabis sativa*)	cotyledon	30–41	28–31
Yellow mustard	(*Sinapis alba*)	cotyledon	38–40	25–35
Sunflower	(*Helianthus annuus*)	cotyledon	20–40	25–40
Cucumber	(*Cucumis sativus*)	cotyledon	38–40	28–30
Safflower	(*Carthamus tinctorius*)	cotyledon	22–40	14–17
Douglas fir	(*Pseudotsuga menziesii*)	megagametophyte	36	32
Lettuce	(*Lactuca sativa*)	endosperm/cotyledon	33–37	24
Cotton	(*Gossypium hirsutum*)	cotyledon	15–33	25–39
Tung	(*Aleurites fordii*)	endosperm	16–33	25
Olive seed	(*Olea europaea*)	endosperm/cotyledon	12–28	c
Yucca	(*Yucca whipplei*)	perisperm	23–27	21–24
Soybean	(*Glycine max*)	cotyledon	13–24	36–38
Tomato	(*Lycopersicon esculentum*)	endosperm/cotyledon	15	c
Maize	(*Zea mays*)	scutellum	4–6[b]	10–12[b]

[a] A range of percent composition of unimbibed seed weight is given for most seeds to include varying values reported in the literature. Data were compiled from the following: Vaughan (1970), Butt and Beevers (1966), Bewley and Black (1978), Sinclair and de Wit (1975), Duffus and Slaughter (1980), Gurr (1980).

[b] Percent of kernel dry wt; the percent oil of the maize embryo (mostly scutellum) is about 55%.

(c) not available.

seed(s) on the plant contribute substantially to variation in lipid content and composition (Section 6.II of Volume 1). Nevertheless, the data show that species can be grouped into three general categories: with low, medium (25%–50%) or high seed oil content. The neutral storage lipids are located within discrete organelles called lipid or oil bodies (Fig. 1A) or wax bodies (Fig. 1B).

A. Structure and Fatty Acid Composition of Triacylglycerols

The fatty acid compositions of the lipids (predominantly triacylglycerols) extracted from seeds of important species are listed in Table II. In most of these lipids saturated fatty acids constitute only a small percentage of the total fatty acid content. Hence these reserve lipids are oils, not fats. Of the unsaturated fatty acids, oleic and linoleic acids are predominant. A notable exception is linseed oil, rich in linolenic acid (18 carbon atoms, three double bonds, all *cis*; 18:3). Other exceptions are species with a high proportion of 'unusual' fatty acids (Sections 6.II and 6.V,C of Volume 1). When present, 'unusual' fatty acids are restricted to the reserve triacylglycerol molecules; they are not components of other acyl lipids even within the same storage cells.

Fatty acids are not esterified haphazardly in triacylglycerols, as emphasized in Section 6.II of Volume 1. Analytical procedures for determining the identities of fatty acids at each of the three positions are described and discussed by Hitchcock and Nichols (1971) and Gurr (1980). Saturated fatty acids usually occupy positions 1 and 3, whereas unsaturated fatty acids (notably oleic, linoleic, and linolenic) are found predominantly in position 2. In most oil seeds, unsaturated fatty acids constitute more than 33% of the total fatty acids, thus unsaturated acids must 'spill over' to occupy one or both of the other positions. There is a slight preference for position 1. Exceptions to these generalizations are found in crambe (*Crambe abyssinica*) and rape, where the characteristic 20:1 and 22:1 fatty acids are found in positions 1 and 3, like palmitic and stearic acids in other seed triacylglycerols.

B. Structure and Composition of Wax Esters

Jojoba (*Simmondsia chinensis*) is the only higher plant known to produce wax esters as a major form of storage lipid (Table I, Fig. 1B). Jojoba is a desert shrub native to the U.S.A. (California and Arizona) and Mexico. Within the last fifteen years, there has been considerable interest in the commercial production of jojoba seeds because the chemical structure of the storage lipid is similar to the oil of sperm whales (*Physeter macrocephalus*). Sperm whale oil has many industrial uses such as additives to shampoos and cosmetics, and as a high quality lubricant. In 1970 the

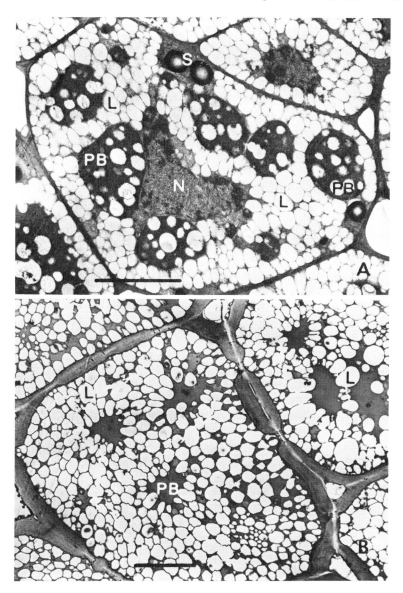

Fig. 1. Electron micrographs of sections through cotyledons of seeds soaked for 2 hours before fixation. (A) cotton cells with numerous triacylglycerol lipid bodies (L) packed around protein bodies (PB). (B) jojoba cells with numerous lipid bodies containing wax esters packed around protein bodies (from Muller et al., 1975). N = nucleus, S = starch. Bars = 5 μm.

Table II. Fatty acid composition of triacylglycerols in selected oil seeds[a]

Plant[b]	8:0	10:0	12:0	14:0	16:0	18:0	20:0	22:0	16:1	18:1	18:2	18:3	20:1	Other acids
Queensland nut	—	—	—	2	8	3	2	—	20	59	2	—	2	—
Brazil nut	—	—	—	2	14	3	—	—	—	58	23	—	—	—
Coconut	10	5	51	19	8	3	—	—	—	5	1	—	—	—
Hazel nut	—	—	—	—	3	3	—	—	—	91	4	—	—	—
Pecan	—	—	—	—	3	2	—	—	—	79	16	—	—	—
Castor bean	—	—	—	—	—	2	—	—	—	7	4	—	—	ricinoleic,88
Sesame	—	1	—	—	8	5	1	—	—	49	38	—	—	—
Watermelon	—	—	1	—	8	6	—	—	—	36	49	—	—	—
Crambe	—	—	—	—	2	1	1	—	—	17	8	6	2	erucic,60
Oil palm	3	3	52	15	8	3	—	2	—	16	1	—	—	—
Peanut	—	—	—	—	9	4	6	—	—	55	27	—	—	—
Pine	—	—	—	—	—	—	—	—	3	43	37	—	—	—
Pumpkin	—	—	—	—	6	7	—	—	—	41	46	—	—	—
Rapeseed (Regina II)	—	—	—	—	4	1	1	1	—	12	16	7	10	erucic,44
Rapeseed (Oro)	—	—	—	—	5	2	1	1	1	48	25	15	2	—
Nigerseed	—	—	—	3	8	5	1	—	—	30	53	—	—	—
Linseed (flax)	—	—	—	—	7	9	—	—	—	13	17	54	—	—
Hemp	—	—	—	—	6	2	—	—	—	6	70	15	—	—
Yellow mustard	—	—	—	—	2	—	1	1	—	22	14	6	—	erucic,44
Sunflower	—	—	—	—	14	—	—	—	—	17	69	—	—	—
Safflower	—	—	—	—	2	2	1	—	—	38	57	—	—	—
Cucumber	—	—	—	—	4	17	—	—	—	39	40	—	—	—
Cotton	—	—	—	1	22	2	—	—	—	31	45	—	—	—
Tung	—	—	—	—	7	—	—	—	—	19	—	—	—	eleostearic,74
Olive seed	—	—	—	—	6	4	—	—	—	83	7	—	—	—
Soybean	—	—	—	—	7	6	—	—	—	23	60	3	—	—
Tomato	—	—	—	—	18	—	—	—	—	24	57	2	—	—
Maize	—	—	—	1	10	3	—	—	2	50	34	—	—	—

[a] Selected data mostly from Hilditch and Williams (1964). Values are the percent of total fatty acids from lipid hydrolysates. For presentation here, values have been rounded to nearest whole number and small quantities of minor acids may have been omitted. Totals, therefore, maybe ± 100%.
[b] Specific scientific names are given in Table I.

importation of sperm whale oil was banned in the U.S.A., hence research on the agricultural production of jojoba seeds has greatly escalated.

Quantitative analyses of jojoba wax esters have been done by Miwa (1971), Spencer *et al.* (1976) and Yermanos and Duncan (1976). Miwa (1971) reported the results of the first quantitative analysis of jojoba oil using gas-liquid chromatography. Table III shows the relative weight percentages of the acids and alcohols. The weight ratio of acid to alcohol is essentially 50:50 (calculated from either the ethanolysis product or the isolated alcohol and acid fractions). Wax esters are composed almost exclusively of unsaturated monoenic acids and alcohols. Eicosenoic acid (20:1) represents 70% of the total acids. Two alcohols, eicosenol (20:1) and docosenol (22:1) represent about 90% of the alcohols and occur in nearly equal amounts. The chain length of the wax esters, as determined by gas-liquid chromatography, is presented in Table IV. They are even numbered and vary from 34 to 48 carbons long, although most are between 40 and 44 carbons. Esters 42 carbons long (C_{42}) are the predominant molecules, representing one-half of the total esters.

Spencer *et al.* (1976) showed that the wax esters from jojoba are not composed of random combinations of acids and alcohols. There is a predominant, non-random combination for *each* chain length (Table V). The main constituents of the C_{42} ester (the most prevalent, Table IV) are docosenol (22:1) and eicosenoic acid (20:1). This indicates that docosenyl eicosenoate is the main wax ester in jojoba wax and is likely to be synthesized exclusively at certain stages of seed maturation (Section 6.IV of Volume 1). Miwa (1971) had assumed that the acids and alcohols in jojoba wax were ω 9-unsaturated. The ω notation is used to locate the double bond position by counting carbons from the methyl terminus. Spencer *et al.* (1976) analysed the wax ester constituents with gas liquid chromatography and mass spectrometry (GLC/MS) and showed that the constituents are *exclusively* $\omega 9$ unsaturated. This indicates that they are related to oleic acid which is important when considering the precursors (Section 6.V,B of Volume 1) or the fate of the products of hydrolysis.

III. DEVELOPMENTAL PATTERNS OF GLYOXYSOMES AND THEIR ENZYMES

A. The Period of Lipid Degradation

Most studies on lipid mobilization have been done with quiescent (non-dormant) seeds. Upon soaking, protein synthesis, respiration, and other metabolic activities begin or increase. Emergence of part of the embryo

Table III. Composition of jojoba acids and alcohols[a]

Number of carbon atoms and double bonds	By ethanolysis and GLC, wt %[b]		By fractionation and GLC, wt %[c]	
	Acids	Alcohols	Acids	Alcohols
14:0	0.1	0.0	0.1	0.0
16:0	0.9	0.3	0.8	0.1
16:1	0.3	0.0	0.2	0.0
18:1	6	0.7	6	0.6
18:2	0.1	0.0	0.1	0.0
20:0	0.1	1	0.0	0.4
20:1	35	22	35	23
20:2	0.1	0.0	0.1	0.0
22:0	0.2	1	0.0	0.5
22:1	7	21	7	21
22:2	0.1	0.0	0.1	0.0
24:0	0.0	0.0	0.0	0.1
24:1	0.5	4	1	4
Total	50.4	50.0	50.4	49.7

[a] From seed oil of Arizona desert sample (Miwa, 1971).
[b] Calculated after ethyl esters were corrected to free acids.
[c] Calculated as if both fractions were remixed as free acids and alcohols.

Table IV. Composition of jojoba oil samples (gas chromatography)

Wax ester chain length	Percent
34	0.1
36	2
38	7
40	30
42	50
44	10
46	0.8
48	0.1
50	—

Data of Spencer et al. (1976).

Table V. Isomer composition for each chain length in jojoba (cv. Apache) wax esters[a]

Wax ester chain length	Alcohol/acid combination	HPLC[b], Sapon.[c] Apache (%)	GC/MS(%)[d] Apache
36	16–20	—	—
	18–18	—	5
	20–16	—	95
38	16–22	4	3
	18–20	21	14
	20–18	70	80
	22–16	5	3
40	16–24	1	1
	18–22	4	5
	20–20	82	81
	22–18	12	12
	24–16	1	1
42	18–24	2	3
	20–22	21	21
	22–20	75	74
	24–18	2	2
44	18–26	—	—
	20–24	12	9
	22–22	28	21
	24–20	60	70
	26–18	—	—
46	20–26	—	—
	22–24	—	—
	24–22	—	—
	26–20	—	—
48	22–26	—	—
	24–24	—	—
	26–22	—	—

[a] Spencer et al. (1976).
[b] HPLC = High pressure liquid chromatography.
[c] HPLC separation and saponification, followed by separate analysis of alcohols and acids.
[d] GC/MS = gas chromatography/mass spectrometry.

(usually the radicle) indicates that the seed has germinated. Nearly all of the lipid and other reserves in oil-rich seeds are mobilized *after* germination, that is, during 'postgerminative growth' of the seedlings. Many authors use the phrase 'during germination' as the description of the time of lipid degradation and assimilation of products. In this chapter, we distinguish between the periods of germination and postgerminative growth.

The time of oil and wax degradation varies with seed species and growth conditions, both in the field and in the laboratory. Laboratory researchers usually have their own preferences for germinating and growing seeds such as using Petri dishes, paper-towelled troughs, flats with soil or other substrates (e.g., vermiculite) at different temperatures, light conditions, and relative humidities. Also, seed coats may have been removed before or after germination to facilitate germination and seedling growth. In most instances, however, changes in enzyme activities are correlated with rates of lipid degradation, allowing comparisons to be made among studies of different (or the same) seeds. In general, the rate of lipid degradation lags for up to 2 days after soaking, then remains constant (at a faster rate) for 3–8 days. Lipid mobilization occurs over a relatively brief period in cotton (see Fig. 9, Section VI,B) compared to cucumber (*Cucumis sativus*) (Fig. 2) and jojoba (Fig. 3). The time taken in jojoba is about one month: seeds germinate after about 5 days, then the linear rate of decrease in wax ester content in the cotyledons lasts about 25 days (Fig. 3). The development of catalase activity [EC 1.11.1.6] is shown for comparison of these three species, since catalase is a marker enzyme for glyoxysomes (Section III,B).

B. Metabolic Pathways: An Overview

The process of lipid mobilization involves numerous enzymes. The activities of many of these enzymes characteristically increase following germination, together with the onset of the linear rate of lipid degradation. Before describing the changes in enzyme activities, we need to discuss briefly which enzymes are involved and what their roles are in lipid degradation and product assimilation.

At least four subcellular compartments are involved. They are (i) storage lipid (oil or wax) bodies, (ii) glyoxysomes, (iii) mitochondria, and (iv) the cytosol. Figure 4 illustrates the relationship among these compartments in the conversion of reserve lipid to carbohydrates (gluconeogenesis), a main pathway occurring in the various storage tissue of all the oil seeds examined. Fatty acids are liberated from triacylglycerols (or wax esters) in the lipid bodies and oxidized to acetyl-CoA through a sequence of β-

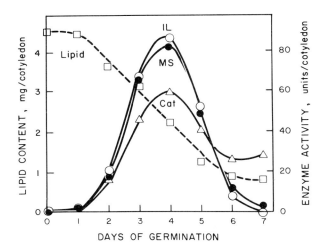

Fig. 2. Changes in lipid and selected glyoxysomal enzymes activities in cucumber cotyledons. IL = isocitrate lyase; MS = malate synthase; Cat = catalase (from Becker et al., 1978).

oxidation enzymes located in glyoxysomes. The metabolism of acetyl-CoA in the glyoxysome via the glyoxylate cycle results in a net synthesis of succinate. Isocitrate lyase [EC 4.1.3.1] and malate synthase [EC 4.1.3.2] are enzymes unique to the glyoxylate cycle; catalase participates in the β-oxidation of fatty acids. Succinate is transported to the mitochondria where it is oxidized to oxaloacetate via three enzymes commonly associated with Krebs cycle metabolism. In the cytosol oxaloacetate is decarboxylated to phosphoenolpyruvate (PEP) which is converted to hexose via reversed glycolysis. In the endosperm of castor beans, there is a quantitative conversion gram to gram of triacylglycerol to sucrose (Beevers, 1980). Sucrose is a main product in most oil seed tissues (including cotyledons), but not the exclusive product as in castor beans. Nevertheless, the processing of fatty acids and production of succinate in glyoxysomes is similar in all oil seed tissues and organs. The sucrose generally is transported to the axis to support seedling growth (Section VII,D).

C. Glyoxysomes

1. Ultrastructure and Distribution

Glyoxysomes were discovered and named in 1967 by Breidenbach and Beevers who isolated them from castor bean endosperm. Before then, it was

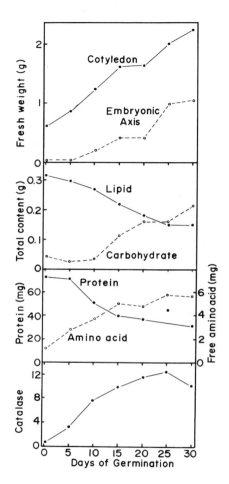

Fig. 3. Compositional changes in cotyledons (five) of jojoba seedlings. Catalase activity is mmol/min (from Moreau and Huang, 1977).

thought that the glyoxylate and Krebs cycles both operated within mitochondria. Since their discovery, glyoxysomes have been isolated in sucrose-density gradients from the storage tissues of 18 oil seed species, and observed with the electron microscope in these and several other oil seeds (Huang et al., 1983). Figure 5A is an electron micrograph illustrating glyoxysomes distributed among, and in close contact with, storage lipid bodies. The characteristic ultrastructure of a glyoxysome is a single boundary membrane enclosing a finely granular or flocculent matrix. They

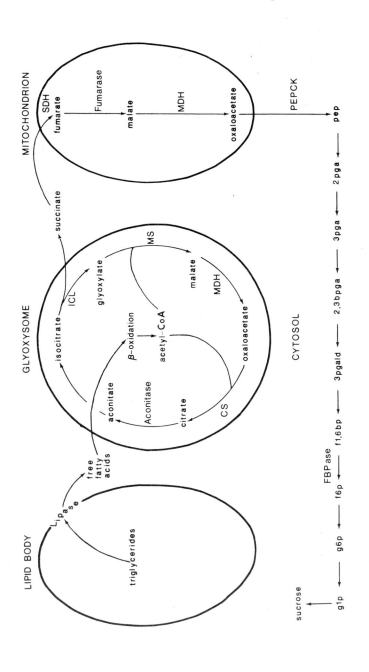

Fig. 4. Diagrammatic illustration of cellular compartments and enzymes involved in gluconeogenesis from reserve lipid in germinated oil seeds. ICL = isocitrate lyase; MS = malate synthase; MDH = malate dehydrogenase; CS = citrate synthase; PEPCK = phosphoenolpyruvate carboxykinase; FBPase = fructose, 1,6-bisphosphatase (diagram courtesy of J. A. Miernyk).

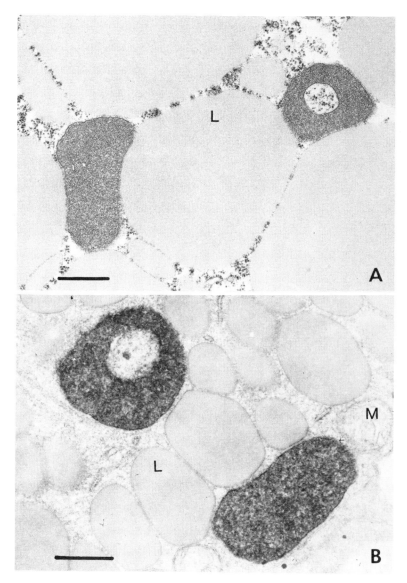

Fig. 5. Electron micrographs of glyoxysomes among lipid bodies (L) in cotyledon cells of germinated seeds. (A) cucumber; tissue prepared in a standard manner. (B) cotton; tissue segments were incubated in H_2O_2 and 3,3'-diaminobenzidine to localize catalase reactivity. Note the electron-dense product throughout the matrix of the glyoxysomes. Pockets (actually invaginations) of cytosolic material within sectioned glyoxysomes are commonly observed (shown in both A and B). M = mitochondrion. Bars = 0.5 μm.

do not possess any internal membrane structure, although amorphous or crystalline proteinaceous inclusions are observed occasionally. Glyoxylate cycle and β-oxidation enzymes are distributed in the organelle matrix or are loosely associated with the boundary membrane. Lipase [EC 3.1.1.3] is the only enzyme known to be an integral membrane protein. The localization of catalase in the matrix of cotton seed glyoxysomes is shown cytochemically in Figure 5B. In higher plants, glyoxysomes have been found only in storage tissues of lipid-rich seeds. They also occur in algal and fungal cells under certain heterotrophic growth conditions. For a more detailed discussion of glyoxysomes, the reader is referred to the monograph by Huang *et al.* (1983).

2. Changes in Enzyme Activities and Glyoxysome Development
In this section we consider only catalase and glyoxylate cycle enzymes; other glyoxysomal enzymes such as lipases and β-oxidation enzymes are discussed later (Sections IV, VI).

The enzyme activities of the glyoxylate cycle show a characteristic rise and fall following oil seed germination (Fig. 2). These activities reach a peak when lipids are being degraded at a constant, maximal rate. It should be noted that considerable lipid remains to be mobilized when the enzyme activities are at their peak. The remaining lipid is mobilized as the glyoxysomal enzyme activities are decreasing (Fig. 2).

Development of the catalase and glyoxylate-cycle enzyme activities has been shown to result from *de novo* synthesis in the seeds of several species (Beevers, 1979). From these results and changes in specific activities of enzymes in isolated glyoxysomes, it was concluded that glyoxysomes and their constituent enzymes were formed *de novo* and increased in number during postgerminative growth. Numerous other studies indicated that glyoxysomes and their enzymes were derived from specialized portions of rough endoplasmic reticulum (ER). This led to the 'ER-vesiculation' model, whereby enzymes are synthesized by ribosomes bound to ER, inserted into the lumen (similar to mechanisms described by the signal hypothesis, Section 3.V,A of Volume 1) then transported to a smooth segment of ER which is 'pinched' off, thereby forming a glyoxysome with its enzymes (Beevers, 1979). Results from recent experiments with several oil seeds indicate that this model is not correct; details and the evidence are given in Huang *et al.* (1983) and Kindl (1982).

Here we will state the recent findings and develop a contemporary concept of glyoxysomal involvement in lipid mobilization. First, of the membrane peripheral proteins and matrix proteins that have been examined, none has been shown to be synthesized on rough ER. In contrast, they have been shown to be synthesized on free ribosomes or the nascent molecules are recovered first in the cytosol. The enzymes are then added post-translationally to pre-existing organelles; the mechanism(s) for this addition have not been documented. Secondly, our laboratory and Kindl's have demonstrated that activities of glyoxysomal enzymes develop during maturation of cotton (see Fig. 9) and cucumber seeds, respectively (Choinski and Trelease, 1978; Miernyk and Trelease, 1981a, b; Kindl, 1982). In addition, both groups have isolated glyoxysomes from maturing or imbibed seeds. Only isocitrate lyase activity was not detected in cotton seed glyoxysomes (but was detected in very low levels in nearly mature cucumber seeds). Miernyk et al. (1979) surveyed dry-seed extracts from several species and varieties of cotton seeds plus 18 other oil seeds (representing 11 different families) for malate synthase and isocitrate lyase activities. They found malate synthase, but not isocitrate lyase activity in all of the dry-seed homogenates. All of these data indicate a universal production of glyoxysomes during maturation of oil seeds.

From these discoveries a new concept is emerging. At least some glyoxysomes with their constituent enzymes are synthesized during seed maturation, very likely in preparation for the massive lipid degradation and product assimilation that follow germination. More glyoxysomal enzymes are synthesized and inserted into already formed or forming glyoxysomes once lipid mobilization begins. It has not been demonstrated unequivocally that newly synthesized enzymes are added to the same organelles made during seed maturation, although Kunce et al. (1984) have shown with a detailed morphometric analysis that the number of glyoxysomes per cell in cotyledons of nearly mature cotton seeds is the same as the number per cell in soaked and early germinated seeds. This indicates that immature-seed glyoxysomes persist through the periods of desiccation and germination. Also, Wanner et al. (1983) and Kunce et al. (1984) have shown using electron microscopy that glyoxysomes are present in soaked watermelon (*Citrullus lanatus*) and cotton seeds, respectively, and Kunce et al. (1984) recorded that the glyoxysomes increase in volume concomitant with the increase in enzyme activities, suggesting that enzymes are added to glyoxysomes already present in dry seeds. The relevance of these data to our discussion is that it appears that certain oil seeds are 'primed' or 'preprogrammed' for lipid mobilization through syntheses (organelle and enzyme) that occur in the cotyledons during seed maturation. This is different from the concept that prevailed only a few years ago: that glyoxysomes and their enzymes are synthesized together from rough ER initiated by some unknown trigger, then increase in number to metabolize the fatty acids released from lipid bodies.

The physiological or molecular mechanisms responsible for the decreases in enzyme activities that occur during lipid mobilization have not been elucidated. In tissues that senesce (e.g., endosperm, scutellum, megagametophyte and perisperm), the decreases seem to result from degradation of glyoxysomes along with the senescing cells. However, similar decreases in enzyme activities also are apparent in cotyledons which do not senesce, but differentiate into photosynthetic organs and persist on the plant well after the storage lipid has been depleted. Several hypotheses describing the fate of glyoxysomes in these organs have been formulated, none of which has gained universal acceptance (see Beevers, 1979; Huang et al., 1983).

A final point needs to be discussed relative to changes in glyoxysomes and their enzymes. The rate of lipid degradation remains relatively constant in all seeds examined during postgerminative growth. Fatty acids apparently do not accumulate (Huang, 1984) hence they must be β-oxidized as they are released from storage triacylglycerols. Yet one consistently finds a rise and fall in enzyme activities during lipid degradation. If the enzyme activities at the beginning of the rise and those near the end of the decline are sufficient to process the fatty acids released at a constant rate, then there must be an excessive production of enzymes during the mid period of lipid degradation. Does this indicate that the numbers of glyoxysomes also are overproduced, or does it indicate the capability of a relatively constant number of glyoxysomes to acquire enzymes that may be needed only under adverse environmental conditions? These aspects of lipid mobilization are not well understood. Most of the work has been done under defined laboratory conditions with seeds that have been bred for high oil content. It would be interesting to repeat many of these studies with wild varieties.

IV. LIPOLYSIS OF TRIACYLGLYCEROLS

A. General Considerations

Our knowledge of lipase action in plants is meager compared to that for most other lipid-related enzyme activities. A likely reason for this is the difficulty in working with enzymes that utilize insoluble substrates. This difficulty has created considerable confusion and misinterpretations of the enzymatic steps involved in lipolysis. Researchers often have resorted to using short-chained water-soluble or other unnatural substrates to assay for 'lipase' activities. True lipases (EC 3.1.1.3, triacylglycerol acylhydrolase) utilize triacylglycerols as their substrates and act only at an oil–water interface. These enzymes also may hydrolyse fatty acids from diacylglycerols and monoacylglycerols, but preferentially attack triacylglycerols. In plant cells there are a number of other esterases and acylhydrolases whose activities can be construed as true lipase activities when improper substrates are used. Another important point is that lipases often are specific, or most active, with oils from the seed in which the lipase functions. Therefore, assays should be done with purified natural triacylglycerols.

Literature on this topic has been reviewed and discussed recently by Huang (1984), Hitchcock (1982) and Galliard (1980), and earlier by Galliard (1975, 1978) and Brockerhoff and Jensen (1974).

B. Time Courses

Numerous time courses are published for lipase activities in germinated seeds. The methods used for preparing and assaying the lipases vary in almost every conceivable aspect, for example, pH, buffer, substrate, assay procedure and enzyme source. A few generalizations can be made from these studies. Usually lipases are not active in dry seeds, but develop and increase in activity concomitant with storage lipid depletion. A decrease in activity after reaching a peak near the mid point of lipid degradation is not a consistent phenomenon. Often, time courses are presented for several lipases at different pH optima in the same seed. It is difficult to determine from these reports whether one or all of these enzymes are responsible for triacylglycerol hydrolysis.

C. Subcellular Compartmentation of Lipases

1. Castor Bean Lipases

Lipid bodies are bounded by a comparatively thin membraneous layer which has been referred to as a 'half-unit' membrane (Yatsu and Jacks, 1972). There is some controversy as to whether this boundary layer actually exists (Gurr, 1980), but recent evidence indicates that proteins and phospholipids form a surface membrane (Section 6.VI of Volume 1). Ory *et al.* (1968) were the first to show (using a cytochemical procedure) that a lipase was active at pH 4.2 and located at the surface of lipid bodies isolated from castor beans. Detailed discussions of additional work done with this acid lipase can be found in reviews cited earlier (Section IV,A). The enzyme is associated with a lipid-body membrane, has a pH optimum of 4.1 and requires both a low molecular weight cofactor and heat-stable glycoprotein for maximal activity. It can hydrolyse triacyglcyerols, diacylglycerols and monoacylglycerols, but preferentially releases fatty acids from triacylglycerols. This acid lipase is in an active state in dry beans, and apparently is incorporated as an active enzyme into the lipid body membrane as the lipid bodies are formed during seed maturation. The mechanism(s) responsible for preventing lipid body autolysis during maturation is not known. Moreau *et al.* (1980) suggested that lower pH may be involved because the enzyme essentially is inactive at pH 7.0.

Muto and Beevers (1974) reported that the activity of castor bean lipases decreased after germination before most of the storage lipid was depleted. They discovered that an alkaline lipase, associated with the glyoxysomal membrane, increased in activity during the later period of lipid degradation. As reported at that time, the glyoxysomal enzyme could hydrolyse only monoacylglycerols. They proposed that the two lipases cooperated in a sequential manner to degrade the lipid. Moreau *et al.* (1980) examined the system further. By altering the homogenization procedure and by correcting for incomplete extraction of lipid and other lipid body components with the aqueous method, they showed that the period of acid lipase activity and period of lipid degradation coincided, so that the acid lipase could be responsible entirely for postgerminative lipolysis. More recently Huang (1984) found that the glyoxysomal enzyme would hydrolyse trilinolein, (indicating it is a true lipase), but that its activity was only a few percent of the lipid body lipase activity (on trilinolein). With the available data, it is difficult to see the need or role of the glyoxysomal alkaline lipase in triacylglycerol lipolysis.

2. Other Lipid Body Lipases

Although considerable information is available for the castor bean system, it should be emphasized that the pattern of lipase activities is not the same for other oil seeds. An active, true acid lipase has not been found associated with lipid bodies in any other *ungerminated* oil seed. However, enzymes active on triacylglycerols or endogenous substrates at an acid pH have been reported for *germinated* seeds (cotton and peanut, St Angelo and Altschul, 1964; cotton, Huang, 1984; mustard, Lin and Huang, 1983; Douglas fir, (*Pseudotsuga menziesii*), Ching, 1968; *Cucumeropsis edulis*, Opute, 1975; soybean, Lin *et al.*, 1982). The subcellular location of the *Cucumeropsis*, peanut and soybean enzymes is unknown. The enzyme from fir is about 80% soluble, although some activity was measured in a 'fat body' fraction.

Lipases associated with lipid bodies and having maximal activities at neutral or acidic pHs have been found in maize, rape, cotton, and mustard (*Sinapis alba*) (Table VI). The *in vitro* rates of lipid body autolysis are sufficient to account for the *in vivo* rate of lipid degradation in these seeds. It is interesting to note that lipid body autolysis does not occur at any pH

Table VI. Subcellular localization and pH optima of true lipases in germinated oilseeds. Only the activity for castor bean acid lipase was found in ungerminated seeds.

Species	Lipid body[a]	pH optimum	Glyoxysome	pH optimum[b]
Castor bean	+[c]	5.0	+	
Maize	+	7.5	ND	
Cotton	+	5.0	ND	
Cotton (glandless)	+	5.0	ND	
Watermelon	ND	ND	+	
Rape	+	6.5	−	
Mustard	+	5.0	ND	
Jojoba	+	9.0[d]	—	
Peanut	−		+	
Soybean	−		+	8.0–9.0
Cucumber	−		+	
Sunflower	−		ND	
Pine (*P. ponderosa*)	−		+	

[a] Autolytic activity of isolated lipid bodies. Data from Huang (1984), Lin and Huang (1983) and Lin et al. (1983).

[b] All species scored with a '+' are activities measured at pH 9 with N-methylindoxylmyristrate as substrate (Huang, 1975). Only the castor bean, peanut and soybean enzymes have been assayed with triacylglycerols; only the soybean enzyme has been shown to have an alkaline *optimum* on triacylglycerol (Lin et al., 1982).

[c] '+' indicates presence of activity, '−' indicates no activity; ND = not determined.

[d] Activity of lipid-body membrane enzyme; optimum was determined with N-methylindoxylmyristrate. The enzyme can hydrolyse wax esters and triacylglycerols, but these activities were not measured at pHs other than 9.0 (Huang et al., 1978).

tested in several of the other common oil seeds shown in Table VI. Lipases have been recovered in membrane 'ghosts' after extracting lipid bodies with diethyl ether or other organic solvents, or in membrane 'appendices' that were presumably dislodged from lipid bodies during cell fractionation (Huang, 1984).

The membrane-bound enzyme from maize has an optimal activity at pH 7.5 with trilinolein and is most active with acylglycerols containing linoleic and oleic acids (Lin et al., 1983). Oil from maize scutellum is composed mostly of these two acids (Table II). The membrane-associated enzyme from mustard seeds has an optimal activity at pH 5.0–5.5 on triolein, trilinolein and trierucin, all three substrates being hydrolysed nearly equally (Lin and Huang, 1983). Theimer's group has studied the rape lipase in a membrane fraction evidently derived from an 'appendix' of lipid body boundary membranes (Rosnitschek and Theimer, 1980; Wanner and

Theimer, 1978; Theimer and Rosnitschek, 1978). The enzyme preparation had optimal activity at pH 9 on sunflower oil. Trierucin was not used in these tests although erucic acid constitutes about 45% of rape oil (Table II). Lin and Huang (1983) obtained contrasting results using a different enzyme preparation from rape. Isolated lipid bodies contained about 50% of the lipase activity in contrast to 90% reported by Theimer's group. Optimal autolytic activity was at pH 6.5, and enzyme from the membrane ghosts was most active on pure trierucin and trilinolein at pH 6.5.

Results with the enzyme associated with soybean lipid body membranes (Lin *et al.*, 1982) were different from those described above. This may be expected because soybean lipid bodies do not undergo autolysis (Table VI). The enzyme hydrolyses monolinolein, but not trilinolein, yet soybean oil is about 60% linoleic acid (Table II). The activity decreases following germination and prior to complete degradation of the neutral lipid. These data indicate that the lipid body lipase is not the main enzyme responsible for lipid degradation; however, soybeans have an active true glyoxysomal lipase (see Section IV.C.*3*).

3. Glyoxysomal Lipases

Huang (1975) compared specific activities of enzymes in glyoxysomes isolated from several common germinated oil seeds. He noted striking similarities in all the enzymes examined, with the exception of the lipases assayed only at pH 9 with N-methylinodoxylmyristrate; specific activities varied up to 10-fold among the different seedlings. All of the species (except soybean) shown to have a glyoxysomal lipase in Table VI are marked on the basis of Huang's survey.

Even though the role of the glyoxysomal lipase is not clear from research on castor beans (Muto and Beevers, 1974; Moreau *et al.*, 1980), the consistent occurrence reported in Huang's survey suggests they may be involved in lipid mobilization. Also, the enzymes from peanut, castor bean and soybean have been shown capable of hydrolysing triacylglycerols, indicating they are true lipases (Lin *et al.*, 1982). However, an alkaline pH *optimum* with triacylglycerols has been determined only for the soybean enzyme.

Besides the glyoxysomal enzyme from castor bean endosperm, only the soybean glyoxysomal lipase has been studied extensively with acylglycerol substrates (Lin *et al.*, 1982). It has a pH optimum between 8 and 9 on trilinolein, but does not exhibit activity with triolein, tripalmitin, or tristearin. The activity is 10-fold higher with monolinolein than trilinolein. This seems odd because linoleic acid comprises 60% of the fatty acid content of soybean oil (Table II).

As discussed above, soybean lipid bodies do not undergo *in vitro* autolysis at any pH tested. The only hydrolase activity found associated with the lipid body membranes was not active on triacylglycerols, but could cleave monolinolein. Interestingly, the glyoxysomal lipase is most active on monolinolein. Lin *et al.* (1982) concluded that the soybean glyoxysomal lipase is a main enzyme responsible for hydrolysis of the soybean storage lipids. This is in direct contrast with conclusions from work with castor bean. The lipid body lipase is seen as the main lipolytic enzyme and the role of the glyoxysomal lipase is not at all clear. A lipase associated with lipid bodies (Huang, 1984) or the lipid body appendices (Rosnitscheck and Theimer, 1980) is believed responsible for the main lipolysis in rape; a glyoxysomal enzyme ostensibly does not exist (Table VI).

4. Lipases in Non-Oil Seeds
Lipases in this category include mostly those in cereal grains (kernels). Some work has been done with these enzymes because of their potential involvement in rancidity during grain or bran storage. Details of research with these enzymes are not given here; see Huang (1984), Hitchcock (1982), and Galliard (1980).

D. Fate of Storage Lipid Bodies

Ultrastructural studies of cells at different stages of postgerminative growth indicate that the number of lipid bodies per cell decreases, rather than all decreasing uniformly in diameter. This suggests that certain lipid bodies, most likely those adjacent to glyoxysomes, are degraded before others. The localization of lipase in the lipid body membrane (Ory *et al.*, 1968; Huang, 1984) indicates that the lipids are hydrolysed inwardly from the periphery until only the membrane is left. Bergfeld *et al.* (1978) reported that lipid bodies in mustard (*S. alba*) cotyledons were degraded centrifugally from the central vacuole(s), and that collapsed lipid-body coats were left behind in glyoxysome invaginations or in the cytosol. These membrane remnants did not accumulate, which suggests that they are degraded by some unknown means.

Similar membrane configurations were described by Theimer's group but their involvement in lipid mobilization was interpreted differently. These structures often were observed attached to lipid bodies and were described as appendices possessing the lipase (Wanner and Theimer, 1978). Released fatty acids are thought to move to glyoxysomes through the appendix or parts of the fluid lipid-body membrane. They also showed structures which were interpreted as membrane remnants, but had the appearance of appendices not attached to lipid bodies (possibly due to section views). It is not clear to us from their micrographs whether appendices are different than remnants. More recently, Wanner *et al.* (1982) concluded that an appendix gives rise to a new glyoxysome which envelops a lipid body and digests its components leaving the membrane remnant within a glyoxysome cavity.

Thus there are several current hypotheses on the mode of lipid hydrolysis, transfer of fatty acids to glyoxysomes, and the fate of the boundary membrane. All of these are based on interpretations of ultrastructural observations made during the period of lipid mobilization. None has received universal acceptance.

V. LIPOLYSIS OF WAX ESTERS

Wax esters located in wax bodies are hydrolysed to alcohols and fatty acids by a wax ester hydrolase in the bounding membrane (Moreau and Huang, 1977; Huang *et al.*, 1979). The enzyme has a pH optimum (on N-methylindoxylmyristrate) near 9. It also is active on monoacylglycerols, commercially prepared wax esters, natural jojoba wax, tripalmitin, and triolein with this order of preference, but the activities were tested only at pH 9. The enzyme may be a true lipase; only its apparent preference for monoacylglycerols casts doubt on such a classification. The listing in Table VI indicates that the optimum at pH 9 for jojoba is different than for other lipid body acylhydrolases, but its activity needs to be assayed at pH 5 or 7 on purified wax esters. Acylhydrolase activity was not detected in glyoxysomes or mitochondria (Table VI).

Moreau and Huang (1979) identified and partially characterized two enzymes, a fatty alcohol oxidase and fatty aldehyde dehydrogenase, which catalyse the conversion of fatty alcohol to fatty acid during postgerminative growth of jojoba seedlings. Both enzymes were localized mainly in the membrane of the wax bodies. The oxidase utilizes molecular oxygen as the electron acceptor and dodecyl alcohol as the preferred substrate and has a pH optimum of 9. The dehydrogenase shows a preference for dodecyl aldehyde and $NADP^+$ at pH 9.

The three enzymes appear to work in concert to mobilize the wax esters. Their activities develop from undetectable levels in dry seeds to a peak when the rate of wax mobilization is greatest (Figs 3, 6). Hence, the patterns of enzyme increases (and declines) are similar to those in other oil seeds (Fig. 2), only extended over a longer time. Fatty acids (including those produced

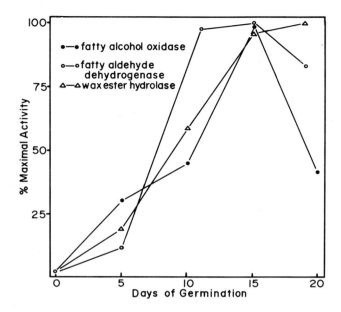

Fig. 6. Changes in activities of wax-ester mobilizing enzymes in cotyledons of jojoba seedlings following seed imbibition. At day 15, activities of the enzymes listed were 16, 635, and 240 nmol/min per seed, respectively (from Moreau and Huang, 1979).

from alcohols) are transported to glyoxysomes by some unknown means (although wax bodies and glyoxysomes are likely to be closely appressed as in other oil seeds) and converted to acetyl-CoA through the β-oxidation system (Section VI). The concomitant rise in soluble carbohydrates and starch and decline in wax esters (Fig. 3) indicate that gluconeogenesis is a main pathway in jojoba seedlings as in other oil seedlings.

VI. β-OXIDATION OF FATTY ACIDS

A. Enzyme Sequence and Compartmentation

Fatty acids released from triacylglycerols or wax esters, and long-chain alcohols converted to fatty acids (Section V), are catabolized to acetyl-CoA via β-oxidation sequences located in glyoxysomes (Fig. 7). Fatty acids are converted first to fatty acyl-CoA by an acyl-CoA synthetase [EC 6.2.1.3]. This enzyme has been shown to be located exclusively in the glyoxysomes

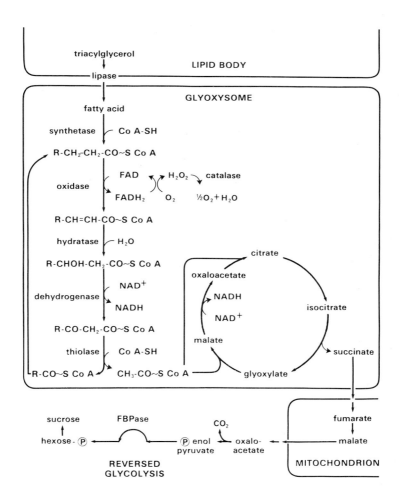

Fig. 7. Pathway of fatty acid metabolism through the β-oxidation sequence in glyoxysomes. Catalase degrades the H_2O_2. Malate is shown leaving the mitochondria prior to its conversion to oxaloacetate (from Huang et al., 1983).

in germinated seeds of castor bean (Cooper, 1971) and cotton (Miernyk and Trelease, 1981a). The β-oxidation sequence consists of four enzymes, two of them in a complex. The first reaction is catalysed by a fatty acyl-CoA oxidase that utilizes molecular oxygen as the electron acceptor and produces hydrogen peroxide which is then degraded by catalase (Fig. 7). The enzyme has not been purified, but indirect evidence suggests it contains a flavin prosthetic group for the transfer of electrons to oxygen. In animal mitochondria that possess β-oxidation sequences, the first enzyme is an acyl-CoA dehydrogenase which is coupled to the electron transport system.

The second and third enzymes (enoyl-CoA hydratase [EC 4.2.1.17] and β-hydroxyacyl-CoA dehydrogenase [EC 1.1.1.35]) have been purified from cucumber cotyledons (Frevert and Kindl, 1980a). Both enzyme activities are associated with the same protein (MW 75 000 daltons, no subunit structure), therefore it often is referred to as β-oxidation multifunctional protein. The chain-length specificity of the enoyl-CoA hydratase activity in cotton is broad; activity has been demonstrated on substrates varying from C_4 to C_{16} (Miernyk and Trelease, 1981c). The last enzyme, 3-oxyacyl-CoA thiolase [EC 2.3.1.9], releases acetyl-CoA which enters the glyoxylate cycle in two separate steps (Fig. 7). The thiolase, purified from cucumber cotyledons, has a molecular weight of 90 000 daltons and is composed of two identical subunits (Frevert and Kindl, 1980b).

Fatty acids in storage lipids typically possess 1 or 2 double bonds (Table II). These fatty acids require some modification before or during the cycles of the β-oxidation sequence. Information on such modifications is meager. The modifications may occur on the triacylglycerols before hydrolysis, or most likely after the fatty acids are released by lipase action. Removal of three successive acetyl-CoAs from oleoyl-CoA leaves 3-*cis*-dodecenoic acid which must be isomerized by 2-*trans*-dodecenoyl-CoA before or during the following round of β-oxidation. Hutton and Stumpf (1971) reported that isolated castor bean glyoxysomes could oxidize [3-*cis*-^{14}C]dodecenoic acid by measuring amounts of released $^{14}CO_2$. Miernyk and Trelease (1981a) showed that the activity of 3-*cis*-2-*trans*-enoyl CoA isomerase [EC 5.3.99.1] was located solely in cotton glyoxysomes. Thus, from the limited data it appears that glyoxysomes are responsible for modifying fatty acids for β-oxidation.

Early studies indicated that β-oxidation activities were located in both mitochondria and glyoxysomes of castor bean endosperm (Hutton and Stumpf, 1969), pine (*Pinus ponderosa*) megagametophyte (Ching, 1970) and maize scutella (Longo and Longo, 1975). Possible contamination of mitochondria by broken glyoxysomes could not be ruled out. More recent work including complete analyses of sucrose gradients indicates that β-oxidation activity (in oil seeds) is restricted to the glyoxysomes in both maturing and germinated (Fig. 8) seeds (Miernyk and Trelease, 1981a; Frevert and Kindl, 1980a). The occurrence of the β-oxidation sequence in jojoba seed glyoxysomes has been shown by the exclusive localization of only one marker activity (β-hydroxyacyl-CoA dehydrogenase) in this organelle (Moreau and Huang, 1977).

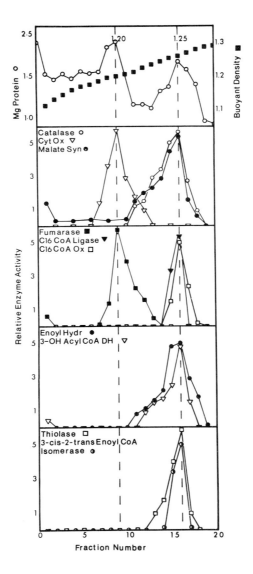

Fig. 8. Distribution of enzyme activities and protein in a sucrose-density gradient following isopycnic centrifugation of a clarified homogenate of cotyledons from forty-five 48-h-old germinated cotton seeds. Glyoxysomes banded at a density of 1.25 g/mL and mitochondria at 1.20 g/mL. Note the exclusive localization of β-oxidation enzymes, malate synthase and catalase in the glyoxysome region (from Miernyk and Trelease, 1981a).

B. Time Courses

Time courses for the activities of β-oxidation enzymes have been reported only for castor beans, cotton, and maize. In cotton and castor beans, activities were measured in maturing seeds. Miernyk and Trelease (1981a) found that activities in cotton (localized in glyoxysomes) begin to increase several days after reserve lipid begins to accumulate, then level off at a peak near seed desiccation (Fig. 9). Thiolase activity is an exception. Hutton and Stumpf (1969) reported a rise and fall in activities during maturation of castor beans; about 10% of the peak activities remained in dry seeds. Both groups noted a disjunct development of thiolase activity, but could not interpret it. The concurrent accumulation of lipid and increase in catabolic enzyme activities seems incongruent. Both groups suggested that the β-oxidation system may be needed during maturation to remove (scavenge) free fatty acids which could inhibit subsequent seed germination. The development of these activities, however, is in line with the concept we discussed in Section III,C,2, that is, that mature seeds are preprogrammed and equipped to initiate lipid mobilization soon after germination.

Following germination, the activities of β-oxidation enzymes rise and fall as do those of several glyoxylate cycle enzymes. In cotton (Fig. 9) and maize (Longo and Longo, 1975), however, the activities peak before the glyoxylate cycle activities (represented by the profile for catalase in Fig. 9). The physiological reason for a rise and fall in enzyme activities during the rapid phase of lipid degradation is not understood, nor is it understood why there are not simultaneous declines in these activities, which presumably are working together in glyoxysomes to metabolize fatty acids released at a constant rate.

VII. METABOLISM OF ACETYL-COENZYME A

A. Glyoxylate Cycle

The elucidation of the biochemical pathway involved in the conversion of acetyl-CoA to C_4 dicarboxylic acids stemmed originally from work with certain microorganisms (i.e. *Escherichia coli*, *Pseudomonas*) that could use acetate both as the sole source of carbon and for energy (for review see Kornberg, 1966). The pathway (Fig. 4) provides for the net synthesis of succinate and NADH for each two molecules of acetyl-CoA which enter the cycle. Isocitrate lyase catalyses the aldol cleavage of isocitrate to succinate and glyoxylate, thus providing a route whereby isocitrate bypasses two oxidative decarboxylating steps in the Krebs cycle. Malate synthase

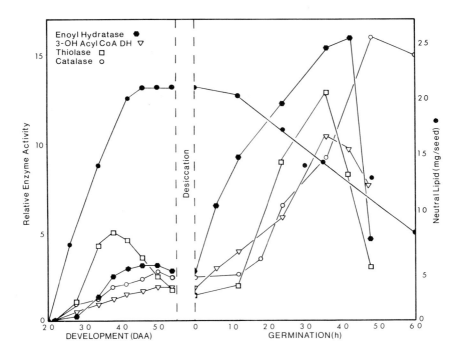

Fig. 9. Changes in β-oxidation enzyme activities and neutral lipid content in cotyledons of maturing and germinated cotton seeds. Not shown are malate synthase and isocitrate lyase activities; the former begins to increase at about 42 days after anthesis (DAA), the latter at about 16 hours postimbibition of mature seeds (from Miernyk and Trelease, 1981a).

catalyses the condensation of the glyoxylate with acetyl-CoA to form malate. The oxidation of malate by malate dehydrogenase [EC 1.1.1.37] produces oxaloacetate which condenses with another molecule of acetyl-CoA to produce citrate.

The demonstration of malate synthase and isocitrate lyase activities in germinated castor bean endosperm led Kornberg and Beevers (1957) to propose a complete sequence for the conversion of lipid to sucrose that involved the glyoxylate pathway. Subsequently, activities of these enzymes have been demonstrated in all other oil seed species in which storage lipid degradation is occurring. An alternative pathway is considered in Section VII,A,*1*. Compartmentation of isocitrate lyase, malate synthase, aconitase [EC 4.2.1.3], malate dehydrogenase and citrate synthase [EC 4.1.3.7] in glyoxysomes has been discussed (Section III). The latter three enzymes are also components of the Krebs cycle in the mitochondria. All glyoxylate cycle enzymes except aconitase have been purified to homogeneity and their properties elucidated (Huang *et al.*, 1983).

Succinate is considered the end product of the glyoxylate cycle. Glyoxysomes isolated from castor bean endosperm were shown to metabolize [5,6-^{14}C]isocitrate to radioactive succinate in the presence of acetyl-CoA (Cooper and Beevers, 1969). When mitochrondria were added to these glyoxysomes, succinate became labeled first, but after a short lag, label began to accumulate in malate. Since both the labeled carbon atoms of [5,6-^{14}C]isocitrate would be lost during strict mitochondrial utilization of this substrate, Cooper and Beevers concluded that the labeling seen in malate had arisen from mitochondrial utilization of succinate produced by isocitrate lyase in the glyoxysomes. How malate escapes oxidation to oxaloacetate *in vivo* is not known.

1. Oxidation of NADH Generated in Glyoxysomes

One problem that requires explanation is the fate of NADH generated in glyoxysomes. For each mol of acetyl-CoA generated during β-oxidation of fatty acids one NADH is produced. Additionally, the conversion of 2 mol of acetate to 1 mol of succinate requires the oxidation of malate to oxaloacetate, and thus the production of 1 mol NADH (Fig. 7). As isolated glyoxysomes do not have the ability to oxidize NADH (Lord and Beevers, 1972) but isolated mitochondria do, it was concluded that glyoxysomal NADH probably is reoxidized in the mitochondria. Glyoxysomes contain only catalytic amounts of NAD$^+$(NADH) (0.22–0.55 nmol/mg protein, Mettler and Beevers, 1980; 0.60 nmol/mg protein, Donaldson, 1982).

There are three possible mechanisms by which NADH could be reoxidized. Although it is unlikely that NADH simply leaves the glyoxysome and is directly oxidized by the mitochondria, this possibility has not been eliminated. Another possibility is that electrons from glyoxysomal NADH may be donated to flavins and cytochromes in the glyoxysomal membrane (Donaldson, 1982; Hicks and Donaldson, 1982; Donaldson *et al.*, 1981), and then discharged to external acceptors. The glyoxysomal membrane has been shown to be capable of NADH oxidation when coupled to cytochrome *c* reduction (Donaldson *et al.*, 1981). The glyoxysomal membrane is rich in flavin, and contains cytochromes *b*5 and P-420, NADH ferricyanide reductase [EC 1.6.99.3] and NADH cytochrome *c* reductase. However, the *in vivo* function of the glyoxysomal membrane redox system is uncertain, as the physiological acceptors of the redox components are not known.

A third possibility involves shuttling of reducing equivalents other than NADH between glyoxysomes and mitochondria. Both the mitochondria and

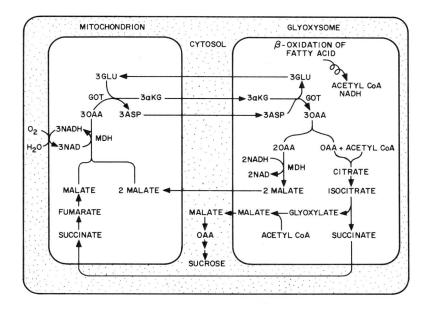

Fig. 10. Proposed malate–aspartate shuttle for the transfer of reducing equivalents between glyoxysomes and mitochondria (from Mettler and Beevers, 1980).

glyoxysomes have active glutamate–oxaloacetate aminotransferases (aspartate aminotransferase [EC 2.6.1.1.]), thus glutamate and α-ketoglutarate may be involved in shuttling between the two organelles. A scheme for these interactions proposed by Mettler and Beevers (1980) is presented in Figure 10. An important feature is the reduction of oxaloacetate via malate dehydrogenase to produce malate and NAD^+. The overall process results in the formation of one unit of malate to be used for sucrose synthesis in the cytosol and the transfer of three equivalents of NADH to the mitochondria for reoxidation. In this scheme the Krebs cycle enzymes in the mitochondria are not involved directly in the flow of carbon in gluconeogenesis; rather they participate in the electron transport system. Further work is needed to detect special transport systems on the glyoxysome membranes.

B. Gluconeogenesis from Oxaloacetate

Oxaloacetate is formed either in the mitochondria (Fig. 4) or in the cytosol from malate which has diffused from either the mitochondria (Fig. 7) or glyoxysomes (Fig. 10). The formation of PEP from oxaloacetate is catalysed by PEP carboxykinase [EC 4.1.1.49] as follows:

$$\text{ATP} + \text{oxaloacetate} \rightarrow \text{ADP} + \text{CO}_2 + \text{PEP} \quad (1)$$

This is the only decarboxylation step in the entire sequence from fatty acids to carbohydrate. The resulting PEP is then converted to fructose-1,6-bisphosphate by a reversal of glycolysis. A cytosolic fructose-1,6-bisphosphatase [EC 3.1.3.11] circumvents the irreversible phosphofructokinase [EC 2.7.1.56] step to give fructose 6-phosphate which is further metabolized to sucrose

$$\text{D-fructose-1,6-bisphosphate} + \text{H}_2\text{O} \rightarrow \text{D-fructose-6-phosphate} + \text{H}_3\text{PO}_4 \quad (2)$$

It has been demonstrated, in both castor bean (Kobr and Beevers, 1971) and in marrow (*Cucurbita pepo*) (Thomas and ap Rees, 1972a) that glycolysis occurs during postgerminative gluconeogenesis. In marrow, the capacity for glycolysis actually increases during postgerminative growth. What arrangements allow the two opposing pathways to operate simultaneously in the same tissue? Clearly, control mechanisms are required to prevent futile cycles. Two mechanisms for regulation have been suggested: (i) compartmental regulation or (ii) regulation of enzyme activities by products acting as effectors.

Nishimura and Beevers (1979) investigated the compartmentation of these enzymes in plastids and the cytosol in endosperm of germinated castor beans. They concluded that the activities of several of the plastid enzymes were too low to account for the *in vivo* rate of gluconeogenesis, whereas those in the cytosol were adequate. Furthermore, PEP carboxykinase, sucrose phosphate synthase [EC 2.4.1.14] and sucrose synthase [EC 2.4.1.13], which catalyse the first and final stages in the conversion of oxaloacetate to sucrose, were found only in the cytosol. In marrow, PEP carboxykinase activity is not associated with any particular organelle fraction (Leegood and ap Rees, 1978a). Nakayama *et al.* (1978) studied the subcellular localization of pyruvate kinase [EC 2.7.1.40] and phosphofructokinase in endosperm of germinated castor bean and found at least 85% of the activities of these enzymes are cytosolic. The available evidence thus suggests that gluconeogenesis and the bulk of glycolysis occur together in the cytosol. In view of the lack of compartmental regulation these reactions must be regulated by the kinetic properties of the enzymes themselves.

Regulation of metabolic reactions can be achieved through control of enzyme activity by factors including substrate and product concentration, cofactor concentration, other metabolites, and by a change in the amount or concentration of an enzyme. The last factor is usually referred to as 'coarse' control as opposed to 'fine' control that is brought about, for instance, by activation or inhibition by a metabolic intermediate. Phosphoenolpyruvate carboxykinase and fructose-1,6-bisphosphatase have been shown to be under coarse control in marrow (Thomas and ap Rees, 1972a, 1972b; Leegood and ap Rees, 1978a) and in castor bean seedlings. Kobr and Beevers (1971) identified phosphofructokinase and pyruvate kinase as major points of control of glycolysis during gluconeogenesis in castor bean endosperm. Regulation of pyruvate kinase would be required to avoid simultaneous operation of reactions involving PEP formation and breakdown. It would seem probable that pyruvate kinase in oil seed tissue is subject to inhibition by ATP, citrate and malate (Turner and Turner, 1980). A substantial increase in ATP concentration occurs during gluconeogenesis in castor bean endosperm (Kobr and Beevers, 1971) and citrate and malate are components of the glyoxylate cycle. Regulation of phosphofructokinase also would be required to prevent futile cycles between fructose-6-phosphate and fructose-1,6-bisphosphate. It is likely that this also could be achieved by fine control. Kobr and Beevers (1971) found a marked increase in the content of PEP and glycerate-3-phosphate, both potent inhibitors of phosphofructokinase. The increased inhibition of phosphofructokinase by these substances is likely to be subject to synergistic enhancement by ATP, and possibly citrate (Turner and Turner, 1980).

After establishing that reactions catalysed by phosphofructokinase, pyruvate carboxylase [EC 6.4.1.1], fructose-1,6-bisphosphatase and PEP carboxykinase were non-equilibrium reactions, Leegood and ap Rees (1978b) attempted to identify which reactions were regulatory. Proof that a non-equilibrium reaction is regulatory can be provided by a demonstration that, *in vivo*, its substrate concentration changes in the direction opposite to the direction of the flux when the latter is varied. To reduce gluconeogenesis, Leegood and ap Rees supplied 3-mercaptopicolinic acid to cotyledons of 5-day-old marrow seedlings. This substance has been shown to reduce the incorporation of acetate into sugars by inhibiting PEP carboxykinase (Leegood and ap Rees, 1978a). This treatment led to significant increases in the amounts of fructose-1,6-bisphosphate and oxaloacetate, indicating that the reactions catalysed by fructose-1,6-bisphosphatase and PEP carboxykinase might be regulatory. Although a wide range of metabolites have been investigated, no regulatory properties have been discovered for PEP carboxykinase in marrow, whereas such properties have been demonstrated for fructose-1,6-bisphosphatase (Turner and Turner, 1980). Based on the higher *in vitro* and *in vivo* activity of fructose-1,6-bisphosphatase activity when compared to phosphofructokinase in marrow and its regulatory properties, Leegood and ap Rees (1978a, b) concluded that this enzyme plays a dominant role in regulating the gluconeogenic flux from PEP to hexose-6-phosphate.

C. Experiments with Radioactive Precursors

1. Experimental Design
In these studies, specifically labeled components of the glyoxylate cycle are supplied to tissues of oil seeds and the products are analysed at intervals. It is generally assumed that the exogenously added substance mixes with the appropriate pool of non-labeled compounds in the tissue. Metabolism of the labeled substance will then reflect normal metabolism assuming there is no enzyme discrimination against the labeled compound. In most studies ^{14}C-acetate is used as a tracer for acetyl-CoA formed via β-oxidation in the glyoxysomes. Results of labeling experiments have been used to confirm glyoxylate-cycle metabolism and to ascertain the ultimate fate of supplied metabolites at various stages of seedling growth.

In many studies, the oil-storing tissue has been excised from the seedling and quartered or sliced before immersion in small quantities of labeling solution (e.g., cotyledons of marrow, Thomas and ap Rees, 1972b; of sunflower and peanut, Bradbeer and Stumpf, 1959; of pumpkin, linseed, watermelon, Sinha and Cossins, 1965; endosperm of castor bean, Canvin and Beevers, 1961). Immersion of cotyledon tissue in solution has been shown to reduce gluconeogenesis in marrow due to reduced availability of oxygen (Leegood and ap Rees, 1978a). Further, excision of the oil-storage tissue from the rest of the seedling can influence reserve mobilization because of removal of the sink (Davies and Slack, 1981; Section VIII). Only one study to date has been done with intact seedlings; this avoided possible complications resulting from sink removal and allowed an estimation of the proportion of labeled material being transported to the roots (Doman *et al.*, 1982).

2. Confirmation of the Gluconeogenic Pathway in Castor Bean Endosperm
Canvin and Beevers (1961) conducted both short-term and long-term labeling experiments with specifically labeled acetate supplied to endosperm slices. They observed the following: (i) early labeling of dicarboxylic acid intermediates of the glyoxylate cycle prior to sucrose labeling; (ii) no formation of $^{14}CO_2$ from [2-^{14}C]acetate; (iii) free sugars were the main

repository of label from [2-^{14}C]acetate; (iv) labeling patterns in degraded malate and hexose were in accord with glyoxylate cycle metabolism. They concluded that the pathway shown in Figure 4 is followed during gluconeogenesis in castor bean endosperm, that acetate is utilized exclusively in the glyoxylate cycle and that none is oxidized in the Krebs cycle. Evidence for this was that none of the C-2 of acetate appeared as CO_2, and glutamate (derived from α-ketoglutarate) was not heavily labeled. Three out of four carbons entering the glyoxylate cycle were recovered in sucrose for an overall efficiency of 75%. Since the observed efficiency of conversion *in vivo* on a carbon basis is 70%, it appears that virtually none of the acetyl-CoA generated *in vivo* from fatty acids is oxidized to CO_2.

D. Fate of Products

The final disposition of glyoxylate cycle intermediates depends on the fate of the storage organ. In castor bean endosperm, a tissue which senesces after reserves have been mobilized, nearly all metabolism is directed toward provision of transport substances for transfer to the axis. The cotyledons rapidly absorb sucrose from the endosperm (Kriedemann and Beevers, 1967a) although some may be stored temporarily in vacuoles (Nishimura and Beevers, 1978). Sucrose is absorbed, transported and unloaded in the axis without hydrolysis. The rate of transfer exceeds 2 mg/h in five-day-old seedlings (Kriedemann and Beevers, 1967a, b). Although the scutellum of maize normally senesces after lipid mobilization, the pattern of ^{14}C distribution from [2-^{14}C]acetate was similar to that found in marrow and cotton cotyledons which persist after reserve mobilization (Table VII). Differences in the physiological functions of scutellum and endosperm might account for the disparity in the fate of oil in these two organs. The persistence and differentiation of cotyledons of epigeic seedlings indicate that some reserve material would be used to support the transition in cotyledon function. Labeling patterns for cotyledons of marrow and cotton supplied with exogenous [2-^{14}C]acetate support this concept (Table VII). Although the cotton seedlings were intact and the marrow cotyledons were detached from their axes, the overall patterns obtained were similar. In the neutral fractions, sucrose was the most heavily labeled sugar, although in marrow, stachyose, another transport sugar, also was labeled. The water-insoluble fractions contained significant amounts of label. Although Leegood and ap Rees suggested that this residue contained ^{14}C-polysaccharides, no analysis was done, whereas the residue in cotton was shown to contain labeled starch, pectins, hemicellulose and cellulose; the distribution of label in these substances after a 3-h pulse and two chase periods is given in Table VIII. Polysaccharides accounted for about 50% of pellet activity,

Table VII. Percent distribution of radioactivity from [2-^{14}C]acetate in various fractions of maize and cotton seedlings and marrow cotyledons

Fraction	Maize[a]	Cotton[b]	Marrow[c]
Period of incubation[d]:	2 h	3 h	4 h
Seedling			
CO_2	ND	3.0	3.2
Cotyledon			
Lipid	5.8	5.3	13.5
Water-solubles	67.1	65.0	46.2
Neutral	16.5	10.8	21.8
Sucrose	7.3	7.0	9.6
Stachyose	ND	0.0	5.6
Glucose	5.1	0.2	1.1
Fructose	4.1	0.3	0.7
Acid (org. acids)	12.3	30.0	10.5
Basic (amino acids)	38.3	24.0	9.0
Ethanol-insolubles	24.5	21.0	30.8
Axis		5.8	ND
Ethanol-solubles	2.7	4.5	ND
Ethanol-insolubles	ND	1.3	ND

[a] Recalculated from Oaks and Beevers (1964).
[b] Doman et al. (1982).
[c] Leegood and ap Rees (1978).
[d] Hours of incubation in [2-^{14}C]acetate.
ND = Not determined.

protein accounted for an additional 25%. Labeling of insolubles also was noted by Bradbeer and Stumpf (1959) in sunflower and peanut cotyledons and by Sinha and Cossins (1965) in pumpkin, linseed and watermelon cotyledons.

Starch synthesis during seed germination and growth has been reported in storage tissues of other oil seeds, including jojoba (Moreau and Huang, 1977) and soybean cotyledons (Adams et al., 1980) and castor bean endosperm (Reibach and Benedict, 1982). These authors postulated that starch was derived from lipid catabolites; Reibach and Benedict (1982) showed incorporation of [2-^{14}C]acetate into starch. In cotton, both lipid and raffinose reserves contribute to starch synthesis (Doman et al., 1982). Starch reserves may serve as a cotyledonary sink for reserve degradation products, thereby limiting the buildup of soluble sugars that may inhibit further degradation (Section VIII).

Table VIII. Distribution of radioactivity from [2-^{14}C]acetate in polymers of cotton cotyledons.[a] Polymers were contained in the ethanol-insoluble pellet shown in Table VII. Values are percent of pellet or total radioactivity recovered.

Fraction	Pulse 3 h		Chase 5 h		Chase 16 h	
	Pellet	Total	Pellet	Total	Pellet	Total
Polysaccharides	55	11.4	50	14.1	47	15.5
Starch[b]	14	3.0	15	4.3	22	7.8
Pectic substances[c]	20	4.2	12	3.4	5	1.5
Hemicellulose[d]	10	2.0	12	3.4	11	3.4
Cellulose[e]	11	2.2	11	3.2	9	2.8
Protein[f]	22	4.3	27	8.0	25	7.7
Unidentified substances[g]	20	4.2	20	5.6	24	7.8

[a] Doman et al. (1982).
[b] [^{14}C]glucose released from pellet by amyloglucosidase digestion.
[c] ^{14}C-acidic compounds released into water by boiling 1 h.
[d] ^{14}C-compounds solubilized from pectin-, starch-, and protein-extracted pellets by cold 5% NaOH (16 h).
[e] ^{14}C-neutral compounds solubilized from hemicellulose-extracted pellets by cellulysin digestion.
[f] ^{14}C-basic compounds released from pectin-extracted pellet by protease digestion.
[g] ^{14}C-compounds (neutral and acidic) released during procedure e.

In both the maize preparation and the cotton seedling only a small percentage of recovered ^{14}C was in the roots (Table VII). After a 16-h chase period in cotton (not shown) only 27% of the label appeared in the roots. Regulation of partitioning of carbohydrates between the axis and cotyledons has not been studied, but it is apparent that a large portion of metabolites from ^{14}C-acetate remains in cotyledons.

VIII. CONTROL OF LIPID MOBILIZATION

A. General Considerations

Knowledge concerning the control of enzyme development and reserve lipid mobilization is limited. Although many studies have been reported, often results are contradictory and inconclusive. This is partly because of experimental design; results obtained from experiments in which excised cotyledons or endosperms are immersed in liquid medium often have been related more to unfavorable incubation conditions than to the factors apparently being studied. For example, Penner and Ashton (1967) reported that isocitrate lyase activity was lower in excised squash (*Cucurbita maxima*) cotyledons than when not excised and attributed this to a lack of hormonal stimulus presumably produced by the embryo axis. The presence of benzyladenine in the culture solution could partially alleviate this depression. Ford *et al.* (1976) subsequently showed that the lowered activity in these cotyledons was a consequence of insufficient oxygen rather than a hormonal deficiency. Similarly, depression of enzyme activity apparently correlated with the absence of the axis in other species could also be the consequence of a general metabolic depression or an injury response to axis or embryo excision. A more detailed critique of experimental techniques is presented by Bewley and Black (1978).

B. Effect of the Axis

Lipid mobilization and seedling growth appear to be synchronized processes; this suggests that the embryo or embryonic axis has a direct influence on reserve mobilization. The effect of axis removal on development of enzyme activities and on rates of lipid degradation has been studied in several species. An apparent lack of axial control has been reported for isocitrate lyase development in peanut (Marcus and Feeley, 1964; Allfrey and Northcote, 1977), soybean (Tester, 1976), castor bean (Huang and Beevers, 1974) and cucumber (Slack *et al.*, 1977; Davies and Chapman, 1979a). Depressions in enzyme activities have been reported for ponderosa pine (Bilderback, 1974) and peanut (Gientka-Rychter and Cherry, 1964). These latter two studies are subject to the above criticisms. Mobilization is inhibited in excised cucumber cotyledons (Davies and Slack, 1981) and in peanut (Allfrey and Northcote, 1977), but not in castor bean endosperm (Marriott and Northcote, 1975).

Two alternative hypotheses can be used to explain how the embryo or embryonic axis controls food mobilization in seeds. The first involves the production of a hormonal stimulus by the axis which stimulates development of hydrolytic enzyme activity in the storage organs. This is discussed in Chapter 7. The second involves the continual operation of a source–sink relationship between the storage organ and the axis during early seedling development (Davies and Slack, 1981). Evidence for the sink effect is discussed here.

The embryo or axis may exert its promotive effect by drawing off the soluble products of mobilization thus maintaining the amounts in the storage organ below certain critical levels. Accumulation of end products could arrest further mobilization by suppression of reactions in the lipid to sugar conversion pathway through the phenomenon of end-product inhibition. In the cotyledons of cotton (Doman *et al.*, 1982), cucumber

6. Lipid Mobilization   239

(Slack et al., 1977), lettuce (*Lactuca sativa*) (Halmer et al., 1978) and marrow (Pinfield and Yousif, 1980), and in the megagametophytes of Douglas fir and pine (*Pinus ponderosa*), (Ching, 1966, 1970) there is no accumulation of soluble sugars while lipid is being utilized. However, sugars do accumulate in castor bean endosperm (Huang and Beevers, 1974) and in jojoba cotyledons (Moreau and Huang, 1977).

Removal of the axis in cucumber resulted in reduced rates of lipid degradation without effecting the development of maximal enzyme activities associated with mobilization (Slack et al., 1977; Davies and Chapman, 1979a, b). Whether the axis is excised immediately after imbibition, or either two or four days after imbibition, an inverse relationship exists between subsequent inhibition of lipid hydrolysis and the accumulation of reducing sugars and sucrose in the cotyledons (Fig. 11). The occurrence of some lipid degradation in the isolated cotyledons suggests that they possess an internal sink. Data obtained from experiments with polyethylene glycol (Davies and Chapman, 1979b) imply that the development of such a sink depends upon the inherent capacity of isolated cotyledons for expansion growth. When expansion is prevented by restricting water uptake, a further reduction in lipid mobilization is observed and is accompanied by an additional increase in sugar levels. These authors suggest that incorporation of sugars into cell wall materials keeps the level of soluble sugars below some critical level.

Other examples in the literature support the contention that the amount of lipid breakdown is controlled by utilization of degradation products. Application of exogenous sucrose to isolated cucumber cotyledons results in a further inhibition of lipid degradation (Slack et al., 1977). The accumulation of starch in isolated cucumber cotyledons (Davies and Slack, 1981), in cotyledons of intact mustard seedlings in which hypocotyl expansion has been inhibited (Bajracharya and Schopfer, 1979), and in cotyledons of intact cotton seedlings before the onset of rapid axial growth (Doman et al., 1982) all demonstrate the existence of internal cotyledon sinks. Further, the application of exogenous glucose partially suppresses isocitrate lyase activity in castor bean endosperm, in squash cotyledons (Lado et al., 1968) and in peanut cotyledons (Longo and Longo, 1970). These examples strongly implicate the sink effect as a major factor in controlling reserve mobilization.

C. Control by the Availability of Oxygen

Ford et al. (1976) showed that the availability of oxygen is an important factor in the development of enzyme activities. Leegood and ap Rees

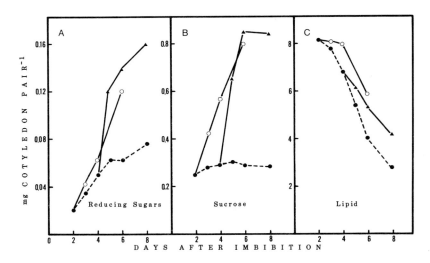

Fig. 11. The effect of axis excision on levels of reducing sugars (A), sucrose (B) and on lipid mobilization (C) in cucumber cotyledons. Axes were not removed (●) or excised either at 2 (○) or 4 (▲) days from the beginning of imbibition. Testas were removed at day 0 (redrawn with permission from Davies and Chapman, 1979b).

(1978a, b) reported that reduced availability of oxygen results in lowered incorporation of [2-^{14}C]acetate into sugars in excised marrow cotyledons. In cucumber, the intact testa substantially inhibits oxygen uptake by cotyledons (Slack *et al.*, 1977). Normally the testa is removed soon after germination by a peg of tissue which develops on the lower surface of the hypocotyl and serves as a lever. When this occurs rapid hydrolysis of lipid reserves begins. This is probably a causal relationship since artificial removal of the testa before axis growth, even from detached cotyledons, promotes enzyme production and lipid breakdown (Fig. 11). Formation of the peg of tissue for testa removal is under hormonal control (Witzum and Gersani, 1975).

IX. SUMMARY AND PERSPECTIVES

Storage lipids are degraded after the seed has germinated at a relatively constant rate (after a brief lag period) over a period of a few to nearly 30 days (jojoba) depending on the species of and growth conditions of seedlings. Conversion of lipid to carbohydrates (gluconeogenesis) is a main pathway in all oil seeds examined, ranging in its predominance from a quantitative conversion in the endosperm of castor beans to about 70% in cotyledons. Sucrose is a main (though not exclusive) product of gluconeneogenesis and typically is transported to the embryo axis to support seedling growth. Other products synthesized from lipid catabolites in cotyledons are retained and used for maintenance and differentiation into photosynthetic organs.

Gluconeogenesis involves four main cellular compartments: (i) lipid bodies; (ii) glyoxysomes, (iii) mitochondria; and (iv) the cytosol. Glyoxysomes, possessing β-oxidation and glyoxylate cycle enzymes, have been studied extensively and shown to metabolize fatty acids and acetyl-CoA similarly in all species (including jojoba) and storage tissues (scutellum, endosperm, cotyledons) examined. Glyoxysomal enzyme activities characteristically increase to peaks when storage lipids are being degraded at their maximal rate, then begin to decline while substantial lipid remains to be mobilized. This suggests there is an excessive production of enzymes during postgerminative growth. Glyoxysomal enzymes are synthesized *de novo* and are added post-translationally to glyoxysomes during the period of lipid degradation. Unimbibed seeds appear to be primed or pre-programmed for lipid degradation by their possession of glyoxysomes synthesized during seed maturation (shown in cotton and cucumber). These glyoxysomes persist through desiccation and seed germination and may be the recipients of enzymes synthesized on cytosolic polysomes following imbibition. Research is continuing to learn whether this is exclusively the case, or whether newly formed glyoxysomes may account for some of the rise in enzyme activities following germination. Essentially no data are available on the mechanisms of enzyme decline; the fate of the glyoxysomes in greening cotyledons is controversial and not resolved.

Although many generalizations can be made for lipid mobilization, the initial patterns of lipolysis seem to differ. Only castor bean possesses an active lipase in dry seeds. This enzyme is localized in the lipid body membrane, has an acidic pH optimum (4.1), and seems to be mainly responsible for postgerminative lipolysis. A glyoxysomal lipase has an undefined role. In cotton, maize, rape, and mustard, lipid body lipases with pH optima between pH 5 and 7 develop activities which can account for lipolysis, and true glyoxysomal lipases may or may not be present. In soybean, a glyoxysomal lipase appears to hydrolyse the lipid and a true lipase does not seem to be associated with the lipid bodies. For other seeds (e.g., peanut, cucumber, sunflower, and pine), the available data do not indicate that lipases in either location are responsible for the main hydrolyses. More research is needed on this critical aspect of mobilization.

The fate of lipid bodies has been investigated primarily by electron microscopy. These studies have generated several current models, notably some that relate to the involvement of lipid body appendices in lipolysis and glyoxysome formation. Further work also is needed here before these models will be accepted. The control of lipid mobilization in seeds has not been studied thoroughly. Source-sink relationships between sugars and gluconeogenesis enzymes seem to be important in that high sugar concentrations tend to slow the mobilization processes. Involvement of the embryo axis as a source of hormones is not clearly established.

ACKNOWLEDGEMENTS

Support by the National Science Foundation during the writing of the chapter is gratefully acknowledged and appreciated. We also thank Ms Diane Mowers for her patience with us during preparation of the manuscript.

REFERENCES

Adams, C. A., Rinne, R. W., and Fjersstad, M. D. (1980). *Ann. Bot.* **45**, 577-582.
Allfrey, J. M., Northcote, D. H. (1977). *New Phytol.* **78**, 547-563.
Appleqvist, L.-Å. (1975). *In* 'Recent Advances in the Chemistry and Biochemistry of Plant Lipids' (T. Galliard and E. I. Mercer, eds), pp. 247-286. Academic Press, New York.
Bajracharya, D., and Schopfer, P. (1979). *Planta* **145**, 181-186.
Barclay, A. S., and Earle, F. R. (1974). *Econ. Botany* **28**, 178-236.
Becker, W. M., Leaver, C. J., Weir, E. M., and Riezman, H. (1978). *Plant Physiol.* **62**, 542-549.
Beevers, H. (1975). *In* 'Recent Advances in the Chemistry and Biochemistry of Plant Lipids' (T. Galliard and E. I. Mercer, eds), pp. 287-299. Academic Press, New York.
Beevers, H. (1979). *Annu. Rev. Plant Physiol.* **30**, 159-193.
Beevers, H. (1980). *In* 'The Biochemistry of Plants' (P. K. Stumpf, ed.), Vol. 4, pp. 117-130. Academic Press, New York.
Bergfeld, R., Hong, Y.-N., Kühnl, T., and Schopfer, P. (1978). *Planta* **143**, 297-307.
Bewley, J. D., and Black, M. (1978). 'Physiology and Biochemistry of Seeds in Relation to Germination', Vol. 1. Springer-Verlag, New York.
Bilderback, D. E. (1974). *Physiol. Plantarum* **31**, 200-203.
Bradbeer, C., and Stumpf, P. K. (1959). *J. Biol. Chem.* **234**, 498-502.
Briedenbach, R. W., and Beevers, H. (1967). *Biochem. Biophys. Res. Commun.* **27**, 462-469.
Brockerhoff, H., and Jensen, R. G. (1974). 'Lipolytic Enzymes.' Academic Press, New York.
Butt, U. S., and Beevers, H. (1966). *In* 'Plant Physiology' (F. C. Steward, ed.), Vol. IVB. Academic Press, New York.

Canvin, D. T., and Beevers, H. (1961). *J. Biol. Chem.* **236**, 988-995.
Ching, T. M. (1966). *Plant Physiol.* **41**, 1313-1319.
Ching, T. M. (1968). *Lipids* **3**, 482-488.
Ching, T. M. (1970). *Plant Physiol.* **46**, 475-482.
Ching, T. M. (1972). *In* 'Seed Biology' (T. T. Kozlowski, ed.), Vol. II, pp. 103-218. Academic Press, New York.
Choinski, J. S., and Trelease, R. N. (1978). *Plant Physiol.* **62**, 141-145.
Cioni, M., Pinzauti, G., and Vanni, P. (1981). *Comp. Biochem. Physiol.* **70B**, 1-26.
Cooper, T. G. (1971). *J. Biol. Chem.* **246**, 3451-3455.
Cooper, T. G., and Beevers, H. (1969). *J. Biol. Chem.* **244**, 3507-3513.
Davies, H. V., and Chapman, J. M. (1979a). *Planta* **146**, 579-584.
Davies, H. V., and Chapman, J. M. (1979b). *Planta* **146**, 585-590.
Davies, H. V., and Slack, P. T. (1981). *New Phytol.* **88**, 41-51.
Doman, D. C., Walker, J. C., Trelease, R. N., and Moore, B. d. (1982). *Planta* **155**, 502-510.
Donaldson, R. P. (1982). *Arch. Biochem. Biophys.* **215**, 274-279.
Donaldson, R. P., Tully, R. E., Young, A. O., and Beevers, H. (1981). *Plant Physiol.* **67**, 21-25.
Duffus, C. M., and Slaughter, J. C. (1980). 'Seeds and Their Uses.' John Wiley and Sons, New York.
Earle, F. R., and Jones, Q. (1962). *Econ. Bot.* **16**, 221-250.
Earle, F. R., Melvin, E. H., Mason, L. H., VanEtten, C. H., and Wolf, I. A. (1959). *J. Amer. Oil Chem. Soc.* **36**, 304-307.
Ford, M. J., Slack, P., Black, M., and Chapman, J. M. (1976). *Planta* **132**, 205-208.
Frevert, J., and Kindl, H. (1980a). *Eur. J. Biochem.* **10**, 79-86.
Frevert, J., and Kindl, H. (1980b). *Hoppe-Seyler's Z. Physiol. Chem.* **361**, 537-542.
Galliard, T. (1975). *In* 'Recent Advances in Chemistry and Biochemistry of Plant Lipids' (T. Galliard and E. I. Mercer, eds), pp. 319-357. Academic Press, New York.
Galliard, T. (1978). *In* 'Biochemistry of Wounded Plant Tissues' (G. Kahl, ed.), pp. 155-201. de Gruyter, Berlin.
Galliard, T. (1980). *In* 'The Biochemistry of Plants' (P. K. Stumpf, ed.), Vol. 4, pp. 85-116. Academic Press, New York.
Galliard, T., and Mercer, E. I. (1975). 'Recent Advances in the Chemistry and Biochemistry of Plant Lipids.' Academic Press, New York.
Gientka-Rychter, A., Cherry, J. H. (1968). *Plant Physiol.* **43**, 653-659.
Gurr, M. I. (1980). *In* 'The Biochemistry of Plants' (P. K. Stumpf, ed.), Vol. 4, pp. 205-248. Academic Press, New York.
Halmer, P., Bewley, J. D., Thorpe, T. A. (1978). *Planta* **139**, 1-8.
Hicks, D. B., and Donaldson, R. P. (1982). *Arch. Biochem. Biophys.* **215**, 280-288.
Hilditch, T. P., and Williams, P. N. (1964). 'The Chemical Constituents of Natural Fats.' John Wiley and Sons, New York.
Hitchcock, C. (1984). *Crit. Rev. Food Sci. Nutr.*, (in press).
Hitchcock, C., and Nichols, B. W. (1971). 'Plant Lipid Biochemistry.' Academic Press, New York.
Huang, A. H. C. (1975). *Plant Physiol.* **55**, 870-874.
Huang, A. H. C. (1984). *In* 'Lipases' (H. L. Brockman and B. Borgstrom, eds), pp. 419-442. Elsevier Press, New York.
Huang, A. H. C., and Beevers, H. (1974). *Plant Physiol.* **54**, 277-279.
Huang, A. H. C., Moreau, R. A., and Lieu, K. D. (1978). *Plant Physiol.* **61**, 339-341.
Huang, A. H. C., Trelease, R. N., and Moore, T. S., Jr. (1983). 'Plant Peroxisomes.' Plant Physiology Monographs. Academic Press, New York.
Hutton, D., and Stumpf, P. K. (1969). *Plant Physiol.* **44**, 508-516.

Hutton, D., and Stumpf, P. K. (1971). *Arch. Biochem. Biophys.* **142**, 48-60.
Kindl, H. (1982). *In* 'International Review of Cytology' (G. H. Bourne and J. F. Danielli, eds), Vol. 80, pp. 193-229. Academic Press, New York.
Kobr, M. J., and Beevers, H. (1971). *Plant Physiol.* **47**, 48-52.
Kornberg, H. L. (1966). *In* 'Essays in Biochemistry' (P. N. Campbell and G. D. Greville, eds), Vol. 2, pp. 1-31. Academic Press, London.
Kornberg, H. L., and Beevers, H. (1957). *Biochim. Biophys. Acta* **26**, 531-537.
Kriedemann, P., and Beevers, H. (1967a). *Plant Physiol.* **42**, 161-173.
Kriedemann, P., and Beevers, H. (1967b). *Plant Physiol.* **42**, 174-180.
Kunce, C., Trelease, R. N., and Doman, D. C. (1984). *Planta* **161**, 156-164.
Lado, P., Schwendimann, and Marré, E. (1968). *Biochim. Biophys. Acta* **157**, 140-148.
Leegood, R. C. and ap Rees, T. (1978a). *Biochim. Biophys. Acta* **524**, 207-218.
Leegood, R. C. and ap Rees, T. (1978b). *Biochim. Biophys. Acta* **542**, 1-11.
Lin, Y.-H., and Huang, A. H. C. (1983). *Arch. Biochem. Biophys.* **225**, 360-369.
Lin, Y.-H., Moreau, R. A., and Huang, A. H. C. (1982). *Plant Physiol.* **70**, 108-112.
Lin, Y.-H., Wimer, L. T., and Huang, A. H. C. (1983). *Plant Physiol.* **73**, 460-463.
Longo, C. P., and Longo, G. P. (1970). *Plant Physiol.* **45**, 249-254.
Longo, C. P., and Longo, G. P. (1975). *Plant Sci. Lett.* **5**, 339-346.
Lord, J. M., and Beevers, H. (1972). *Plant Physiol.* **49**, 249-251.
Marcus, A., and Feeley, J. (1964). *Biochim. Biophys. Acta* **89**, 170-171.
Marriott, K. M., and Northcote, D. H. (1975). *Biochem. J.* **148**, 139-144.
Mayer, A. M., and Poljakoff-Mayber, A. (1982). 'The Germination of Seeds.' Pergamon Press, New York.
Mayer, A. M., and Shain, Y. (1974). *Annu. Rev. Plant Physiol.* **25**, 167-193.
Mazlaik, P., Benveniste, P., Costes, C., and Douce, R. (1980). 'Biogenesis and Function of Plant Lipids.' Elsevier Press, New York.
Mettler, I. J., and Beevers, H. (1980). *Plant Physiol.* **66**, 555-560.
Miernyk, J. A., and Trelease, R. N. (1981a). *Plant Physiol.* **67**, 341-346.
Miernyk, J. A., and Trelease, R. N. (1981b). *Plant Physiol.* **67**, 875-881.
Miernyk, J. A., and Trelease, R. N. (1981c). *FEBS Lett.* **129**, 139-144.
Miernyk, J. A., Trelease, R. N., and Choinski, J. S. (1979). *Plant Physiol.* **63**, 1068-1071.
Miwa, T. K. (1971). *J. Amer. Oil Chem. Soc.* **48**, 259-264.
Moreau, R. A., and Huang, A. H. C. (1977). *Plant Physiol.* **60**, 329-333.
Moreau, R. A., and Huang, A. H. C. (1979). *Arch. Biochem. Biophys.* **194**, 422-430.
Moreau, R. A., Liu, K. D. F., and Huang, A. H. C. (1980). *Plant Physiol.* **65**, 1176-1180.
Muller, L. L., Hensarling, T. P., and Jacks, T. J. (1975). *J. Amer. Oil Chem. Soc.* **52**, 164-165.
Muto, S., and Beevers, H. (1974). *Plant Physiol.* **54**, 23-28.
Nakayama, H., Fujii, M., and Miura, K. (1978). *Plant Cell Physiol.* **19**, 777-784.
Nishimura, M., and Beevers, H. (1978). *Plant Physiol.* **62**, 44-48.
Nishimura, M., and Beevers, H. (1979). *Plant Physiol.* **64**, 31-37.
Oaks, A., and Beevers, H. (1964). *Plant Physiol.* **39**, 431-434.
Opute, F. I. (1975). *J. Exp. Bot.* **26**, 379-386.
Ory, R. L., Yatsu, L. Y., and Kircher, H. W. (1968). *Arch. Biochem. Biophys.* **123**, 255-268.
Penner, D., and Ashton, F. M. (1967). *Plant Physiol.* **42**, 791-796.
Pinfield, N. J., and Yousif, O. A. F. (1980). *Z. Pflanzenphysiol.* **99**, 215-224.
Reibach, P. H., and Benedict, C. R. (1982). *Plant Physiol.* **70**, 252-256.
Rosnitschek, I., and Theimer, R. R. (1980). *Planta* **148**, 193-198.

Sinclair, T. R., and de Wit, C. T. (1975). *Science* **189**, 565–567.
Sinha, S. K., and Cossins, E. A. (1965). *Can. J. Biochem.* **43**, 1531–1541.
Slack, P. T., Black, M., and Chapman, J. M. (1977). *J. Exp. Bot.* **28**, 569–577.
Spencer, G. F., Plattner, R. D., and Miwa, T. (1976). *J. Amer. Oil Chem. Soc.* **54**, 187–189.
St Angelo, A. J., and Altschul, A. M. (1964). *Plant Physiol.* **39**, 880–883.
Stumpf, T. K. (1980). 'The Biochemistry of Plants,' Vol. 4. Academic Press, New York.
Tester, C. F. (1976). *Plant Sci. Lett.* **6**, 325–333.
Theimer, R. R., and Rosnitschek, I. (1978). *Planta* **139**, 249–256.
Thomas, S. M., and ap Rees, T. (1972a). *Phytochemistry* **11**, 2187–2194.
Thomas, S. M., and ap Rees, T. (1972b). *Phytochemistry* **11**, 2177–2185.
Turner, J. F., and Turner, D. H. (1980). *In* 'The Biochemistry of Plants' (D. D. Davies, ed.), Vol. 3, pp. 279–316. Academic Press, New York.
Vaughan, J. G. (1970). 'The Structure and Utilization of Seeds.' Chapman and Hall, London.
Wanner, G. R., and Theimer, R. (1978). *Planta* **140**, 163–169.
Wanner, G. R., Vigil, E. L., and Theimer, R. R. (1982). *Planta* **156**, 314–325.
Witzum, A., and Gersani, M. (1975). *Bot. Gaz.* **136**, 5–16.
Yatsu, L. Y., and Jacks, T. J. (1972). *Plant Physiol.* **49**, 937–943.
Yermanos, D. M., and Duncan, C. C. (1976). *J. Amer. Oil Chem. Soc.* **53**, 80–82.

CHAPTER 7

Axis–Cotyledon Relationships during Reserve Mobilization

DAVID R. MURRAY

I.	The Influence of Cotyledons on Axis Development	247
	A. The Critical Period for Survival of the Axis	248
	B. Reserve Materials of the Axis	249
	C. Do the Cotyledons Transmit Growth Substances to the Axis?	260
II.	Regulation of Reserve Mobilization from Cotyledons by the Axis	261
	A. Identification of Enzymes Synthesized *de novo* in Cotyledons	263
	B. Does the Axis Control the Enzyme Activities of Cotyledons?	265
	C. Effects of Light and the Role of Cytokinins	268
	D. Are Cytokinins Transmitted by the Axis?	272
III.	Conclusions	276
	References	277

I. THE INFLUENCE OF COTYLEDONS ON AXIS DEVELOPMENT

The resumption of growth in the embryo leading to germination and the successful establishment of the seedling has been the theme of this Volume. This final Chapter considers the functions of the cotyledons and their interaction with the embryonic axis, concentrating on those species with cotyledons adapted for storage.

The single cotyledon of cereals, the scutellum, is often an organ adapted primarily for nutrient absorption and transmission, as substrates are released from adjacent endosperm (Chapters 4 and 5). A storage function is also evident in species like maize (*Zea mays*), where the scutellum stores oil utilized early following imbibition (Dure, 1960) and phytin is also stored predominantly in the embryo (Chapter 5; Chapter 4 of Volume 1).

In dicotyledons, leaf-like cotyledons can function first in the absorption of sucrose and other nutrients from surrounding endosperm, then emerge and expand to function as the first effective leaves, for example, castor bean (*Ricinus communis*) (Chapter 6). Alternatively, many species possess cotyledons adapted for the storage of reserves, but the cotyledons are also able to undertake the transition to a photosynthetic organ, for example, cucumber (*Cucumis sativus*) (Becker *et al.*, 1982) and *Acacia* (*A. iteaphylla*, Ashcroft and Murray, 1979; *A. dealbata*, Smith, 1981). Ultimately, the commitment of cotyledons to the storage and mobilization of reserves is absolute: they fail to develop stomates and so display no useful photosynthetic activity, as demonstrated with $^{14}CO_2$ for the cotyledons of pea (*Pisum sativum*), common bean (*Phaseolus vulgaris*) and runner bean (*P. coccineus*) by Lovell and Moore (1970, 1971).

Dicotyledons belonging to these latter categories have frequently been the subjects of experiments designed to determine the ways that cells in one region of a seedling can influence the course of continued differentiation and development in other regions. Although it is obvious that the axis depends upon the delivery of substrates and other metabolites from the cotyledons to become established, it is less obvious that many changes in cotyledon cells necessary for the mobilization of reserves depend upon attachment of the cotyledons to the axis. The interdependent development of axis and cotyledons might involve gradients of growth substances (phytohormones) comparable to those believed to regulate leaf development and senescence. However, the identities of the chemical regulators that effect communication in either direction have been established in very few instances. Some would argue that they have not been established at all.

A. The Critical Period for Survival of the Axis

Axes excised from pea seedlings after five days (at the end of 'Phase 1', Bain and Mercer, 1966b) show no further growth in the absence of a carbon source and eventually die. Removal of cotyledons from pea seedlings after about two weeks, when the transmission of reserves from the cotyledons is substantially completed, has no discernible effect on the subsequent growth of the seedlings (Killeen and Larson, 1968). Between these extremes, premature removal of cotyledons can slow the growth rate of the seedling and permanently disadvantage it (Section I,C).

In species like pea, with cotyledons adapted completely for storage, the early growth rate of the axis depends upon the size of the cotyledons. Trimming the cotyledons of cowpea (*Vigna unguiculata*) to remove about one-third and two-thirds has the effect of producing seedlings similar in weight and leaf area to the seedlings produced by intact seeds of the corresponding initial weights (Table I).

No satisfactory *in vitro* substitute for the cotyledons has yet been devised. Axes removed from unimbibed embryos of bean (*P. vulgaris*) can be cultured in several media (Monnier, 1982), but growth is very slow, equivalent to the retention of only 3% of the reserves normally available to the axis from attached cotyledons. The provision of sucrose in mineral medium (optimal concentration 5%, about 0.15 M) allows growth of excised pea embryonic axes to continue for at least three weeks (Davis, 1983a). In studies like these, attention should now be paid to determining whether a more suitable source of nitrogen can be found, for example, this could be homoserine (Pate, 1977) in the case of pea.

B. Reserve Materials of the Axis

1. The Myth of Early Independence

There is abundant evidence that the kinds of reserve material accumulated in the cotyledons are present also in the storage parenchyma of the axis. In pea, protein bodies and starch grains are present in mature axis cells (Bain and Mercer, 1966a, b; Yoo, 1970). Protein bodies and oil bodies (spherosomes) are the major reserves in the axis of *Bidens cernua* (Simola, 1971) and hazel (*Corylus avellana*, see Fig. 3 of Chapter 2). These reserves may certainly be degraded (e.g., see Fig. 3B of Chapter 2). However, this information is not sufficient to support the widely held view that the early growth of the axis leading to germination is independent of the main reserves stored in the cotyledons.

An increase in the dry matter content of the axis is the most obvious measure of whether export of metabolites from the cotyledons has begun. In reviewing earlier studies on pea, Sutcliffe and Bryant (1977) considered that growth of the pea axis, defined as dry weight increase, did not occur until 'after emergence'. According to Murray *et al.* (1979), an increase in the dry matter content of the radicle accompanies expansion growth leading to emergence (Fig. 1). In this study of the germinating pea seed, a lag of about 24 h at 22°C was observed before any increase in the dry matter content of the axis occurred, specifically in the radicle (Fig. 1). An increase in the dry matter content of the epicotyl plus plumule was not detected until after two days. In the germinating common bean seed, Simon and Meany (1965) recorded a lag of 20 hours at 24°C before simultaneous increases in total nitrogen and dry matter contents of the axis became detectable (Fig. 2). As in the pea embryo, the earliest gains in dry matter content in the bean axis are confined to the radicle (Simon and Meany, 1965).

Table 1. The effects of seed size on seedling development in cowpea (*Vigna unguiculata*) ten days after sowing

Weight class	Mean seed weight (mg)	Plant weight (mg)	Primary leaf area (cm^2)
1	69.4	189.5	25.0
2	88.8	233.9	31.2
3	112.1	293.9	37.1
4	128.9	310.1	38.3
5	148.5	352.1	41.6
6	174.1	388.7	43.3
7	195.2	435.3	49.9
8	219.4	451.8	48.4
(1/3 × 7)^a	69.8	172.3	25.3
(2/3 × 7)^b	120.8	319.9	37.0

^{a} Seeds of class 7 with only one third of cotyledon weight retained.
^{b} Seeds of class 7 with two thirds of cotyledon weight retained.
Data for cowpea (*V. unguiculata* subsp. *unguiculata*) accession no. TVu 4552 reproduced with permission from W. M. Lush and H. C. Wien (1980): 'The importance of seed size in early growth of wild and domesticated cowpeas', *J. Agric. Sci.* **94**, 177–182, Cambridge University Press.

In contrast to these observations, Bain and Mercer (1966a) reported that no change in the dry matter content of the pea axis occurred between 24 hours and 48 hours from the beginning of imbibition. They suggested that the growth of the axis by cell expansion took place at the expense of local reserves alone for 5 days under ambient winter (glasshouse) conditions, until the radicle had reached a length of about 4 cm. This was the period corresponding to 'Phase 1' (Bain and Mercer, 1966a, b). This conclusion is widely cited, but is at odds with some of the data actually presented, for example, Figure 3 of Bain and Mercer (1966a) indicates that the dry matter content of the axis more than doubled between 2 and 5 days.

The data presented here (Figs 1, 2) suggest that any period of independence on the part of the axis could not extend beyond about 20–24 hours, according to temperature and species.

2. Starch Mobilization within the Axis

Changes in the activities of starch-degrading enzymes in the pea axis and their relationship to net starch breakdown have been studied by Davis (1977, 1979). Initially, β-amylase [EC 3.2.1.2] predominates, and this activity begins to increase further after a lag of about one day (Davis, 1977). After a lag of about two days, α-amylase [EC 3.2.1.1] activity also increases. In this study, the α-amylase was characterized by its sensitivity

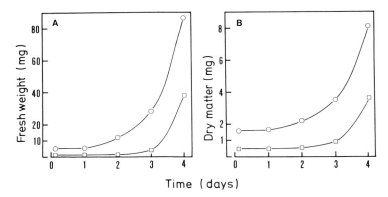

Fig. 1. Changes in the fresh weight (A) and dry matter contents (B) of the radicle (O) and plumule plus epicotyl (□) following imbibition of pea seeds (*Pisum sativum* cv. Greenfeast) at 22°C (from Murray et al., 1979).

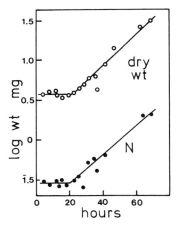

Fig. 2. Increases in total nitrogen (N) and dry matter content of the embryonic axis in seeds and seedlings of *Phaseolus vulgaris* cv. Belfast New Stringless maintained at 24°C in darkness (from Simon and Meany, 1965).

to ethylene diamine tetraacetic acid (EDTA) and its ability to degrade limit dextrins, not shared with the exoamylase. In contrast to the general distribution of β-amylase in the axis tissues, the α-amylase proved to be localized mainly in the hypocotyl (Fig. 3). This specific development of activity in the hypocotyl is associated with the onset of transmission of material from the cotyledons to the epicotyl (Fig. 1). However, it has been shown that sugars such as sucrose, glucose and fructose that might increase

252 David R. Murray

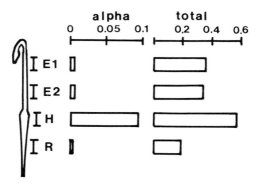

Fig. 3. Distribution of α-amylase and total amylase activities in 5 mm segments of the axis of 5-day old etiolated pea seedlings (*Pisum sativum* cv. Early Alaska). The total activity is expressed as mg maltose equivalents per min; α-amylase activity as the decrease in absorbance at 700 nm per h, accompanying the loss of iodine from β-limit dextrin substrate (from Davis, 1977).

in concentration in the axis following import of sucrose from the cotyledons do not have the ability to derepress the α-amylase activities of the axis (Davis, 1983b).

The starch reserves of the axis are actually degraded, in pea to the extent of 50% in the first 3 days at 22°C (Davis, 1979). This early breakdown is accompanied by the accumulation of sugars, mainly sucrose (Collins and Wilson, 1972). Provided the cotyledons remain attached to the axis, the remaining starch in the axis is spared, and the amount subsequently increases. By comparison, the culture of excised axes with no available carbon source promotes an earlier increase in α-amylase activity and continued depletion of their starch reserves (Davis, 1979).

3. Nitrogen Mobilization within the Axis

The protein content of the axis is similar to or greater than that of adjacent storage cotyledons. In determining the extent to which local protein reserves might be utilized, Murray et al. (1979) could not detect a net decline in the protein content of the pea radicle (Fig. 4). Although polypeptides of estimated molecular weight (MW) 13 000, 30 000 and 84 000 daltons were almost completely degraded after 2, 4 and 4 days respectively, most of the polypeptides of the pea radicle appeared remarkably stable (Murray, 1979b). The rapid utilization of an albumin of MW 12 000 daltons in the axis of germinating mung bean (*Vigna radiata*) has been described by Manickam and Carlier (1980). This polypeptide evidently corresponds to that of MW 13 000 daltons from the axis of pea (Murray, 1979b). It is not

7. *Axis–Cotyledon Interaction* 253

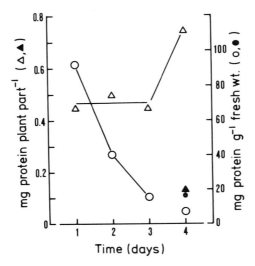

Fig. 4. Soluble protein content of the pea radicle on a fresh weight basis (○) and per radicle (△). The closed symbols indicate values for the 1 cm tip of the root at four days. Growth conditions as for Fig. 1 (from Murray *et al.*, 1979).

certain whether this albumin is located in the cytoplasm (Manickam and Carlier, 1980) or whether it is localized in protein bodies. A pellet obtained after centrifuging extracts of pea radicles imbibed for only 3 hours was enriched in a number of polypeptides, including one (or more) of MW 13 000 daltons (Murray, 1979b).

It has been suggested that many of the stable polypeptides of the pea axis could belong to mitochondrial enzymes rather than to reserve proteins (Murray, 1979b). The major disulphide-linked polypeptides of legumin, one of the major reserve globulins from pea cotyledons (Chapter 3 of Volume 1), were absent from the gel profiles of axis polypeptides (Murray, 1979b). In agreement with these observations, Guldager (1978) could not detect legumin immunochemically in extracts from the axis of pea. Subsequent studies of common bean and cowpea have also shown that the major reserve globulins of the cotyledons are not prominent constituents of the axis (Vidovic and Murray, 1984). Many proteins of the axis are, however, held in common with the cotyledons, as shown for common bean (Fig. 5). Lectins (Manen and Pusztai, 1982) and proteinase inhibitors (Xavier-Filho and do Monte Negreiros, 1979) are included among these, and are utilized as a reserve of nitrogen both in the axis and in the cotyledons.

Early protein breakdown in the axis is more extensive in common bean (Fig. 6) and cowpea (Fig. 7) than in pea. At least seven polypeptides ranging

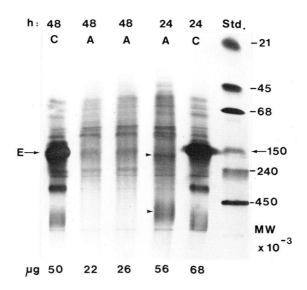

Fig. 5. Electrophoretic separation in gradient polyacrylamide gel ('Gradipore', 2.7%–27%) at pH 8.2 of protein components of the axis (A) and cotyledons (C) of bean seeds (*Phaseolus vulgaris*) germinating in vermiculite at 22°C. E = euphaseolin (Murray and Crump, 1979). MW × 10^{-3} daltons for standard or marker proteins (Std, RH lane). The arrows in the centre of the gel indicate two axis protein bands present after 24 h but not detected after 48 h (from Vidovic and Murray, 1984).

in MW from 72 000 to 11 000 daltons show the first evidence of decline, between 24 and 48 hours after the beginning of imbibition (Figs 6, 7). However, the net effect on protein content of the axis is marginal (Table II), as protein breakdown is apparently offset by the synthesis of other polypeptides (Figs 6, 7). Protein synthesis is known to commence within an hour of the beginning of imbibition in the axis of *P. vulgaris* and related species (Section 3.III,B).

Proteolytic enzyme activities of the axis have been measured following imbibition and compared with the maximum values attained during seed development. Aminopeptidases AP1 and AP2 of the pea radicle recover quickly to values 20%–25% of the maxima observed during development (Collier and Murray, 1977; Murray *et al.*, 1979). After 24 hours imbibition, the spectrum of endopeptidases shifts away from the acid proteinases [EC 3.4.23] predominant at maturity in favour of both serine [EC 3.4.21] and thiol [EC 3.4.22] dependent types (Murray *et al.*, 1979). An increase in carboxypepidase activity [EC 3.4.12] occurs between 24 and 48 hours after the beginning of imbibition in the pea radicle, an increase that accompanies

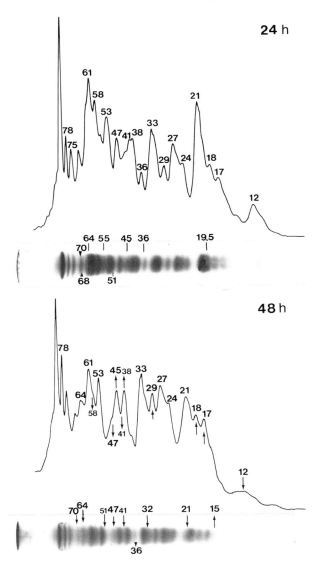

Fig. 6. Polypeptide composition of the axis of *Phaseolus vulgaris* embryos after 24 h (upper gel and densitometer scan) and after 48 h (lower gel and scan). The times given are from the beginning of imbibition at 22°C. Gels were run in the presence of sodium dodecyl sulphate (SDS) and 2-mercaptoethanol and each was loaded with 50 μg of protein. Upward and downward arrows indicate increase or decline in relative content. MW × 10^{-3} daltons (from Vidovic and Murray, 1984).

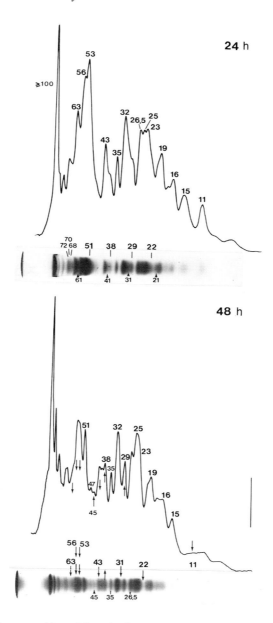

Fig. 7. Polypeptide composition of the axis of cowpea (*Vigna unguiculata* subsp. *unguiculata*, cv. Vita 3) after 24 h (upper gel and scan) and after 48 h (MW × 10^{-3} daltons). Other conditions as for Fig. 6 (from Vidovic and Murray, 1984).

the earliest detected selective breakdown of polypeptides. Whether similar changes in proteolytic enzymes occur in the axis of bean or cowpea has not been determined.

Protein breakdown in the axis is evidently balanced by new protein synthesis, so the increase that is observed in free amino nitrogen in the axis between 24 and 48 hours from the beginning of imbibition (Table II) clearly indicates the importation of nitrogenous solutes from the cotyledons. In bean and cowpea axes (Table II), these increases are accompanied by increases in total dry matter content and exceed the amounts of amino nitrogen theoretically available from the slight net declines in axis protein content in the same interval. Allowing 8 μmol constituent amino acids per mg of protein, increments of 1.72 and 1.26 μmol amino nitrogen per axis (Table II) require the degradation of 0.22 mg and 0.16 mg protein respectively. These amounts exceed the observed net decline in the protein content of the axis under these conditions (0.08 mg for each species). Comparison of these data with information on the pea axis is possible by summing the amounts of individual amino acids recorded by Lawrene and Grant (1963): the free amino nitrogen content of the pea axis is 0.64 μmol after 24 hours and this amount approximately doubles in the next 24 hours (Beevers and Guernsey, 1966).

The cotyledons themselves show marked increases in free amino nitrogen content in the interval 24–48 hours following the beginning of imbibition: from 7.25 to 14.9 μmol per cotyledon for common bean, and from 6.44 to 12.4 μmol per cotyledon for cowpea, under the conditions outlined in Table II. These increases are again comparable with those reported earlier for pea cotyledons (Lawrence and Grant, 1963; Bain and Mercer, 1966a; Beevers and Guernsey, 1966). Although there may be a lag of some days while newly synthesized endopeptidase is acquired by protein

Table II. Changes in the embryonic axis between 24 and 48 h from the beginning of imbibition (22°C)

Parameter	Phaseolus vulgaris		Vigna unguiculata	
	24 h	48 h	24 h	48 h
Fresh weight, mg	14.2	43.1	10.4	28.7
Dry matter, mg	5.59	7.15	3.70	4.50
Protein, mg	1.14	1.06	0.90	0.82
Amino N, μmoles	0.80	2.52	0.79	2.05

Data of Vidovic and Murray (1984). Seeds from batches showing 100% viability were selected by weight in the ranges 407 ± 9 mg (bean) and 190 ± 10 mg (cowpea). Extracts were prepared from 20 axes at each stage.

bodies in the cotyledons of these species (Section IIA,), there are already exopeptidases and peptide hydrolases present that could act in the interval 24-48 hours (Yomo and Taylor, 1973; Prisco et al., 1975; Crump and Murray, 1979). If fully expressed, the proteinase activity of one day-imbibed pea cotyledons (Guardiola and Sutcliffe, 1971a) could generate 15 µmol of free amino nitrogen per cotyledon in the next 24 h. Similar calculations applied to the data of Yomo and Taylor (1973) and Feller (1979) indicate that the proteolytic activity of one day-imbibed bean cotyledons is more than sufficient to account for the observed increases in the amino nitrogen content of the cotyledons plus the axis in the interval from 24 to 48 hours.

Because of sampling variation and other experimental errors, it is more difficult to demonstrate a decline in the protein content of the cotyledons in this same interval. Nevertheless, such a decline has sometimes been detected (Bain and Mercer, 1966a; Basha and Beevers, 1975; Fig. 8). Attempts have been made to identify those polypeptides that are the first to be degraded in the cotyledons. It has been suggested that cytoplasmic albumins might fulfil this role in pea cotyledons, before the protein bodies containing the major globulins become autolytic (Murray, 1979a), and that cytoplasmic lectins might be preferentially degraded in common bean cotyledons (Murray, 1982). In the cotyledons of germinating cowpea seeds, the polypeptides showing the earliest evidence of net breakdown *in vivo* are two albumins of MW 43 000 and 32 000 daltons (Fig. 9). Albumins also

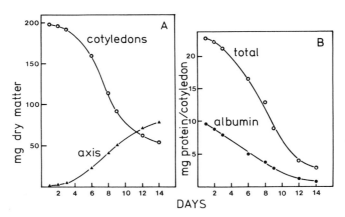

Fig. 8. (A) Changes in the distribution of dry matter between cotyledons and the axis following imbibition of pea seeds (*Pisum sativum* cv. Greenfeast) and growth at 21°C. (B) Changes in the amounts of protein recovered in the albumin and total soluble protein fractions per cotyledon. (From Murray, 1979a).

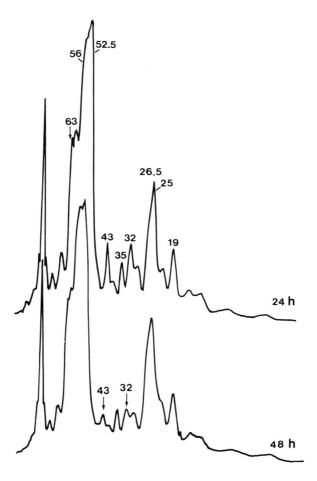

Fig. 9. Densitometer scans at 570 nm of 10% polyacrylamide disc gels stained with Coomassie Brilliant Blue following electrophoretic separation of polypeptides from cowpea (*Vigna unguiculata* subsp. *unguiculata*) cotyledons after 24 h and 48 h respectively (imbibition at 22°C, as for Fig. 7) (from Vidovic and Murray, 1984).

show the earliest signs of breakdown in *Acacia* cotyledons (Ashcroft and Murray, 1979), but the main period of protein breakdown is delayed while selected non-protein amino acids such as albizziine are first transferred to the axis (A. Bradke and D. R. Murray, unpublished data).

Raising the temperature from 22°C to 25°C or 28°C accelerates the transmission of nitrogen from cotyledons to axis in cowpea so that an increase in the protein content of the axis is observed between one and two days following imbibition, rather than a slight decline (Prisco et al., 1975; Vidovic and Murray, 1984). These observations also support the conclusion that the utilization of local nitrogenous reserves in the axis is not confined to an initial lengthy period of independence, but instead this very selective process overlaps with the reception of nitrogenous solutes imported from the cotyledons at an accelerating rate. It appears that the initial restriction on transport of solutes from the cotyledons to the radicle is the time taken to complete the differentiation of functional phloem cells. This process requires about 12 hours or more in pea (Smith and Flinn, 1967). Prior to phloem loading, the pathway for amino acid and amide transfer out of the cotyledons will often involve the intercellular spaces, which provide passage around cells adjacent to vein endings when these cells are among the last to begin breakdown of starch grains and protein bodies (Öpik, 1966; Murray and Knox, 1977).

C. Do the Cotyledons Transmit Growth Substances to the Axis?

In some species exhibiting dormancy, the axis is the source of gibberellins for the cotyledons (Section 2.III,B). In contrast, there is evidence that in pea the cotyledons are the source of gibberellins for the axis as it develops following germination. Shininger (1972) found that the removal of cotyledons from 8 day old light-grown pea seedlings reduced the rate of stem elongation to 50% of the rate for intact seedlings. This removal of the cotyledons lowered both the rate of node formation and the rate of internode elongation. Applied gibberellic acid (GA_3) did not restore the rate of node formation, but the normal rate of internode elongation was soon resumed after a general growth inhibition of 120 h that probably resulted from the loss of about half the cotyledonary reserves of carbon, nitrogen and other nutrients.

As Shininger (1972) has pointed out, the results of this study do not imply that the cotyledons are actively synthesizing gibberellins or gibberellin precursors. Some of his experiments involved the application of the growth retardant 2-isopropyl-4-dimethylamino-5-methyl phenyl-1-piperidine carboxylate methyl chloride (Amo-1618), believed to act by inhibiting new synthesis of gibberellins. The similar responses of shoots with and without cotyledons to the application of Amo-1618 indicated that conversion of inactive to active forms of gibberellin was taking place in the shoot itself.

Bean cotyledons also contain gibberellins, where their presence is masked by inhibitors (Van Onckelen et al., 1977). Provision of gibberellins to the axis in the early stages of germination might stimulate the production of ribonucleases [EC 3.1.4], an effect of exogenous gibberellin demonstrated with excised pea embryonic axes (Takaiwa and Tanifuji, 1978). Increased ribonuclease activity could in turn promote RNA turnover linked to the renewal of net synthesis of RNA in the second day after the beginning of imbibition (Sutcliffe and Bryant, 1977; Section 3.III,C).

II. REGULATION OF RESERVE MOBILIZATION FROM COTYLEDONS BY THE AXIS

A decline in the dry matter content of pea cotyledons can be detected between one and two days from the beginning of imbibition (Fig. 8) and the rate of this decline gradually accelerates (Collier and Murray, 1977; Murray, 1979a). Similar patterns of removal from the cotyledons are followed by total nitrogen, phosphorus, magnesium and potassium (Guardiola and Sutcliffe, 1971a, b, 1972; Sutcliffe and Bryant, 1977). A high proportion of the carbon received initially by the axis is retained, but growing root and shoot systems progressively lose higher proportions of the carbon received in respiration (Fig. 10). By the end of the period of most rapid transmission of reserves from cotyledons to axis, about 40% of the dry matter originally present has been lost as CO_2, in common bean (Fig. 10) and pea alike (Fig. 8). In both these studies, low light intensity was provided in a 12 h cycle, so that the seedlings could not become autotrophic.

Clearly the axis tissues contribute the major proportion of these respiratory losses, because the respiration rates of pea and bean cotyledons do not continue to increase as these become senescent. On the contrary, the respiration rates of cotyledons decline from maximum values with about half the original dry matter still present (*P. vulgaris*, Öpik, 1966) or when half the total nitrogen remains to be mobilized (pea, Guardiola and Sutcliffe, 1971a).

The continued presence of axis-derived organs as sinks for substances loaded into phloem is obviously one factor that will govern the rate of solute transfer in the phloem. If the rate of seedling growth is stimulated by the provision of exogenous calcium ions, then the rates of mobilization of the major cotyledonary reserves are increased (Ferguson and Bollard, 1976). To what extent though is the mobilization of reserves from cotyledons a predetermined activity, and to what extent is it controlled or regulated by some early stimulus from the axis? Is it true that key hydrolytic enzyme activities will fail to develop in the cotyledons if they are deprived of some substance or substances they would normally receive from the axis in the first few days following the beginning of imbibition?

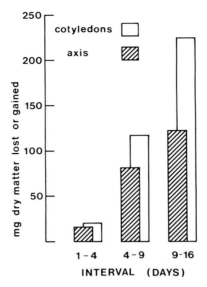

Fig. 10. Disappearance of dry matter from the cotyledons compared with the acquisition of dry matter by the axis in bean seedlings (*P. vulgaris* cv. Hawkesbury Wonder) growing under low light (1 klx 12 h photoperiod) at 19°C as described by Crump and Murray (1979). Data are expressed per seedling.

It is worthwhile asking these questions, as there are many conflicting factors to be taken into account. One is experimental design. In many earlier studies, excised cotyledons were incubated in Petri dishes under conditions that might have: (i) restricted the availability of oxygen; (ii) allowed the accumulation of volatile compounds such as ethylene (Sutcliffe and Bryant, 1977).

When aeration is improved by incubating excised pea cotyledons in large culture vessels, the extent of their increases in acid phosphatase [EC 3.1.3.2] and acid ribonuclease [EC 3.1.4] activities is greater than the extent of the increases normally observed in the cotyledons of intact seedlings, and not less, as would have occurred during incubation of excised cotyledons in Petri dishes (Bryant and Haczycki, 1976). Poor aeration can inhibit the development of isocitrate lyase [EC 4.1.3.1] activity in excised cotyledons of cucumber and squash (*Cucurbita maxima*) (Ford *et al.*, 1976).

A major difficulty with the hypothesis that there is a transmitted chemical stimulus from the axis to the cotyledons is one raised by Öpik (1966), namely the incongruity of the cells furthest from the vein endings in bean cotyledons responding first. The cells closest to the phloem retain their reserves for the longest time in this species and many others, whereas

7. Axis–Cotyledon Interaction   263

these cells might be expected to mobilize their reserves first if the process were entirely directed by some compound from the axis arriving in the phloem. Eight distinct tissue patterns of reserve mobilization have been described for legume cotyledons by Smith (1981). The programme for the initiation of reserve breakdown appears to be vested in the cotyledons, and not the axis.

A. Identification of Enzymes Synthesized *de novo* in Cotyledons

Sucrose is the predominant form of transport carbohydrate in the majority of plants. The supply of sucrose to phloem cells in cotyledons is derived initially from the reserves of free sugars, which are often substantial as in the cotyledons of wrinkle-seeded pea cultivars (e.g. Murray, 1979c). Subsequently, this supply of sucrose for export can be maintained through lipid breakdown (Chapter 6), oligosaccharide mobilization (Section 5.II,B of Volume 1) or polysaccharide breakdown. In pea cotyledons, the initial breakdown of starch must involve starch phosphorylase [EC 2.4.1.1] alone, since β-amylase can act only on soluble substrates (not intact starch grains) and α-amylase is absent (Juliano and Varner, 1969; Abbott and Matheson, 1972). There is a residual inorganic phosphate content of about 4 μmol in the mature pea seed (Murray, 1979c) that would allow starch phosphorylase to act following imbibition. The activity of starch phosphorylase in pea cotyledons increases after one day and reaches a maximum value after 6 days at 30°C (Matheson and Richardson, 1976). The initial activity is attributed to two isoenzymes, I and II, and the increase in activity is attributed entirely to isoenzyme I (Matheson and Richardson, 1976, 1978). Isoenzyme I corresponds to the 'fast' form detected electrophoretically (Yomo and Varner, 1973).

The most rapid breakdown of starch in cotyledon cells appears to depend upon α-amylase (Nolan and Davis, 1984), although the increases in both α-amylase and β-amylase activities are often reported to be similar (Yomo and Varner, 1973; Van Onckelen *et al.*, 1977). The increases in total amylase activities in pea cotyledons involve the synthesis of specific isoenzymes: one ('fast') for β-amylase, and two for α-amylase (Yomo and Varner, 1973). It is often presumed that these hydrolases make their way to the eroding surfaces and fissures of the starch grains, but Matile (1968) has observed that both β-amylase and α-glucosidase [EC 3.2.1.20] are localized in autolytic protein bodies reverting to vacuoles (see Section 3.V,B of Volume 1).

These and other enzymes whose increasing activities in cotyledons depend upon *de novo* protein synthesis (or at least appear to do so) are listed in Table III.

Table III. Enzyme activities concerned with the mobilization of reserves in cotyledons that show increases probably resulting from *de novo* synthesis following imbibition and germination

Enzyme activity	Species	References
endopeptidase		
(thiol)	*Pisum sativum*	Matile (1968)
		Guardiola and Sutcliffe (1971a)
	Phaseolus vulgaris	Yomo and Taylor (1973)
		Feller (1979)
	Vigna radiata	Baumgartner *et al.* (1978)
	Cucurbita maxima	Penner and Ashton (1967)
	Cucurbita sp.	Hara and Matsubara (1980)
aminopeptidase		
'fast' isoenzyme	*Arachis hypogaea*	Basha and Cherry (1978)
chelator sensitive	*Phaseolus vulgaris*	Crump and Murray (1979)
total	*Cucurbita* sp.	Hara and Matsubara (1980)
carboxypeptidase		
N-CBZ-L-Phe-Ala	*Phaseolus vulgaris*	Feller (1979)
	Gossypium hirsutum	Ihle and Dure (1969, 1972a, b)
peptidases (*esterases*)		
BTEEase	*Phaseolus vulgaris*	Crump and Murray (1979)
N-CBZ-L-Tyr-pNP	*Phaseolus vulgaris*	Crump and Murray (1979)
BAPAase	*Cucurbita* sp.	Hara and Matsubara (1980)
asparagine synthetase		
	Glycine max	Duke *et al.* (1978)
	Vigna radiata	Kern and Chrispeels (1978)
	Gossypium hirsutum	Dilworth and Dure (1978)
glutamine synthetase		
total	*Glycine max*	Duke *et al.* (1978)
total	*Vigna radiata*	Kern and Chrispeels (1978)
isoenzyme 2	*Arachis hypogaea*	Winter *et al.* (1982)
total	*Gossypium hirsutum*	Dilworth and Dure (1978)
cytoplasmic form	*Cucurbita* sp.	Nishimura *et al.* (1982)
glutamate dehydrogenase		
	Glycine max	Duke *et al.* (1978)
	Gossypium hirsutum	Dilworth and Dure (1978)
arginine decarboxylase		
	Cucumis sativus	Suresh *et al.* (1978)
acid phosphatase		
	Pisum sativum	Varner *et al.* (1963)
		Johnson *et al.* (1973)
	Phaseolus vulgaris	Lai *et al.* (1971)
		Crump and Murray (1979)
acid ribonuclease		
	Pisum sativum	Matile (1968)
		Bryant and Haczycki (1976)
peroxidase	*Vigna radiata*	Dendsay and Sachar (1978)
α-amylase		
2 isoenzymes	*Pisum sativum*	Yomo and Varner (1973)

Table III—*continued*

Enzyme activity	Species	References
total	*Pisum sativum*	Morohashi and Ueno (1980)
total	*Phaseolus vulgaris* Von Onckelen *et al.* (1977)	
1 isoenzyme	*Vigna mungo*	Koshiba and Minamikawa (1983)
β-amylase		
'fast' isoenzyme	*Pisum sativum*	Yomo and Varner (1973)
total	*Phaseolus vulgaris*	Van Onckelen *et al.* (1977)
starch phosphorylase		
'fast' isoenzyme	*Pisum sativum*	Yomo and Varner (1973)
isoenzyme I	*Pisum sativum*	Matheson and Richardson (1976)
sucrose: UDP glucosyltransferase	*Vicia faba*	Pridham *et al.* (1969)
lipase	oil-rich species	Section 6.IV,C
isocitrate lyase	oil-rich species	Section 6.III,C
malate synthetase	oil-rich species	Section 6.III,C

abbreviations: BAPA, α-N-benzoyl-D,L-arginine-*p*-nitroanilide; BTEE, N-benzoyl-L-tyrosine ethyl ester; N-CBZ-L-Phe-Ala, N-carbobenzoxy-L-phenylalanine-L-alanine; N-CBZ-L-Tyr-*p*NP, N-carbobenzoxy-L-tyrosine-*p*-nitrophenol.

B. Does the Axis Control the Enzyme Activities of Cotyledons?

The simple answer to this question is yes, if comparisons are made between cotyledons excised from the axis for varying periods and cotyledons from intact seedlings of corresponding ages. However, the incubation conditions chosen for the excised cotyledons can affect the outcome of the comparison, as noted above (Section II) and in Section 6.VIII. Species with cotyledons performing both storage and photosynthetic functions have provided confusing results, those most frequently studied being cucumber, squash, pumpkin (*Cucurbita pepo*) and radish (*Raphanus sativus*). Poor aeration inhibits the development of glyoxysomal enzyme activities in excised cotyledons, but when excised cucumber cotyledons are incubated on moist paper towels in open seed trays rather than in closed Petri dishes, they show a development of lipolytic and glyoxysomal enzyme activities that is essentially the same as in the cotyledons of intact seedlings (Slack *et al.*, 1977; Davies and Chapman, 1979a).

1. Proteolytic Enzymes

An interesting feature of many of these species is that the protein bodies in the cotyledon storage cells already contain an endopeptidase responsible for the initial (post-imbibition) cleavage of the larger subunit of the major reserve globulin (Hara-Nishimura et al., 1982). This endopeptidase is controlled during seed development by virtue of its location in the matrix portion of the protein body, whereas the potential substrate forms the crystalloid interior of the protein body (Hara-Nishimura et al., 1982; for similar inclusions in protein bodies of castor bean, see Fig. 9 of Chapter 3, Volume 1). The major reserve globulin is converted quantitatively to water-soluble polypeptides over an initial period of about 3 days in cucumber (Davies and Chapman, 1980) or 4 days in pumpkin (Reilly et al., 1978). There is no further synthesis of the endopeptidase responsible for this limited proteolysis, and its activity declines after 2 days (Reilly et al., 1978). The action of this enzyme might be described as delayed processing (Section 3.VI,B of Volume 1). As such it could be automatic, and there could be no possibility for hormonal regulation of the activity of this enzyme.

This conclusion was reached earlier by Penner and Ashton (1967), who determined that there were two major endopeptidases present in squash cotyledons, one that was unresponsive to experimental conditions (precipitating above 40% saturation with ammonium sulphate) and another, whose synthesis could be inhibited (precipitating between 30%-40% saturation with ammonium sulphate). This second proteinase activity develops over several days, and is evidently the one responsible for subsequent endopeptidase attack on accumulated water-soluble polypeptides derived from the major reserve globulin (Reilly et al., 1978; Davies and Chapman, 1979a; Hara and Matsubara, 1980).

A slower rate of reserve protein breakdown is observed in excised cucumber cotyledons compared to those of intact seedlings (Davies and Chapman, 1979a, b; 1980). However, this is not predisposed by any lack in the development of proteolytic enzyme activities (Davies and Chapman, 1979a). It thus seems unlikely that any hormonal regulation of the synthesis of enzyme proteins is involved. Certain peptides, for example, the tripeptide leupeptin, are potent inhibitors of the thiol endopeptidases active in reserve protein degradation (Alpi and Beevers, 1981). Feedback inhibition of both phloem loading of amino compounds and of total proteinase activity should be expected once export of metabolites to the axis is prevented by excision. This is the probable explanation of the slower rate of protein breakdown observed in excised cucumber cotyledons (Davies and Slack, 1981).

In the cotyledons of mung bean, newly synthesized thiol endopeptidase is transported to the vacuolar protein bodies (Baumgartner et al., 1978). The maintenance of normal endopeptidase activity in the cotyledons then depends upon their continued attachment to the axis (Kern and Chrispeels, 1978). However, the removal of the axis from the beginning of imbibition does not completely prevent an increase in the thiol endopeptidase activity from taking place (Fig. 11). In contrast, cycloheximide-sensitive increases

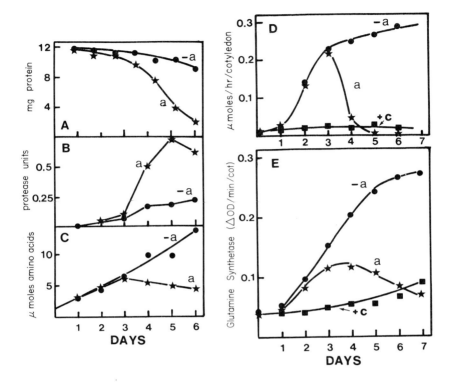

Fig. 11. Changes in protein content (A), thiol endopeptidase activity (B), free amino acid content (C), asparagine synthetase activity (D) and glutamine synthetase activity (E) in the cotyledons of mung bean seedlings (*Vigna radiata*), maintained in the absence (−a) or the presence (a) of the axis. Values are expressed per cotyledon (D, E) or per seedling (A, B, C). The treatment +c indicates results obtained for excised cotyledons incubated on sand moistened with 5 μM cycloheximide solution instead of water (from Kern and Chrispeels, 1978).

in glutamine synthetase [EC 6.3.1.2] and asparagine synthetase [EC 6.3.5.4] activities occur in the cotyledons regardless of attachment or excision (Fig. 11D, E). In excised cotyledons, the activities of both synthetases continue to increase, well beyond the time when they would begin to decline in the cotyledons of intact seedlings.

The protein bodies that can be isolated intact from the cotyledons of common bean initially lack the thiol-dependent endopeptidase necessary for rapid protein degradation (Pusztai *et al.*, 1977). The increases in this enzyme activity that have been reported to occur in excised bean cotyledons may exceed those of attached cotyledons (Yomo and Srinivasan, 1973) or the activities may fall behind (Feller, 1979; Gepstein and Ilan, 1980). Excision of bean cotyledons has no effect on their development of carboxypeptidase activity, which increases to a maximum value after 6 days at 23°C. In the same study, excision was found to delay the normal decline shown in total aminopeptidase activity of the cotyledons, and to promote the activity of a serine-dependent endopeptidase (Feller, 1979).

2. Reassembly of Mitochondria

The presence of the axis is necessary throughout the first 48 h for a number of reorganizational events involving membranes to take place in pea cotyledons (Varner *et al.*, 1963; Bain and Mercer, 1966b). These events include the reassembly of the internal membranes (cristae) of the mitochondria. It is well known that the recovery of respiratory activities of mitochondria in cotyledon cells is not uniform (Section 3.III,A). The capacity of mitochondria isolated from pea cotyledons to utilize succinate develops faster than their capacity to utilize malate or α-oxoglutarate (James and Spencer, 1979; Morohashi and Bewley, 1980). Initially, all of these pathways are cyanide sensitive, but subsequently a significant proportion of the total respiration involving these substrates develops cyanide-insensitivity (James and Spencer, 1979).

These developmental changes in pea probably do not require the *de novo* synthesis of any mitochondrial enzymes or their subunits on cytoplasmic ribosomes (Table IV), but rather an increasing level of organization within the inner membranes, with the possible contribution of some products of translation on mitochondrial (70S) ribosomes (Malhotra and Spencer, 1973).

Morohashi and Bewley (1980) have shown that excised pea cotyledons develop differently from attached cotyledons in some manner that affects the recovery of respiratory activities in the mitochondria. The development of the initial capacity of mitochondria to utilize succinate or exogenous NADH is unaffected by prior removal of the axis, but the capacity of mitochondria to utilize malate is seriously impaired, the degree of impairment depending on the time of excision (Table V).

C. Effects of Light and the Role of Cytokinins

Among the phytochrome-mediated effects of light on seedling growth and morphology are some that may be promoted by cytokinins, including the expansion of cotyledon or leaf cells. In the absence of light, cytokinin-enhanced cell expansion in cotyledons undertaking the conversion from

Table IV. Effects of cycloheximide (50 μM) on development of mitochondrial activities in pea cotyledons attached to the axis

Substrate	Treatment	O_2 uptake	Respiratory control R.C. ratio	ADP/O ratio
succinate	water	100	2.04	1.21
	+ CH	102	2.10	1.21
malate	water	100	2.71	1.50
	+ CH	95	2.80	1.45

Mitochondria were isolated and assayed after incubation of single cotyledons for 18 hours at 25°C in closed Petri dishes (9 cm diameter, 10 mL water or cycloheximide (CH) solution, 20 cotyledons per dish). Data of Morohashi and Bewley (1980).

Table V. Effects of time of excision on the capacity of mitochondria isolated from pea cotyledons to utilize malate

Treatment	O_2 uptake nmol/min per cotyledon
Attached 24 hours	4.6
Excised at t_0	1.5
Excised after 4 h	1.7
Excised after 8 h	1.7
Excised after 12 h	2.2
Excised after 18 h	3.3

Cotyledons were allowed to imbibe water for 24 h at 25°C with the axis removed at the times indicated. Data of Morohashi and Bewley (1980).

storage to photosynthetic organ has been found to provide a convenient bioassay for this class of phytohormone (Letham, 1971, 1978; Letham and Palni, 1983). Cytokinin-stimulated growth of cotyledons may be enhanced by red light, or cytokinins may interact synergistically with potassium ions, for example, in cotyledons of cucumber (Knypl, 1970; Knypl and Chylinska, 1974) and radish (Gordon and Letham, 1975). Potassium ions may stimulate the rate of protein synthesis without greatly affecting the rate of RNA synthesis, whereas treatment with cytokinins can stimulate the rate of RNA synthesis.

The effects of applied cytokinins or their analogues vary widely according to species. In excised cotyledons of sunflower (*Helianthus annuus*) maintained in darkness, the activities of microbody enzymes concerned with photosynthesis are enhanced by benzyladenine to a greater extent than those concerned with lipid mobilization, isocitrate lyase and malate synthetase [EC 4.1.3.2] (Servettaz et al., 1976). Some mitochondrial enzyme activities are also considerably elevated. In excised cotyledons of watermelon (*Citrullus lanatus*) maintained in darkness, mobilization of lipids, the synthesis of carotenoids and elevation of peroxisomal hydroxypyruvate reductase activity [EC 1.1.1.81] are all promoted by brief exposure to benzyladenine (Longo et al., 1978). Specific elevation of the cytochrome a_3 content of mitochondria, and of mitochondrial cytochrome oxidase activity [EC 1.9.3.1] are also observed in response to benzyladenine treatment of excised watermelon cotyledons (Longo et al., 1979). In cucumber cotyledons detached after 2 days imbition in the dark and maintained subsequently in the light, the activity of arginine decarboxylase [EC 4.1.1.19] and the concentration of putrescine (a polyamine) are both increased in response to kinetin and benzyladenine (Suresh et al., 1978).

The effects of benzyladenine on the accumulation of chlorophyll precursors in darkness have been reviewed by Dei (1982). Although benzyladenine was reported to elevate the activity of ribulose bisphosphate carboxylase [EC 4.1.1.39] in excised cucumber cotyledons kept in darkness (Harvey et al., 1974), no increase in the amount of enzyme protein can be detected following electrophoresis of extracts from excised radish cotyledons treated with kinetin in darkness for up to 4 days and compared with excised cotyledons maintained with no added kinetin in darkness (D. R. Murray, unpublished data). Cytokinins cannot completely substitute for red light, and according to Thomas et al. (1980) do not appear to accelerate the conversion of radish cotyledons from storage to photosynthetic organs. Nevertheless, in both radish and cucumber, the initial effects of applied cytokinins are consistent with their promotion of a faster rate of production of sucrose, with the consequent water uptake leading to a faster rate of cell expansion (Rayle et al., 1982).

In cotyledons adapted entirely for the storage and remobilization of reserves, the retention of products of reserve catabolism for the synthesis of chloroplasts and peroxisomes does not have the high priority it would in cotyledons performing both storage and photosynthetic functions, for example, in those of cotton (*Gossypium hirsutum*) (Doman et al., 1982; Tables VII, VIII of Chapter 6). Furthermore, as the storage cells are fully expanded during seed development (before final desiccation), cytokinins are not required for cell expansion following imbition. Even when storage cotyledons are epigeal, as in common bean, the conversion of proplastids to chloroplasts is obvious but vestigial, confined to cells of the hypodermis and bundle peripheries (Öpik, 1966; Lovell and Moore, 1970). Any effects of light on the rate of mobilization of reserves from such cotyledons are likely to reflect shifting 'sink' demands, as the shoot undergoes photomorphogenesis and then grows at a faster rate. Illumination can immediately promote the rate of export from already depleted cotyledons on removal of seedlings from darkness (Murray et al., 1973). Common bean seedlings grown entirely in darkness utilize the reserves of the cotyledons at a detectably slower rate than do seedlings exposed to a light–dark cycle under otherwise identical conditions (Figs 12, 13). The extent of the increase in α-amylase activity in the cotyledons of the light-grown plants is greater than the increase in this activity in the cotyledons of plants grown in darkness (Fig. 13). These results contrast with those of Van Onckelen et al.. (1977), but these authors used continuous illumination, which evidently prevented any further increase in the α-amylase activity in intact seedlings after only five days. There are no major differences in the patterns of development of proteolytic enzymes in cotyledons from dark-grown bean seedlings compared to those grown with a diurnal light–dark cycle (Feller, 1979).

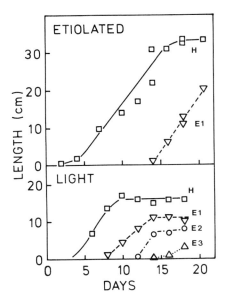

Fig. 12. Differences in the development of hypocotyl (H) and epicotyl (E) internodes in bean seedlings (*P. vulgaris* cv. Kentucky Wonder) grown in darkness at 25°C or in the light. The latter plants were placed under a 12 h photoperiod (8 klx) with temperatures of 23°C ('day') or 20°C ('night') after the first four days in darkness (data of B. D. Davis).

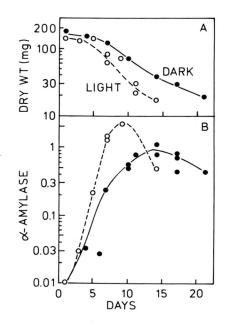

Fig. 13. Changes in (A) dry matter, and (B) α-amylase activity per cotyledon in light (O) or dark (●) grown bean (*Phaseolus vulgaris*) seedlings (for conditions, see Fig. 12) (data of B. D. Davis).

D. Are Cytokinins Transmitted by the Axis?

The possibility that the axis might influence the mobilization of reserves in the cotyledons by transmitting some chemical stimulus has been entertained for more than 20 years (Section II,A; II,B). Support for the identification of cytokinins as a transmitted regulatory factor comes from the studies of the responses of excised cotyledons to cytokinins and other phytohormones, usually tested at a range of concentrations with the exposures varied from intervals as brief as one minute to continuous (Long et al., 1978). Assessment of the possible role of each substance is based upon how closely the hormone-treated cotyledons match the cotyledons of intact seedlings of the same chronological age. Interpretations of the meaning of the results obtained from such studies are many and there is no concensus. Halmer and Bewley (1982) concluded that 'the degree to which the axis controls the cotyledonary reserves (of carbohydrate) is uncertain'. Davies and Slack (1981) have emphasized evidence in favour of feedback regulation, whereas Ilan and Gepstein (1980/1981) have argued in favour of photohormone transmission, taking into account the necessary qualifications about restricted aeration of cotyledons excised and maintained in closed Petri dishes.

The role of abscisic acid (ABA, or 'dormin') is usually inhibitory (Milborrow, 1974; King, 1982; Section 2.III,A). Its relationship to the regulation of growth in germinating non-dormant embryos is relevant here. In the immature embryonic axis of developing common bean seeds, ABA is believed to restrict cell enlargement, and to suppress precocious germination (Morris, 1978; Long et al., 1981). In early stages after the beginning of imbibition, common bean cotyledons display a pronounced capacity for breakdown of ABA (Van Onckelen et al., 1981). This capacity persists while the activities of α-amylase and β-amylase are increasing (Van Onckelen et al., 1977), but then declines. In attached common bean cotyledons entering the senescent phase, increasing levels of endogenous ABA are believed to be instrumental in bringing about the declines then observed in amylase activities (Van Onckelen et al., 1981). Removal of the apical shoot at this late stage defers both the increase in endogenous ABA level and the decrease in amylase activity (Fig. 14). These results imply that the cotyledons are responsive to shifts in the cytokinin:auxin ratio, where the source of auxins is the shoot apex and the source of cytokinins the root apex. If the cotyledons are receiving cytokinins at this stage (Fig. 14), then how soon after the beginning of imbibition does this reception begin?

Cytokinins and gibberellins are well known antagonists of ABA and so it is not surprising to find that the best evidence for hormonal replacement of the axis concerns cytokinin acting alone, or in concert with gibberellin to reinstate α-amylase and β-amylase activities in excised cotyledons of pea and bean (Gepstein and Ilan, 1979, 1979; Locker and Ilan, 1975; Ilan and Gepstein, 1980/1981). In support of the hypothesis that the axis begins to transmit cytokinin to the cotyledons in the first two days is the observation of Gepstein and Ilan (Table VI) that the radicle rather than the whole axis is identifiable as the source of transmitted factor. This finding is doubly significant: (i) the root apex is known to be the major source of cytokinins in the young pea seedling (Torrey and Zobel, 1977); and (ii) the first effective phloem connections to be established are those between the radicle and the cotyledons (Fig. 1).

So we may accept the feasibility of the hypothesis that cytokinins are delivered to the cotyledons from about 12 hours following the beginning of imbibition and may then promote the development of α-amylase (Gepstein and Ilan, 1979), β-amylase (Van Onckelen et al., 1977) or thiol endopeptidase activities, by enhancing their rates of mRNA synthesis. However, we need to be aware that this kind of interaction with transcription and translation is neither substantiated nor disproved.

274 David R. Murray

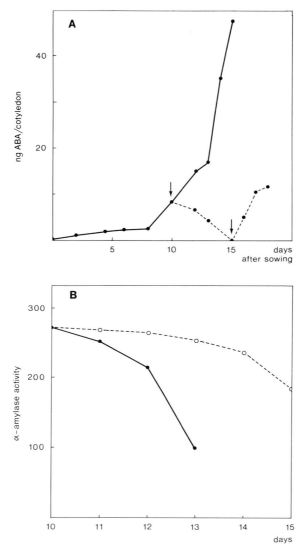

Fig. 14. (A) Changes in the content of endogenous abscisic acid in the cotyledons of etiolated bean seedlings (*Phaseolus vulgaris* cv. Limburg) grown in darkness at 21°C. Dashed lines connect the values for seedlings from which the apical shoot was removed at 10 d (1st arrow). Axillary bud growth commenced in these seedlings at 15 d (2nd arrow). (B) changes in α-amylase activity (as iodine-dextrin colour units at 50°C) in the cotyledons of intact (●) or 'decapitated' (○) etiolated bean seedlings. (From Van Onckelen et al., 1981).

Table VI. The development of amylase activity in bean[a] cotyledons as affected by the presence or absence of components of the attached embryonic axis

Treatment	Amylase activity[b]
Embryonic axis intact	0.36 ± 0.04
Minus whole axis	0.20 ± 0.03
Minus plumule	0.37 ± 0.05
Minus hypocotyl and radicle	0.23 ± 0.02
Minus radicle	0.20 ± 0.02

[a] *Phaseolus vulgaris* cv. Brittle Wax.
[b] Expressed as mg maltose per cotyledon. Standard errors also shown. Samples of 10 cotyledons were incubated at 25°C in darkness for 2 days in Petri dishes containing 3 sheets of filter paper and 6 mL of 5 mM potassium phosphate buffer, pH 6.9. Reproduced with permission from Ilan and Gepstein (1980/1981).

The initial stages of starch and protein breakdown in storage cotyledons are independent of attachment to the axis (Section II,A). Starch breakdown begins without any α-amylase being present. Increases in the activities of both α-amylase and β-amylase result from the *de novo* synthesis of enzyme molecules, both in excised cotyledons, and in the cotyledons of intact seedlings (e.g., Yomo and Varner, 1973; Koshiba and Minamikawa, 1983). However, major differences in the rates of accumulation of newly synthesized active enzyme molecules are soon evident when excised and attached cotyledons are compared. This is true for the thiol endopeptidase (Kern and Chrispeels, 1978; Gepstein and Ilan, 1980) as well as for both amylases.

The increases in enzyme activity in detached cotyledons are not always less than the increases observed for cotyledons from intact seedlings. In one study of common bean, increases in the proteolytic activity of detached cotyledons continued for 12 days in darkness at 22°C, although the rate of protein breakdown in these cotyledons progressively declined (Yomo and Srinivasan, 1973). The development of α-amylase activity in excised cotyledons of peanut (*Arachis hypogaea*) is identical with the pattern of development shown by attached cotyledons (Allfrey and Northcote, 1977). In excised cotyledons of black gram (*Vigna mungo*), α-amylase activity increases to a higher value than that attained in attached cotyledons (Koshiba and Minamikawa, 1983). This increase was determined by immunoassay to be related to an increased amount of enzyme protein. The failure of many enzyme activities to decline in detached cotyledons (e.g., Fig. 11D, E) suggests that access of these enzyme molecules to proteolytic enzymes is somehow prevented, or that insufficient endopeptidase activity develops (e.g., Fig. 11B).

III. CONCLUSIONS

Is it necessary to postulate the early transmission of cytokinins from the axis in order to account for differences in development of enzyme activities in excised cotyledons compared to cotyledons of intact seedlings? Certainly restoration of the endopeptidase activity of excised cotyledons by application of cytokinin is consistent with a hormonal regulation hypothesis. However, such results cannot disprove an alternative hypothesis, that lower endopeptidase activity develops in excised cotyledons because of feedback effects, with accumulating peptides slowing the rate of reserve protein breakdown through their inhibitory action on existing enzyme molecules (Davies and Chapman, 1980). A second element of feedback inhibition then requires that the accumulating peptides also interact with the programme for transcription of mRNA species corresponding to endopeptidases and possibly other proteolytic enzymes. Similar arguments may be applied to studies of the development of amylase activities, but here the onus is on the proponents of a feedback hypothesis to identify the components of feedback inhibition more precisely (Gepstein and Ilan, 1981). The accumulation of sugars in cotyledons following excision (e.g., Fig. 11 of Chapter 6) has no readily identifiable effect on feedback, but the possibilities should be investigated further.

To conclude, let me reiterate two points made earlier. First, the radicle is the source of a transmitted stimulus that might well be zeatin or a related cytokinin responsible for maintaining the rate of increase in amylase activities in common bean cotyledons (Table VI). Secondly, we require an explanation for the still mysterious effect of the axis in promoting the reorganization of intracellular membrane structure, especially within mitochondria (Table V). These effects are certainly within the ambit of those attributed to cytokinins in cotyledons that become photosynthetic (Section II,C).

The apparent impasse between the two main schools of thought, namely transmitted hormonal stimulus versus feedback inhibition, presents an excellent opportunity for the integrated application of a range of analytical techniques. It should be possible to determine transport and turnover rates for hormones, mRNA and enzyme molecules, and to link such studies with immunocytochemical procedures for localizing macromolecules of interest in serial sections of the cotyledons and the axis. Sensitive antibody techniques are now being applied to the measurement of the concentrations of individual cytokinins (e.g., Constantinidou *et al.*, 1978; Jayabaskaran and Jacob, 1982). In the latter study, cucumber cotyledons were compared with embryonic axes after 4 days growth in darkness. The cotyledons were shown to possess a similar spectrum of cytokinins to the axis. With extension, this approach may provide the answer to a key question: whether the cytokinins of the cotyledons arise mainly by synthesis *in situ*, by RNA breakdown *in situ*, or by translocation from the radicle. Perhaps all of these possible sources vary in importance with species or cultivar, and with stage of development.

REFERENCES

Abbott, I. R., and Matheson, N. K. (1972). *Phytochemistry* **11**, 1261-1272.
Allfrey, J. M., and Northcote, D. H. (1977). *New Phytol.* **78**, 547-563.
Alpi, A., and Beevers, H. (1981). *Plant Physiol.* **68**, 851-853.
Ashcroft, W. J., and Murray, D. R. (1979). *Aust. J. Bot.* **27**, 343-352.
Bain, J. M., and Mercer, F. V. (1966a). *Aust. J. Biol. Sci.* **19**, 69-84.
Bain, J. M., and Mercer, F. V. (1966b). *Aust. J. Biol. Sci.* **19**, 85-96.
Basha, S. M. M., and Beevers, L. (1975). *Planta* **124**, 77-87.
Basha, S. M. M., and Cherry, J. P. (1978). *J. Agric. Food Chem.* **26**, 229-234.
Baumgartner, B., Tokuyasu, K. T., and Chrispeels, M. J. (1978). *J. Cell Biol.* **79**, 10-19.
Becker, W. M., Riezman, H., Weir, E. M., Titus, D. E., and Leaver, C. J. (1982). *Ann. N.Y. Acad. Sci.* **386**, 329-349.
Beevers, L., and Guernsey, F. S. (1966). *Plant Physiol.* **41**, 1455-1458.
Bryant, J. A., and Haczycki, S. J. (1976). *New Phytol.* **77**, 757-760.
Collier, M. D., and Murray, D. R. (1977). *Aust. J. Plant Physiol.* **4**, 571-582.
Collins, D. M., and Wilson, A. T. (1972). *Phytochemistry* **11**, 1931-1935.
Constantinidou, H. A., Steele, J. A., Kozlowski, T. T., and Upper, C. D. (1978). *Plant Physiol.* **62**, 968-974.
Crump, J. A., and Murray, D. R. (1979). *Aust. J. Plant Physiol.* **6**, 467-474.
Davies, H. V., and Chapman, J. M. (1979a). *Planta* **146**, 579-584.
Davies, H. V., and Chapman, J. M. (1979b). *Planta* **146**, 585-590.
Davies, H. V., and Chapman, J. M. (1980). *Planta* **149**, 288-291.
Davies, H. V., and Slack, P. T. (1981a). *New Phytol.* **88**, 41-51.
Davis, B. D. (1977). *Plant Physiol.* **60**, 513-517.
Davis, B. D. (1979). *Plant Physiol.* **63**, 1010-1015.
Davis, B. D. (1983a). *Am. J. Bot.* **70**, 816-820.
Davis, B. D. (1983b). *Am. J. Bot.* **70**, 821-826.
Dei, M. (1982). *Physiol. Plant.* **56**, 407-414.
Dendsay, J. P. S., and Sachar, R. C. (1978). *Phytochemistry* **17**, 1017-1019.
Dilworth, M. F., and Dure, L. S. III. (1978). *Plant Physiol.* **61**, 698-702.
Doman, D. C., Walker, J. C., Trelease, R. N., and Moore, B. D. (1982). *Planta* **155**, 502-510.
Duke, S. H., Schrader, L. E., Miller, M. G., and Niece, R. L. (1978). *Plant Physiol.* **62**, 642-647.

Dure, L. S. (1960). *Plant Physiol.* **35**, 919-925.
Feller, U. (1979). *Z. Pflanzenphysiol.* **95**, 413-422.
Ferguson, I. B., and Bollard, E. G. (1976). *Ann. Bot.* **40**, 1047-1055.
Ford, M. J., Slack, P., Black, M., and Chapman, J. M. (1976). *Planta* **132**, 205-208.
Gepstein, S., and Ilan, I. (1970). *Plant Cell Physiol.* **11**, 819-822.
Gepstein, S., and Ilan, I. (1979). *Plant Cell Physiol.* **20**, 1603-1607.
Gepstein, S., and Ilan, I. (1980). *Plant Cell Physiol.* **21**, 57-63.
Gepstein, S., and Ilan, I. (1981). *Physiol. Plant.* **53**, 188-190.
Gordon, M. E., and Letham, D. S. (1975). *Aust. J. Plant Physiol.* **2**, 129-154.
Guardiola, J. L., and Sutcliffe, J. F. (1971a). *Ann. Bot.* **35**, 791-807.
Guardiola, J. L., and Sutcliffe, J. F. (1971b). *Ann. Bot.* **35**, 809-823.
Guardiola, J. L., and Sutcliffe, J. F. (1972). *J. Exp. Bot.* **23**, 322-337.
Guldager, P. (1978). *Theor. Appl. Genet.* **53**, 241-250.
Halmer, P., and Bewley, J. D. (1982). *In* 'Encyclopedia of Plant Physiology' (F. A. Loewus and W. Tanner, eds), New Series, Vol. 13A, pp. 748-793. Springer-Verlag, Berlin.
Hara, I., and Matsubara, H. (1980). *Plant Cell Physiol.* **21**, 233-245.
Hara-Nishimura, I., Nishimura, M., Matsubara, H., and Akazawa, T. (1982). *Plant Physiol.* **70**, 699-703.
Harvey, B. M. R., Lu, B. C., and Fletcher, R. A. (1974). *Can. J. Bot.* **52**, 2581-2586.
Ihle, J. N., and Dure, L. S. III (1969). *Biochem. Biophys. Res. Commun.* **36**, 705-710.
Ihle, J. N., and Dure, L. S. III (1972a). *J. Biol. Chem.* **247**, 5034-5040.
Ihle, J. N., and Dure, L. S. III (1972b). *J. Biol. Chem.* **247**, 5041-5047.
Ilan, I., and Gepstein, S. (1980/1981). *Isr. J. Bot.* **29**, 193-206.
James, T. W., and Spencer, M. S. (1979). *Plant Physiol.* **64**, 431-434.
Jayabaskaran, C., and Jacob, T. M. (1982). *Plant Physiol.* **70**, 1396-1400.
Johnson, C. B., Holloway, B. R., Smith H., and Grierson, D. (1973). *Planta* **115**, 1-10.
Juliano, B. O., and Varner, J. E. (1969). *Plant Physiol.* **44**, 886-892.
Kern, R., and Chrispeels, M. J. (1978). *Plant Physiol.* **62**, 815-819.
Killeen, L. A., and Larson, L. A. (1968). *Am. J. Bot.* **55**, 961-965.
King, R. W. (1982). *In* 'The Physiology and Biochemistry of Seed Development, Dormancy and Germination' (A. A. Khan, ed.), pp. 157-181. Elsevier Biomedical Press, The Hague.
Knypl, J. S. (1970). *Current Sci.* **39**, 534-535.
Knypl, J. S., and Chylinska, K. M. (1974). *Biochem. Physiol. Pflanzen.* **166**, 345-350.
Koshiba, T., and Minamikawa, T. (1983). *Plant Physiol.* **71**, 173-176.
Lai, Y. F., Thompson, J. E., and Barrell, R. W. (1971). *Phytochemistry* **10**, 41-49.
Lawrence, J. M., and Grant, D. R. (1963). *Plant Physiol.* **38**, 561-566.
Letham, D. S. (1971). *Physiol. Plant.* **25**, 391-396.
Letham, D. S. (1978). *In* 'Phytohormones and Related Compounds' (D. S. Letham, P. B. Goodwin and T. J. V. Higgins, eds), Vol. 1, pp. 205-263. Elsevier/North Holland, Amsterdam.
Letham, D. S., and Palni, L. M. S. (1983). *Annu. Rev. Plant Physiol.* **34**, 163-197.
Locker, A., and Ilan, I. (1975). *Plant Cell Physiol.* **16**, 449-454.
Long, S. R., Dale, R. M. K., and Sussex, I. M. (1981). *Planta* **153**, 405-415.
Longo, G. P., Olginati, M., Rossi, G., Valente, L., and Longo, C. P. (1978). *Plant Cell Environ.* **1**, 39-43.
Longo, G. P., Pedretti, M., Rossi, G., and Longo, C. P. (1979). *Plant Sci. Lett.* **14**, 213-223.
Lovell, P. H., and Moore, K. G. (1970). *J. Exp. Bot.* **21**, 1017-1030.

Lovell, P. H., and Moore, K. G. (1971). *J. Exp. Bot.* **22**, 153–162.
Lush, W. M., and Wien, H. C. (1980). *J. Agric. Sci.* **94**, 177–182.
Malhotra, S. S., and Spencer, M. (1973). *Plant Physiol.* **52**, 575–579.
Manen, J. F., and Pusztai, A. (1982). *Planta* **155**, 328–334.
Manickam, A., and Carlier, A. R. (1980). *Planta* **149**, 234–240.
Matheson, N. K., and Richardson, R. H. (1976). *Phytochemistry* **15**, 887–892.
Matheson, N. K., and Richardson, R. H. (1978). *Phytochemistry* **17**, 195–200.
Matile, Ph. (1968). *Z. Pflanzenphysiol.* **58**, 365–368.
Milborrow, B. V. (1974). *Annu. Rev. Plant Physiol.* **25**, 259–307.
Monnier, M. (1982). *Am. J. Bot.* **69**, 896–903.
Morohashi, Y., and Bewley, J. D. (1980). *Plant Physiol.* **66**, 70–73.
Morohashi, Y., and Ueno, K. (1980). *Z. Pflanzenphysiol.* **96**, 303–310.
Morris, D. A. (1978). *Z. Pflanzenphysiol.* **86**, 433–441.
Murray, D. R. (1979a). *Plant, Cell Environ.* **2**, 221–226.
Murray, D. R. (1979b). *Planta* **147**, 117–121.
Murray, D. R. (1979c). *Plant Physiol.* **64**, 763–769.
Murray, D. R. (1982). *Z. Pflanzenphysiol.* **108**, 17–25.
Murray, D. R., and Crump, J. A. (1979). *Z. Pflanzenphysiol.* **94**, 339–350.
Murray, D. R., and Knox, R. B. (1977). *J. Cell Sci.* **26**, 9–18.
Murray, D. R., Wara-Aswapatti, O., Ireland, H. M. M., and Bradbeer, J. W. (1973). *J. Exp. Bot.* **24**, 175–184.
Murray, D. R., Peoples, M. B., and Waters, S. P. (1979). *Planta* **147**, 111–116.
Nishimura, M., Bhusawang, P., Strzalka, K., and Akazawa, T. (1982). *Plant Physiol.* **70**, 353–356.
Nolan, M. C., and Davis, B. D. (1984). *Am. J. Bot.* **71**, 137–141.
Öpik, H. (1966). *J. Exp. Bot.* **17**, 427–439.
Pate, J. S. (1977). *In* 'The Physiology of the Garden Pea' (J. F. Sutcliffe and J. S. Pate, eds), pp. 349–383. Academic Press, London.
Penner, D., and Ashton, F. M. (1967). *Plant Physiol.* **42**, 791–796.
Pridham, J. B., Walter, M. W., and Worth, H. G. J. (1969). *J. Exp. Bot.* **20**, 317–324.
Prisco, J. T., Ainouz, I. L., and Melo, S. de C. (1975). *Physiol. Plant.* **33**, 18–21.
Pusztai, A., Croy, R. R. D., Grant, G., and Watt, W. B. (1977). *New Phytol.* **79**, 61–71.
Rayle, D. L., Ross, C. W., and Robinson, N. (1982). *Plant Physiol.* **70**, 1634–1636.
Reilly, C. C., O'Kennedy, T. T., Titus, J. S., and Splittstoesser, W. E. (1978). *Plant Cell Physiol.* **19**, 1235–1246.
Servettaz, O., Cortesi, F., and Longo, C. P. (1976). *Plant Physiol.* **58**, 569–572.
Shininger, T. L. (1972). *Plant Physiol.* **49**, 341–344.
Simola, L. K. (1971). *Physiol. Plant.* **25**, 98–105.
Simon, E. W., and Meany, A. (1965). *Plant Physiol.* **40**, 1136–1139.
Slack, P. T., Black, M., and Chapman, J. M. (1977). *J. Exp. Bot.* **28**, 569–577.
Smith, D. L. (1981). *In* 'Advances in Legume Systematics' (R. M. Polhill and P. H. Raven, eds), Part 2, pp. 927–940. Royal Botanic Gardens, Kew, U.K.
Smith, D. L., and Flinn, A. M. (1967). *Planta* **74**, 72–85.
Suresh, M. R., Ramakrishna, S., and Adiga, P. R. (1978). *Phytochemistry* **17**, 57–63.
Sutcliffe, J. F., and Bryant, J. A. (1977). *In* 'The Physiology of the Garden Pea' (J. F. Sutcliffe and J. S. Pate, eds), pp. 45–82. Academic Press, London.
Takaiwa, F., and Tanifuji, S. (1978). *Plant Cell Physiol.* **19**, 1507–1518.
Thomas, J., Kugrens, P., and Ross, C. W. (1980). *Am. J. Bot.* **67**, 456–464.
Torrey, J. G., and Zobel, R. (1977). *In* 'The Physiology of the Garden Pea' (J. F. Sutcliffe and J. S. Pate, eds), pp. 119–152. Academic Press, London.

Van Onckelen, H. A., Caubergs, R., and De Greef, J. A. (1977). *Plant Cell Physiol.* **18**, 1029-1040.
Van Onckelen, H. A., Horemans, S., and De Greef, J. A. (1981). *Plant Cell Physiol.* **22**, 507-515.
Varner, J. E., Balce, L. V., and Huang, R. C. (1963). *Plant Physiol.* **38**, 89-92.
Vidovic, M., and Murray, D. R. (1984). *Z. Pflanzenphysiol.* **113**, 117-128.
Winter, H. C., Powell, G. K., and Dekker, E. E. (1982). *Plant Physiol.* **69**, 41-47.
Xavier-Filho, J., and do Monte Negreiros, A. N. (1979). *Z. Pflanzenphysiol.* **95**, 423-430.
Yomo, H., and Srinivasan, K. (1973). *Plant Physiol.* **52**, 671-673.
Yomo, H., and Taylor, M. P. (1973). *Planta* **112**, 35-43.
Yomo, H., and Varner, J. E. (1973). *Plant Physiol.* **51**, 708-713.
Yoo, B. Y. (1970). *J. Cell Biol.* **45**, 168-171.

Plant Species Index

A

Abutilon, 12
   theophrasti, 19, 22, 32, 34
Acacia, 9, 21, 24, 26, 29–32, 35, 38
   accola, 16
   albida, 34
   aneura, 32
   baileyana, 16
   bidwillii, 32
   brachybotrya, 28
   buxifolia, 28
   calamifolia, 28
   cyclopis, 34
   dealbata, 248
   farnesiana, 13, 16, 25, 29
   giraffae, 34
   iteaphylla, 248, 259
   karoo, 32
   longifolia, 26, 28, 29, 37
   mearnsii, 3
   melanoxylon, 6, 28
   pycnantha, 28
   raddiana, 34
   retinodes, 28
   saligna, 16
   sclerosperma, 16
   sophorae, 26, 27, 28, 29
   suaveolens, 10
   tortilis, subsp. *spirocarpa*, 34
Acer
   platanoides, 46, 52, 60, 68, 72
   pseudoplatanus, 49, 50, 52, 56
   saccharum, 52, 55, 56
Agropyron desertorum, 95
Agrostemma githago, 47

Albizzia
   julibrissin, 3, 4, 6, 32
   lophantha, 17, 18, 29, 32
Aleurites fordii, 202, 203, 206
Allium cepa, 95
Amaranthus
   palmeri, 34
   retroflexus, 57, 58
Annonaceae, 147
annual ryegrass, *see Lolium rigidum*
Apium graveolens, 80, 148, 155, 156
Arabidopsis thaliana, 50
Arachis hypogaea, 56, 84, 86, 91, 99, 202, 203, 206, 219, 220, 221, 234, 236, 238, 239, 241, 264, 275
Aristolochia, 9
Asarum, 9
ash, European, *see Fraxinus excelsior*
Avena
   fatua, 58, 61, 65, 66
   sativa, 60, 61, 92, 102, 103, 109, 119, 182, 183, 190

B

barley, *see Hordeum vulgare*
bean, common, *see Phaseolus vulgaris*
beech, European, *see Fagus sylvatica*
Bertholletia excelsa, 203, 206
Bidens cernua, 249
Brassica napus, 78, 104, 106, 202, 203, 204, 206, 219, 220, 221, 222, 241
Brazil nut, *see Bertholletia excelsa*
broad bean, *see Vicia faba*
Bromus mollis, 146
butternut, *see Juglans cinerea*

C

Caesalpiniaceae, 9
Camellia sinensis, 83
Cannabis sativa, 203, 206
Canna compacta, 7
caraway, *see Carum carvi*
carob, *see Ceratonia siliqua*
Carthamus tinctorius, 202, 203, 206
Carum carvi, 147, 155
Carya
 illinoensis, 203, 206
 ovata, 24
Cassia multijuga, 6
castor bean, *see Ricinus communis*
celery, *see Apium graveolens*
Ceratonia siliqua, 150, 153
charlock, *see Sinapis arvensis*
Chenopodiaceae, 6
Chenopodium
 album, 6, 7
 murale, 6
cherry, *see Prunus avium*
chick pea, *see Cicer arietinum*
Cicer arietinum, 30, 105
Citrullus lanatus (vulgaris), 203, 206, 216, 220, 234, 236, 270
cocklebur, *see Xanthium pennsylvanicum*
coconut, *see Cocos nucifera*
Cocos nucifera, 202, 203, 206
Coffea arabica, 147
coffee bean, *see Coffea arabica*
Convolvulaceae, 147
Coronilla varia, 13, 24, 25, 32, 36
Corylus avellana, 28, 46, 49, 50, 51, 53–55, 65, 68, 69–71, 203, 206, 249
cotton, *see Gossypium hirsutum*
cowpea, *see Vigna unguiculata*
crambe, *see Crambe abyssinica*
Crambe abyssinica, 203, 204, 206
Crotalaria, 19, 32
cucumber, *see Cucumis sativus*
Cucumeropsis edulis, 219
Cucumis sativus, 79, 109, 203, 206, 210, 211, 214, 216, 220, 226, 234, 238, 239, 240, 241, 248, 262, 264, 265, 266, 269, 270, 277
Cucurbita, 264
 maxima, 237, 238, 239, 262, 264, 265, 266
 pepo, 203, 206, 232, 233, 234, 235, 236, 239, 240, 266
Cuscuta campestris, 12
Cyamopsis tetragonolobus, 150, 151

D

date palm, *see Phoenix dactylifera*
Douglas fir, *see Pseudotsuga menziesii*

E

Ebenaceae, 147
Elaeis guineensis, 202, 203, 206

F

Fagus sylvatica, 52
fenugreek, *see Trigonella foenum-graecum*
flax, *see Linum usitatissimum*
Fraxinus excelsior, 48, 49, 50, 67

G

Gleditsia triacanthos, 13, 25, 150
Glycine max, 79, 82, 84, 85, 86, 90, 92, 93, 94, 96, 110, 202, 203, 206, 219, 220, 221, 222, 236, 238, 241, 264
Gossypium, 12, 20, 32
 hirsutum, 3, 22, 50, 90, 92, 93, 106, 107, 184, 202, 203, 205, 206, 210, 214, 216, 219, 220, 225, 226, 227, 228, 229, 234, 235, 236, 237, 238, 239, 241, 264, 270
guar, *see Cyamopsis tetragonolobus*
Guizotia abyssinica, 203, 206

H

Haplopappus gracilis, 110
hazel nut, *see Corylus avellana*
Helianthus annuus, 91, 202, 203, 206, 220, 234, 236, 241, 269
hemp, *see Cannabis sativa*
hickory nut, *see Carya ovata*
honey locust, *see Gleditsia triacanthos*
Hordeum
 leporinum (murinum), 6
 spontaneum, 129
 vulgare, 50, 90, 96, 102, 118, 120, 121, 122, 123, 124, 125, 126, 127–129, 131, 133, 135–137, 138, 140, 142, 143, 144, 146, 156, 164, 165, 169, 170, 174–176, 177, 178–181, 184, 185, 186, 188, 191, 192, 193, 194

Plant Species Index

J
jojoba bean, *see Simmondsia chinensis*
Juglans
  cinerea, 24
  nigra, 24, 28
  regia, 56
Juncus
  bufonius, 6
  tenuis, 6

L
Lactuca sativa, 7, 47, 62, 63, 90, 92, 96, 100, 108, 148, 154, 155, 184, 203, 239
Leguminosae, 2, 5, 6, 7, 8, 38, 147
lettuce, *see Lactuca sativa*
Leucadendron daphnoides, 56
Lima bean, *see Phaseolus lunatus*
lime, large-leafed, *see Tilia platyphyllos*
Linum usitatissimum, 202, 203, 206, 234, 236
loblolly pine, *see Pinus taeda*
Loganiaceae, 147
Lolium rigidum, 109
Lotus corniculatus, 84
lucerne, *see Medicago sativa*
Lupinus
  angustifolius, 30
  arcticus, 6, 7
  cosentinii, 36
Lycopersicon esculentum, 203, 206

M
Macadamia ternifolia, 202, 203, 206
maize, *see Zea mays*
Malus sylvestris, 52, 56, 68
Malvaceae, 2, 5, 6, 7, 8, 19, 22, 38
Malva parviflora, 6
maple
  Norway, *see Acer platanoides*
  sugar, *see Acer saccharum*
marrow, *see Cucurbita pepo*
Medicago
  polymorpha (hispida), 6
  sativa, 150
Melilotus, 13, 17
  indicus, 6
Mimosaceae, 9, 19, 22, 39
Mucuna urens, 12
mung bean, *see Vigna radiata*
mustard, *see Sinapis alba*

N
Nelumbo nucifera, 6, 7
nigerseed, *see Guizotia abyssinica*

O
oak, *see Quercus robur*
oats, *see Avena sativa*
Olea europaea, 202, 203, 206
onion, *see Allium cepa*
Oryza sativa, 5, 92, 142, 143, 144, 168, 169, 171, 184, 185, 186, 187

P
palm, oil, *see Elaeis guineensis*
Palmae, 147
Papilionaceae, 6, 9, 11, 14, 15, 19, 32
Parkia, 22
pea, *see Pisum sativum*
peanut, *see Arachis hypogaea*
pear, *see Pyrus communis*
pearl millet, *see Pennisetum americanum*
Pennisetum americanum, 119, 120
Peperomia, 9
Phaseolus
  coccineus, 248
  lunatus, 25, 85, 92, 102, 107
  vulgaris, 13, 17, 59, 82, 85, 91, 92, 93, 95, 96, 97, 98, 101, 104, 110, 248, 249, 251, 253, 254, 255, 257, 258, 260, 261, 262, 264, 265, 267, 268, 270–275, 276
Phoenix dactylifera, 147
Physostigma venenosum, 12
Picea sitchensis, 53, 62
Pinus, 28, 241
  lambertiana, 53, 66
  pinea, 203, 206
  ponderosa, 220, 226, 238, 239
  sylvestris, 69, 168, 169, 170, 171, 177, 181, 196
  taeda, 53, 60
Pisum
  elatius, 23
  sativum, 49, 79–81, 82, 84, 85, 86, 87, 90, 91, 92, 93, 94, 95, 97, 98, 102, 104, 110, 144, 189, 248, 249, 250–253, 254, 257, 258, 260, 261, 262, 263, 264, 265, 268, 269, 273
Poaceae, 6
potato, *see Solanum tuberosum*

Potentilla norvegica, 58
Pothos longipes, 7
Prosopis farcata, 12, 13, 25
Protea compacta, 56
Prunus
    avium, 52
    cerasus, 63, 65, 67
Pseudotsuga menziesii, 53, 56, 68, 203, 219, 239
pumpkin, *see Cucurbita pepo*
Pyrus communis, 66, 72

Q

Queensland nut, *see Macadamia ternifolia*
Quercus robur, 52

R

radish, *see Raphanus sativus*
rape, *see Brassica napus*
Raphanus sativus, 100, 104, 265, 269, 270
redroot pigweed, *see Amaranthus retroflexus*
Rhus
    glabra, 34
    ovata, 32
rice, *see Oryza sativa*
Ricinus communis, 9, 28, 97, 98, 170, 171, 172, 173, 177, 181, 182, 184, 196, 202, 203, 206, 211, 218, 219, 220, 221, 225, 226, 229, 230, 231, 232, 233, 234, 235, 236, 238, 239, 241, 248, 266
Robinia pseudo-acacia, 3, 6
Rubiaceae, 147
rye, *see Secale cereale*

S

safflower, *see Carthamus tinctorius*
Schefflera actinophylla, 7
Scots pine, *see Pinus sylvestris*
Secale cereale, 7, 102, 104, 106, 109
sesame, *see Sesamum indicum*
Sesamum indicum, 202, 203, 206
Sida spinosa, 12, 22, 23
Simmondsia chinensis, 203, 204, 205, 207–209, 210, 212, 220, 223, 224, 226, 236, 239, 240, 241
Sinapis
    alba, 96, 203, 206, 219, 220, 222, 239, 241
    arvensis, 95
sitka spruce, *see Picea sitchensis*

Sium circutaefolium, 6
Solanum tuberosum, 83
sorghum, *see Sorghum bicolor*
Sorghum bicolor, 146, 174, 192
sour cherry, *see Prunus cerasus*
soybean, *see Glycine max*
Spergula arvensis, 6, 7, 59, 65
stone pine, *see Pinus pinea*
Striga asiatica, 56
subclover, *see Trifolium subterraneum*
sugar pine, *see Pinus lambertiana*
sunflower, *see Helianthus annuus*
sycamore, *see Acer pseudoplatanus*

T

Tilia platyphyllos, 52
tomato, *see Lycopersicon esculentum*
Trifolium, 6, 13
    subterraneum, 18, 20, 23, 25
Trigonella foenum-graecum, 148, 149, 150, 151, 153, 154
Triticale, 187
Triticum aestivum, 49, 50, 78, 102, 103, 104, 105, 106, 110, 118, 119, 123, 124, 127, 130, 135, 137, 138, 140, 142, 143, 144, 146, 156, 164, 169, 170, 176, 184, 185, 187, 188, 191, 192
tung, *see Aleurites fordii*

U

Umbelliferae, 155

V

Vaccaria pyramidata, 47
velvetleaf, *see Abutilon theophrasti*
Viburnum blattaria, 7
Vicia faba, 91, 92, 98, 106, 107, 265
Vigna
    mungo, 265, 275
    radiata, 96, 105, 185, 187, 252, 264, 266, 267
    unguiculata
        subsp. *dekindtiana*, 15
        subsp. *mensensis*, 14
        subsp. *unguiculata*, 12, 90, 248, 250, 253, 254, 256, 257, 258, 259, 260

W

walnut, *see Juglans*
watermelon, *see Citrullus lanatus*

wheat, *see Triticum aestivum*
wild barley, *see Hordeum spontaneum*
wild oat, *see Avena fatua*
witch-weed, *see Striga asiatica*

X
Xanthium
  pennsylvanicum, 57, 65, 72
  strumarium, 78

Y
yucca, *see Yucca whipplei*
Yucca whipplei, 203

Z
Zea mays, 3, 78, 92, 93, 107, 110, 146, 165, 166, 170, 172, 173, 178, 184, 186, 187, 189, 191, 202, 203, 206, 219, 220, 235, 236, 241, 247

Subject Index

A

abscisic acid (ABA), 179, 180
    effects on aleurone cells, 125, 126, 153, 155
    in dormant seeds, 48, 49, 50, 52
    as germination inhibitor, 55, 273
    in non-dormant seeds, 49, 50, 274
    effects on translation, 50
[^{14}C]acetate, metabolism following germination, 234–237
acetyl-CoA, 210, 211, 213, 224, 228, 229, 230, 234, 241, *see also* β-oxidation of fatty acids
acid phosphatase
    of aleurone cells
        relationship to phytase, 185, 188
        secretion, 129, 130, 131, 132, 134–137
        synthesis, 129, 134, 188
    of cotyledons, 262, 264
aconitase, 213, 225, 229, 230
ADP, 100, 232
Aepyceros melampus, 33, 34
albizziine, 259
albumins, as reserve proteins, 252, 253, 258, 259
aleurone cells
    cell wall hydrolysis, 133, 135–140
    sensitivity of response to GA, 110, 146
aleurone grains, *see* protein bodies
alleles
    affecting ABA levels, 50
    affecting seed protein composition, 165, 184

α-amanatin, 104, 105
amides, 257
amino acids, 47, 68, 257
aminoacyl-tRNA synthetases, 72, 102
aminopeptidase
    activity, 170, 171, 172, 173, 174, 175, 177, 254, 264, 268
    localization, 169, 170
α-amylase, 120
    of aleurone cells
        properties, 123, 124, 127–129, 133
        secretion, 122, 128, 130, 131, 132–135
        synthesis, 101, 110, 121, 122, 125, 126, 133, 145, 146
    of cotyledons, 142, 143, 263, 265, 271, 272, 273, 274, 275
    of embryo axis, 250, 251, 252
    of scutellum, 142, 143
β-amylase, 120
    of cotyledons, 263, 265, 273, 275
    of embryo axis, 250, 251, 252
    of endosperm, 127, 143, 144
amylopectin, *see* starch
amylose, *see* starch
anaerobiosis, on imbibition, 92, 99, 100
Aphelenchus avenae, 90
apical dominance, 273, 274
α-arabinofuranosidase, secretion by aleurone cells, 139
arabinoxylans, 138
arginine, in proteins, 164
arginine decarboxylase, 264, 270
asparagine synthetase, 264, 267

aspartate aminotransferase, 231
ATP, 232, 233
  synthesis
    in dormant seeds, 66
    on imbibition, 63, 95, 100, 103
  see also energy charge, changes on imbibition
ATPase, of plasmalemma, 109
auxins, 60, 125, 273
axis, embryonic, 247-277
  cell division, 95, 107
  cell expansion, 82, 89, 107, 108, 109, 110, 250
  DNA synthesis, 107, 109
  excised, 63, 82, 85, 86, 96, 102, 248, 249, 252, 261
  growth in dormant seeds, 63, 108
  imbibition, 82, 86, 93
  metabolism, 51, 69, 70, 95, 261, 262
  phloem differentiation, 140, 260
  protein synthesis, 108, 254-257, 260
  reserves, 67, 69, 70, 72, 105, 249, 252, 253
  RNA metabolism, 46, 47, 50, 72, 104, 108, 261

B
benzyladenine, 63, 238, 270
bruchid beetles, role in germination, 34

C
calcium
  as cofactor, 128
  growth promotion, 261
  in phytin, 184
carboxypeptidase, 106, 264
  activity, 167, 169, 170, 171, 172, 173, 174, 254, 264, 268
  localization, 169, 171, 174, 176, 254, 264
  properties, 167, 174
  release from aleurone cells, 176
catalase, 64, 173, 211, 212, 214, 215, 225, 227
cation exchange, in aleurone cell walls, 130, 132
cattle, seed ingestion, 33, 34
cellulase, 140
cell wall
  breakdown
    in aleurone, 133, 135-140
    in endosperm, 133, 141, 146-156
  loosening, 109, 110
  synthesis, 185, 235-237

chalaza, 19
chalazal cap, 20, 21
chalazal discontinuity, 8, 19, 20, 30, 32, 33, 36, 39
2-chloroethyltrimethylammonium chloride (CCC), 51, 180
p-chloromercuribenzoic acid, 171, 172
citrate, 225, 229, 233
citrate synthase, 213, 225, 229
CO_2
  in gas phase of seeds, 83
  requirement for ethylene action, 57, 59
cordycepin, 104, 105, 176
cotyledon, 47, 59, 67, 68, 93, 95, 181, 184, 185, 189, 202, 203, 212, 227, 247-277
  cell expansion, 239, 268
  development of photosynthetic capacity, 217, 235-237, 239, 241, 248, 265, 269, 270, 276
  excised, 51, 63, 82, 235, 237, 239, 240, 262, 265-270, 273, 275
  reserve mobilization, see mobilization of seed reserves
cuticle
  within hilum, 16
  of seed coat palisade cells, 2, 12, 22
cyclic-3,5-AMP, 125
cycloheximide, 57, 97, 99, 110, 122, 153, 176, 189, 267, 268, 269
cytochrome oxidase, 64, 66, 85, 97, 98, 173, 227, 270
cytokinins, 55, 56, 62, 63, 268, 269, 270, 276, 277
cytoplasm, 85, 87, 89, 178
cytosol, as metabolic compartment, 210, 211, 213, 216, 231, 232, 241

D
debranching enzymes, see limit dextrinase
decarboxylation, in reserve lipid mobilization, 213, 225, 232
dicarboxylic acids, 213, 228, 234, see also malate; oxaloacetate; succinate
dikdik, see Rhynchotragus kirkii
dinitrophenol, 178, 179, 180
dipeptidases, 167, 168, 171, 174, 175, 177
dipeptides
  model substrates, 174, 176, 180
  see also glycylsarcosine
dithiothreitol, 191
DNA
  hybridization, 124
  polymerase, 106
  synthesis, 106, 107

Subject Index

docosenol, 207-209
Dorcas gazelle, see *Gazella dorcas*
dormancy of seeds
   advantages, 3, 45, 46
   alleviation of coat-imposed dormancy
      by freezing, 36
      by heat, 19, 20, 21, 29, 35, 36, 37
      by impaction, 8, 18, 19, 20, 35, 36
      by ingestion, 32, 33, 34
      by microwave irradiation, 29, 35, 37, 38
      by natural means, 31, 32, 33, 34
      by solvents, 3, 4, 20, 35
   alleviation of embryo dormancy
      cold stratification, 46, 47, 49, 51, 52, 53-55, 56, 60, 63, 65, 66, 67, 68, 69, 71, 72
      by fatty acids, 61
      phytochrome, 47, 53, 58, 59, 61-63, 155
      phytohormones, 45-73
      respiratory inhibitors, 64, 65, 66
   definition of dormancy, 2-5, 45, 50
dormin, see abscisic acid

E

eicosenoate
   in triacylglycerols, 204, 206
   in wax esters, 207-209
eland, see *Taurotragus oryx*
elephant, African, see *Loxodonta africana*
endopeptidases, 167, 168, 170, 258, 265, 266
   acid, 168, 170, 171, 173, 174, 254
   serine, 254, 268
   thiol, 168, 170, 171, 172, 254, 264, 266, 267, 268, 273, 275, 276
endoplasmic reticulum, 59, 67, 125, 126, 127, 132, 133, 134, 135, 215
endosperm, see mobilization of seed reserves
endosperm cell wall hydrolysis, 133, 141, 146-156, see also aleurone cells
endoxylanase, of aleurone cells
   secretion, 130, 131, 139
   synthesis, 130, 131
energy charge, changes on imbibition, 100
enoyl-CoA hydratase, 225, 226, 227
enoyl-CoA isomerase, 226, 227
epihilum, 9, 15
erucic acid, in triacylglycerols, 204, 206, 221
Escherichia coli, 228
esterase, of aleurone cells
   secretion, 129, 130, 131, 132, 135-137
   synthesis, 129

ethanol, and soaking damage, 92
ethylene, 153
   binding sites, 57, 59
   breaking dormancy, role in, 56-60, 65
   as germination inhibitor, 58, 59, 262
   xylanase release, 139, 140
N-ethylmaleimide, 171
euphaseolin, 254
Evans blue, test for cell damage, 86
exopeptidases, 167, 168, 170, see also aminopeptidase; carboxypeptidase; dipeptidases; iminopeptidase

F

fatty acids
   degradation, see β-oxidation
   release from oil or wax bodies, 210, 213, 217-223
   short chain, 61
   in triacylglycerols, 204-206
fatty acyl-CoA ligase, see fatty acyl-CoA synthetase
fatty acyl-CoA oxidase, 225, 226, 227
fatty acyl-CoA synthetase, 224, 225, 227
fatty alcohol oxidase, 223, 224
fatty alcohols, 207-209
   breakdown, 224
   release from wax esters, 223, 224
fatty aldehyde dehydrogenase, 223, 224
ferulic acid
   in aleurone cell walls, 138
   in scutellum cell walls, 143
fluoride, as phytase inhibitor, 186
fructose-1,6-bisphosphatase, 173, 213, 225, 232, 233, 234
fructose-1,6-bisphosphate, 186, 213, 232, 233
fructose-6-phosphate, 232, 233
fumarase, 85, 213, 225, 227
funiculus, 9, 16, 17
fusicoccin, 109, 110
Fusicoccum amygdali, 109

G

galactomannans, 147-150
galactose, 151
   uptake by embryo, 150, 152, 153
α-galactosidase, 62, 150, 151, 152, 153, 154
gas, displacement from seeds, 82, 83
Gazella
   dorcas, 33, 34
   thompsonii, 33, 34
germination of seeds
   effects of temperature, 78, 79, 92-94
   choice of experimental conditions, 210

germination of seeds—*continued*
  terminology, 209, 210
  see also dormancy of seeds; imbibition of water
gibberellic acid (GA$_3$), 179, 180
  effects on aleurone cells, 126, 127
    enzyme release, 122, 129, 132-140, 176, 191, 193, 196
    enzyme synthesis, 110, 121-125, 129, 131, 132, 133, 134, 191
    glutamine release, 188
  as growth stimulant, 49, 54, 55, 62, 69
  sites of storage, 145
  sites of synthesis, 144, 145
  translocation, 63, 145
gibberellins, 51-55
  content in seeds, 51, 52, 53, 62, 128, 144, 145, 146
  effects on aleurone cells, 125
  effects on endosperm cells, 155
  as growth stimulants, 47, 48, 51, 52, 53-55, 108, 273
  inhibitors, 53, 260
  sites of storage, 51, 145, 260
  sites of synthesis, 51, 52, 144, 145, 260
  translocation, 51, 145, 260
  see also gibberellic acid
gliadins, 164, 165, 172
globulin, 165, 166
β-glucanase, of aleurone cells
  secretion, 129, 130, 131, 140
  synthesis, 129, 130
gluconeogenesis, 210, 213, 224, 225, 231-235, 241, *see also* PEP carboxykinase
glucose-6-phosphate dehydrogenase, 65, 84
α-glucosidase
  of aleurone cells, 120, 121
    secretion, 129
    synthesis, 129
  of cotyledons, 263
β-glucosidase, 171
glutamate, 231, 235
glutamate dehydrogenase, 84, 264
glutamine, 180
  in protein, 164
  transport, 178, 181, 182, 188
glutamine synthetase, 264, 267
glutelins, 164, 165, 166, 169, 170, 172, 174
glutenins, 144, 164, 165, 169, 170
glycerate-3-phosphate, 233
glycine, 179
glycolysis, 100, 101, 150, 213
  regulation during gluconeogenesis, 232, 233

glycylsarcosine, 179
glyoxylate, 213, 225, 228
glyoxylate cycle
  in microorganisms, 228
  in oil seeds, 211, 213, 215, 225, 228-231, 234, 235, 241
  see also glyoxysomes
glyoxysomes
  constituent enzymes, 213, 214, 215-217, 220, 221, 224-231
  development, 207, 210-217, 265
  isolation, 212, 215, 227
  membrane constituents, 215, 219, 230, 231
  ultrastructure, 212, 214, 215

H

hemicellulose
  in seed coats, 12, 22, 23
  synthesis following germination, 235-237
hilar groove, 14, 15
hilum, 9, 10, 12, 14-17, 32, 33, 35
  as hygroscopic valve, 14-17, 39
Homarus americanus, see lobster, muscle lipids
homoserine, 249
hordein, 165, 174
hordenin, 165
hormone sensitivity, 48, 49, 55, 110, 146, 152, 156, 157
β-hydroxyacyl-CoA dehydrogenase, 173, 225, 226, 227
hydroxylamine, 189

I

imbibition of water
  cell damage, 82, 83-91
  chilling injury, 92-94
  lag phase, 78-81, 109
  rate, 79-81
  soaking injury, 91, 92
iminopeptidase, 167
immunoassay, 253, 275
immunocytochemistry, 23, 276, 277
immunoelectrophoresis, 253
impala, *see Aepyceros melampus*
internode, 260, 271
iodoacetamide, 171
myo-inositol, 185
  phosphate esters, 185, 186
isocitrate, 213, 228, 230

Subject Index

isocitrate lyase
  activity, 68, 69, 71, 173, 211, 229, 239, 262, 265, 270
  localization, 211, 213, 229
  properties, 228
isoelectric focusing, 129
isoenzymes
  acid phosphatase, 185
  aminopeptidase, 264
  α-amylase, 127–129, 134, 263
  β-amylase, 263
  carboxypeptidase, 169, 174
  endopeptidase, 174
  β-glucanase, 140
  glutamine synthetase, 264
  starch phosphorylase, 265

J

jojoba oil, *see* wax esters

K

α-ketoglutarate, 231
kinetin, 125, 270

L

leakage, solute loss from seeds, 81, 83–91, 92, 93
lectin, mobilization, 253, 255, 257, 258
legumin, 253
lens
  of seed coats, 9, 10, 11, 17, 18
    role in water entry, 8, 16, 17, 29, 30, 32, 33, 35, 36, 37, 38, 39
    role in water loss, 16
leupeptin, 171, 172, 173, 266
light
  effects on reserve mobilization, 270, 271, 272
  effects on seedling development, 268, 269, 270–272
  requirement for germination, 108
limit dextrinase, 120, 121, 129
linoleate, in triacylglycerols, 204, 206, 220, 221, 222
linolenate, in triacylglycerols, 204, 206
lipase, 217, 265
  in glyoxysomes, 219, 221, 222, 241
  in oil bodies, 213, 218, 219, 220, 221, 222, 241
  pH optima, 68, 218, 219, 220, 221, 241
  substrate specificity, 218, 219, 220, 221
  temperature optima, 68

lobster, muscle lipids, 90
longevity, 4, 5
  of dry seeds, 6, 7
  of imbibed seeds, 6, 7
Loxodonta africana, 33, 34
lysophosphatidylcholine
  degradation, 192, 193
  occurrence, 192
lysophospholipase, 192, 193, 194
  regulation, 193

M

macrosclereids, of seed coats, 8, 10, 11, 12, 14, 15, 16, 18, 19, 36, 38
magnesium, in phytin, 184
malate, 233, 268
  as oxaloacetate precursor, 213, 225, 230, 231, 269
  synthesis in glyoxysomes, 213, 225, 230, 231, 235
malate dehydrogenase
  glyoxysomal, 213, 225, 229, 231
  mitochondrial, 97, 98, 213, 231
malate synthase
  activity, 211, 265, 270
  localization, 213, 225, 227
  properties, 228, 229
Malpighian cells, *see* macrosclereids, of seed coats
maltose, 120
mannans, 147
  degradation, 118, 148–155
β-mannase, 150, 151, 152, 154
mannobiose, 150, 151
mannose, 150, 151, 152
β-mannosidase, 150, 151, 152
mannotriose, 150, 151
megagametophyte, 168, 169, 170, 171, 203, 226, 239
membrane
  hydration, 87, 88, 89, 99
  phospholipids, 86, 87, 88, 89, 94
  proteins, 86, 90, 91, 99
  repair, 86, 87, 88, 90
  rupture, 86, 87, 88, 90, 94
  structure, 72, 86–91, 98, 99
  synthesis, 126, 127
  see also plasmalemma
2-mercaptoethanol, 172, 187
3-mercaptopicolinic acid, as PEP carboxykinase inhibitor, 233
methionine, as ethylene precursor, 59
N-methyglycine, *see* sarcosine
6-methylpurine, 176

micropyle, 9, 10, 39
  function during imbibition, 17
mitochondrion
  biogenesis, 99, 143
  electron transport chain, 96, 231
  isolation, 96, 97, 98, 100
  Krebs cycle, 96, 231
  reassembly, 97, 98, 268, 269, 276
  role in lipid mobilization, 210, 211, 212, 213, 225, 226, 230, 231, 241
  structure, 81, 253
  see also respiration rate
mobilization of seed reserves
  in aleurone cells, 123, 153, 165, 184–190, 194, 196
  from cotyledons, 249, 260, 261–277
    lipids, 63, 67, 68, 210–242
    oligosaccharides, 96, 149, 154
    phytin, 261
    proteins, 63, 106, 212, 253, 254, 258, 259
    starch, 263
  in dormant seeds, 66–72
  in the embryo axis, 68, 69, 70, 72, 249, 250, 251, 252–260
  from endosperm
    lipids, 210–242
    nucleic acids, 190–192
    oligosaccharides, 147, 153
    phospholipids, 192–194, 195
    phytin, 184–190, 195, 196
    polysaccharides, see galactomannans; starch
    proteins, 164–182, 190, 195, 196
  from megagametophyte
    lipids, 68, 226, 239
    proteins, 168, 169, 170, 171

N
NAD, 225, 230, 231
NADH, 225, 228, 230, 231, 268
NADH cytochrome c reductase, 98, 230
NADH ferricyanide reductase, 230
NADP, 64, 223
NADPH, 64
nematodes, rehydration, 90
p-nitrophenyl phosphate, 185, 186
3′-nucleotidase, 190, 191

O
oil bodies
  of aleurone cells, 126, 146
  of axis cells, 70, 249
  of cotyledon cells, 214, 221, 222
  degradation, 213, 219, 222, 223, 241, 242
  of endosperm cells, 155, 218
  membrane-bound enzymes, 213, 218, 219, 220, 221
oleate, in triacylglycerols, 204, 206, 220
oligosaccharide mobilization, 147, 149, 153, 154, 236, 263
osteosclereids of seed coats, 10, 11, 13, 16
oxaloacetate, 211, 213, 225, 230, 231, 232, 233
β-oxidation of fatty acids
  component enzymes, 224–229
  glyoxysomal localization, 210, 211, 215, 224–229, 230, 241
oxygen
  in lipid metabolism, 223, 225, 226, 234, 240
  see also anaerobiosis, on imbibition

P
palisade cells, see macrosclereids, of seed coats; seed coats, impermeability
palmitate
  release from phospholipids, 194
  in triacylglycerols, 204, 206
pentose phosphate pathway
  in dormant seeds, 59, 64, 65
PEP (phosphoenolpyruvate), 211, 213, 225, 232, 233, 234
PEP carboxykinase, 213, 232, 233
pepstatin-A, 168, 169, 170, 171, 172, 177, 196
peptide hydrolases, 264
  classification, 166–168
peroxidase, 64
  of aleurone cells
    secretion, 130, 131, 132, 135–137
    synthesis, 130
  of cotyledons, 264
  of seed coats, 23
peroxisome, 217, 241, 270
phenylmethylsulfonyl fluoride, 176
phloem, 261
  differentiation, 140, 260, 273
phosphate, inorganic
  exogenous, 188
  as inhibitor of phytase, 187, 188, 190
  phosphorylase requirement, 263
  from phytin, 183, 189
phosphatidylcholine, 88
phosphatidylethanolamine, 88
phosphodiesterase, 171

Subject Index   293

phosphofructokinase, 232, 233, 234
6-phosphogluconate dehydrogenase, 65
phospholipid
    breakdown, 192, 193, 222
    synthesis, 126
phosphomannoisomerase, 150
Physeter macrocephalus, 204, 207
phytase
    function, 171, 189, 190, 196
    properties, 185–190
phytic acid, 182
phytin, 247
    degradation, 182–190
phytochrome, 268
plasmalemma, 88, 132, 135, 188, 190
    ATPase, 109
    expansion, 89
plasmodesmata, between aleurone cells, 137, 138
pleurogram, 35
polyamines, 108, 270
polypeptide, processing, 124, 142, 266
polyphenol oxidase, of seed coats, 23
potassium
    loss on imbibition, 81, 84, 91
    in phytin, 184
    uptake on imbibition, 109
prolamin, 164, 165, 166, 172, 174
proline
    in protein, 164, 174
    specific carboxypeptidase, 174
    see also iminopeptidase
protein
    accumulation in seeds, 101, 102, 203
    composition, 164, 174
    degradation, *see* mobilization of seed reserves
    synthesis, 47, 48, 57, 63, 69, 72, 92, 94, 101, 102, 103, 106, 109, 123–126, 130, 133, 176, 263, 269
proteinase (protease)
    activity, 68, 69, 71, 169, 170
    classification, 167, 168
    localization, 59, 72, 129, 171
    release from aleurone cells, 130, 131, 144, 174, 176
proteinase inhibitors, 69
    of endopeptidases, 169, 170, 177
    of exopeptidases, 176
    mobilization, 169, 178, 253
protein bodies
    aleurone grains, 153, 164, 165
    autolysis, 69, 153, 165, 265, 266
    hydrolases, 72, 171, 258

    inclusions
        crystalloid proteins, 171, 266
        globoid crystals, 164, 184, 205
        matrix proteins, 70, 171, 266
    membranes, 70, 72
        binding sites for ethylene, 59
Pseudomonas, 228
putrescine, 270
pyruvate carboxylase, 233
pyruvate kinase, 232, 233

R

radicle
    damage, 84, 93, 101
    expansive force, 28, 33, 35
    growth, 49, 56, 60, 62, 69, 77, 78, 79, 82, 106, 108, 149, 150, 154, 210, 250, 251
    membrane structure, 90
    source of cytokinins, 273, 275
raffinose, conversion to starch, 236
raphe, 17
respiration rate, 63
    cyanide-insensitive, 57, 64, 65, 66, 268
    of dry seeds, 95
    following germination, 97
    prior to germination, 57, 64, 65, 66, 68, 69, 84, 95, 96, 97
Rhizoctonia, effects on seed coats, 32
Rhynchotragus kirkii, 34
ribonuclease
    of aleurone cells
        secretion, 129, 130, 131, 132
        synthesis, 129, 130, 132
    in embryo, 191, 261, 262
    in endosperm, 171, 190, 195, 196
ribosomes, 72, 123, 127, 153, 215, 241, 268
    retention in mature seeds, 101, 102
ribulose bisphosphate carboxylase, 270
RNA
    breakdown, 105, 106, 190, 191, 195, 196
    mRNA, 47, 50, 102, 103, 104, 105, 106, 121, 123, 124, 125, 129, 273, 276
    poly(A)-rich RNA, 46, 50, 103, 104, 105, 123
    rRNA, 46, 47, 72, 102, 105, 127
    synthesis, 46, 47, 50, 67, 92, 105, 106, 109, 121, 123, 124, 125, 127, 190, 191, 269
    tRNA, 102
    turnover, 103, 261

S

salicylhydroxamic acid, 57, 65, 96
sarcosine, 179, 180
scutellum, 247
   amino acid uptake, 164, 177, 178, 179, 195, 196
   aminopeptidase activity, 175, 177, 196
   dipeptidase activity, 175, 177, 196
   enzyme secretion, 141-143, 146, 147, 192
   peptide uptake, 177, 179-181, 195, 196
   source of GA, 142, 145
   starch synthesis, 142
   sucrose
      synthesis, 140, 141
      transmission, 140, 141
seed coats
   hormone content, 49, 50
   impermeability, 3, 12, 13, 22-25, 27, 28, 240
   light screening properties, 62
   mechanical properties, 24, 26-31
      compression, 13, 24, 26, 27
      deformation energy, 27, 28, 29
      deformation histograms, 24, 37
      elastic properties (L/D ratio), 26, 27, 28
      tension, 26, 27, 28, 29
      thin shell theory, 29-31
   moderation of water inflow, 79-81, 82, 85, 86, 91, 94
   morphology, 10, 11, 15, 20, 21, 22
   mucilage, 12
   pigmentation, 17, 18, 22, 23
   structure, 8-24
   vascular bundles, 16, 17
   water entry, 13, 17, 19, 22, 29, 30, 32, 36, 37, 38, see also chalazal discontinuity; lens
seed storage
   in adobe bricks, 6
   ambient conditions, 5, 6, 7
   frozen, 6
   in soil, 6, 7, 28
   in water, 6
seed viability, 5, 7, 78
signal hypothesis, 124, 134, 142
sperm whale oil, see *Physeter macrocephalus*
spherosomes, see oil bodies, wax bodies
stachyose, 235, 236
starch
   breakdown
      in embryos, 250-252, 260
      in endosperm, 118-121, 174
   synthesis
      during dormancy, 67, 68
      following germination, 148, 150, 212, 236, 239
starch granules
   of embryo axis, 249
   of endosperm, 118, 119
   patterns of breakdown, 118, 119
   phospholipid content, 192, 193
   size heterogeneity, 118, 119, 192
starch phosphorylase, 120, 121
   of cotyledons, 263
starling, see *Sturnus vulgaris*
stearate, in triacylglycerols, 204, 206
strophiole
   definition, 9
   term applied to lens, 8, 9
Sturnus vulgaris, 34
succinate, 268
   production by glyoxysomes, 211, 213, 225, 228, 230
succinate dehydrogenase, 97
sucrose
   accumulation, 211, 251, 252, 276
   as inhibitor, 239, 240, 276
   synthesis, 211, 213, 229, 232, 235, 236, 241, 270
   translocation, 211, 241, 251, 252
   uptake, 248, 249
sucrose phosphate synthase, 232
sucrose UDP glucosyl transferase, 265

T

Taurotragus oryx, 34
tetrazolium, 86
thin shell theory
   hoop stresses, 30, 31
   meridian stresses, 30, 31
thiolase, 225, 226, 227, 228, 229
Thompson's gazelle, see *Gazella thompsonii*
tracheid bar, 11, 14, 15, 16
triacylglycerols
   acyl composition, 204, 206
   content, 202, 203
   conversion to sucrose, 210, 211, 213
   rate of breakdown, 210, 211, 215, 217
   regulation of breakdown, 222, 237
      by embryonic axis, 234, 238, 239, 242
      by oxygen, 234, 237, 238, 239, 240

U

unsaturated fatty acids, see individual compounds

V

vacuoles, 69, 70, 235

W

water content
 of cotyledons, 98
 of organelles, 85
 of seeds, 78, 93
wax bodies
 membrane-bound enzymes, 223
 occurrence, 204, 205
wax esters
 composition, 207–209
 degradation, 210, 212, 223, 224
 hydrolases, 220, 223
 storage, *see* wax bodies

X

Xenopus laevis oocytes, 124
β-xylopyranosidase, secretion by aleurone cells, 139

Z

zeatin, 276
zein, 165, 166, 172